CW00690833

# High-Power Laser–Plasma Interaction

With the rise in laser power there has been phenomenal growth in the field of high-power laser–plasma interaction, with diverse applications ranging from laser fusion and laser acceleration of electrons and ions, to laser ablation of materials and laser coupling to graphene plasmonics. The laser power has increased from 100 watts in 1960 (the first laser) to terawatts in 1985 through the invention of chirped pulse amplification proposed by Mourou and Strickland, to 100 petawatt and beyond today. Presently, lasers constitute the most powerful coherent source of radiation on the Earth. Also, we have advanced our understanding of the most fascinating but complex state of high-temperature matter, plasma, in which matter is ionized to contain free electrons and ions. Plasma supports many collective oscillations, such as electron plasma waves and ion acoustic waves, and the electromagnetic wave is also strongly modified by plasma. Parametric coupling between lasers and plasma waves and quasi-modes gives rise to stimulated Raman, Brillouin and Compton scattering, two-plasmon decay, and four-wave processes including filamentation, modulational and oscillating two-stream instabilities. Nonlinear refraction leads to self-focusing and self-guiding of the laser over long distances in plasma and air, offsetting diffraction divergence. Laser interaction with metallic surfaces gives rise to surface-enhanced Raman scattering from adsorbed molecules and mode conversion to surface plasma waves with application to ablation and thin film deposition. With the rapid development of laser technology and plasma understanding, we can expect many exciting new frontiers of research to develop in coming years.

This book builds a systematic theoretical understanding and analytic treatment of laser–plasma interaction. It can serve as a textbook for graduate students in plasma physics and as a reference book for researchers in laser and plasma communities, with diverse interests in laser–plasma interaction, free electron lasers and laser ablation of materials, and related areas.

**Chuan Sheng Liu** is a theoretical physicist active in research on basic plasma physics and controlled fusion. He carried out pioneering work on parametric instabilities in laser–plasma interaction at the Institute for Advanced Studies at Princeton. He is now a Professor Emeritus at the University of Maryland, College Park, USA. Between 2003 and 2006, he served as President of the National Central University (NCU) in Taiwan, where he led the effort in building a 100 terawatt laser system.

**Vipin K. Tripathi** is a Professor of Physics at the Indian Institute of Technology Delhi, India. He established a plasma group with the focus on beam-plasma systems, radio frequency heating and current-drive in tokamaks, high-power laser–plasma interaction, free electron lasers, gyrotrons, and surface plasmonics. He has taught a wide range of courses, including electrodynamics, classical mechanics, quantum mechanics, thermodynamics, plasma physics, laser physics, microwaves, and high-power laser–plasma interaction.

**Bengt Eliasson** is an Associate Professor (Reader) at the University of Strathclyde in Glasgow, United Kingdom, where he teaches courses in plasma physics and numerical methods. He is an expert in numerical and theoretical plasma physics and has made significant contributions in the fields of space plasma, quantum plasma physics, and laser–plasma interaction.

# High-Power Laser–Plasma Interaction

C. S. Liu
V. K. Tripathi
Bengt Eliasson

**CAMBRIDGE**
**UNIVERSITY PRESS**

University Printing House, Cambridge CB2 8BS, United Kingdom

One Liberty Plaza, 20th Floor, New York, NY 10006, USA

477 Williamstown Road, Port Melbourne, VIC 3207, Australia

314 to 321, 3rd Floor, Plot No.3, Splendor Forum, Jasola District Centre, New Delhi 110025, India

79 Anson Road, #06–04/06, Singapore 079906

Cambridge University Press is part of the University of Cambridge.

It furthers the University's mission by disseminating knowledge in the pursuit of education, learning and research at the highest international levels of excellence.

www.cambridge.org
Information on this title: www.cambridge.org/9781108480635



Printed in India by Nutech Print Services, New Delhi 110020

*A catalogue record for this publication is available from the British Library*

ISBN 978-1-108-48063-5 Hardback

Additional resources for this publication at www.cambridge.org/9781108480635

To

Marshall Rosenbluth and Roald Sagdeev,
two pioneers in laser–plasma physics.

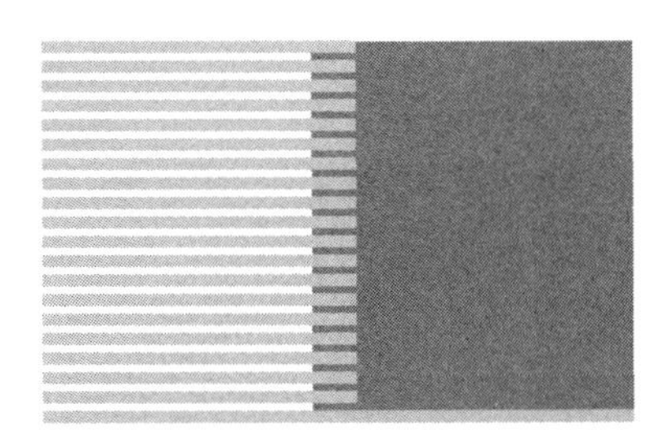

# Contents

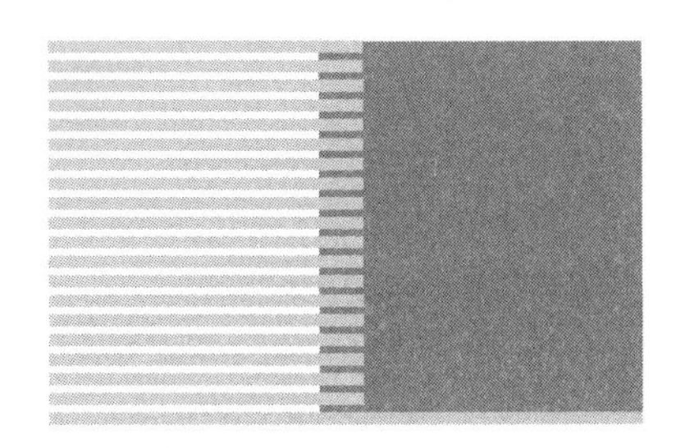

# FIGURES

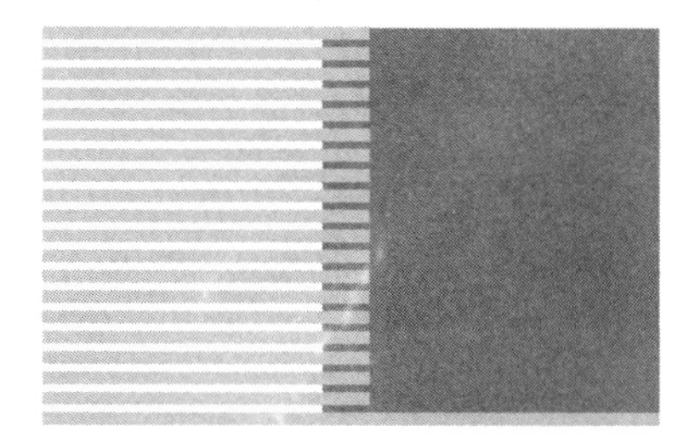

# PREFACE

The laser, with its coherent, monochromatic, and well collimated character, has been a most remarkable discovery of the twentieth century. Along with semiconductors, its multifaceted applications have broadly touched and greatly improved our lives – it has made an indelible mark in the field of sensing, printing, barcode scanning, surgery, communications, and so on. It has also become a major tool for scientific research. For example, Thomson scattering and laser induced fluorescence are important tools for plasma diagnostics. Lasers have been used successfully for cooling of atoms and heating of plasmas.

The laser peak power has increased about a 1000 fold every decade since its invention. Starting from hundred watts in the 1960s, table top terawatt Ti: sapphire lasers became available in the 1990s following the discovery of the chirped pulse amplification (CPA) by Mourou and Strickland in 1985. These lasers do not only have high power but also very short pulses of a few femtoseconds, opening a new field of ultra-short pulse lasers and their interactions with matter, such as electron dynamics in molecules. In the past few years, we have seen worldwide efforts to build high power laser infrastructures. The Extreme Light Infrastructure (ELI) has been approved to construct three petawat laser facilities in Eastern Europe. Similar efforts are being made in Korea, Japan and China.

With the rise in laser power, there has been a phenomenal growth in the field of high power laser–plasma interaction with diverse applications, ranging from laser driven fusion and laser acceleration of charged particles to laser ablation of materials. The field has revealed a rich variety of fascinating new phenomena. Parametric coupling between lasers and plasma eigenmodes and quasi-modes gives rise to stimulated Raman, Brillouin, and Compton scattering, two-plasmon decay, and four-wave processes of filamentation, modulational, and oscillating two-stream instabilities of the laser. Nonlinear refraction gives rise to self-focusing and self-guiding of lasers over long distances in plasma and air, offsetting diffraction divergence. Laser interaction with rough metallic surfaces reveals surface-enhanced Raman scattering (SERS) where Raman scattered power from adsorbed molecules rises a million times due to surface plasmon resonance. Laser mode conversion to surface plasma waves (SPWs) on metallic surfaces has been shown to enhance the ablation yield and thin film deposition rates by orders of magnitude, making pulsed laser deposition a very attractive scheme. Surface plasma waves can focus light to sub-wavelength dimensions and holds the key to the development of nano-devices. Experiments with gas jet targets, embedded with clusters, has revealed the phenomenon of ion Coulomb explosion, providing a high yield source of neutrons.

In 2004, the remarkable acceleration of quasi mono energetic electrons to 200 MeV with laser wake-fields in plasma was reported in *Nature* by three independent groups. The cover story of the journal was the "dream beam" of electron acceleration! The same year, J.P. Wang reported the first direct measurement of an acceleration gradient of GeV/m in laser electron acceleration. Recent experiments have achieved multi GeV electrons.

With the completion of the creation of 192 laser beams with a total energy of 1.8 megajoules at the National Ignition Facility (NIF) at Livermore a few years ago, there was a great optimism for its successful demonstration. But early experiments failed to show ignition. Two reasons were given for this setback: the lack of symmetry of the deuterium-tritium (D–T) pellet and a high level of Raman backscattering with a reflectivity of the laser beams by the hohlraum plasma of up to 30 percent. Both indicated the need for further in-depth study of laser–plasma interaction.

While the laser is the most coherent source of radiation, plasma is the most chaotic state of matter. Hence, the nonlinear physics of laser–plasma interaction is indeed a most fascinating subject to study – not only for its importance to projects like laser fusion, but also for the new fields of research opening up such as relativistic nonlinear physics, collective QED as we approach the Schwinger field.

We hope this book can provide graduate students an overview of the field of high-power laser–plasma interaction, with an in-depth discussion of the basic concepts, systematic knowledge, and tools developed in the last forty years, so that they can embark on their creative career in these exciting fields. The book can be used as a reference and guide for plasma experimentalists and simulation experts, and should also benefit researchers in materials science.

Chuan Sheng Liu
Vipin K. Tripathi
Bengt Eliasson

# 1 INTRODUCTION

Plasma is an ionized state of matter whose main constituents are electrons and ions. It is the most predominant state of visible matter in the universe. There are various parameters that define the characteristics of a plasma such as characteristic frequency and Debye length. A characteristic frequency of plasma oscillations is the electron plasma frequency $\omega_p = (ne^2/m\varepsilon_0)^{1/2}$, where $n$, $-e$, and $m$ are the electron number density, charge, and mass, respectively and $\varepsilon_0$ is the vacuum electric permitivity. According to linear theory, an electromagnetic wave of frequency $\omega$ normally incident on a plasma slab is reflected at the critical layer where $\omega = \omega_p$. For waves of higher frequency, $\omega > \omega_p$, the plasma behaves as an optically transparent medium with the refractive index $\eta = \left(1 - \omega_p^2 / \omega^2\right)^{1/2}$. Electron and ion collisions cause the wave to be damped.

There also exists a characteristic length in a plasma, the electron Debye length $\lambda_D = v_{\text{th}}/\omega_p$, where $v_{\text{th}} = \sqrt{T_e / m}$ is the electron thermal speed, and $T_e$ is the electron temperature in energy units. The electron Debye length characterizes the length over which the electric potential of a charged particle is screened by the electrons and ions of the plasma. An ionized gas of size much greater than $\lambda_D$ can be called a plasma. The screening of the potential is crucial for the description of the dynamics of charged particles, otherwise the electric and magnetic fields produced by each charged particle would influence the motion of all the other particles (billions and billions in number) and each one would move in the fields of all others. The Debye screening localizes the micro-fields of the individual particles to within a Debye sphere, containing a large number of particles $N \approx n\lambda_D^3 >> 1$, where $N$ is called the plasma parameter.

At a distance of a Debye length, the potential energy of a particle equals its kinetic energy, and beyond that the latter prevails. The quantity $1/N$, provides us an expansion parameter, valid to the zeroth order, in which one can neglect the correlations and collisions of charged particles. This is called the Vlasov approximation. Binary collisions are an effect to the next order.

## 1.1 Laser Produced Plasma

A high power laser is an effective means of producing high density plasma on a nanosecond, pico-second, or even femto-second time scale with a temperature greater than 100 electronvolts (eV). Most laser–plasma experiments employ Nd:glass lasers (1.06 μm) or Ti:Sapphire lasers (0.8 μm) on targets that include gas jets or thin foils or spherical pellets. Experiments with $CO_2$ lasers (10.6 μm) have also been conducted.

### 1.1.1 Tunnel ionization

A striking feature in using lasers at high intensities (> $10^{14}$ W/cm$^2$) is the process of tunnel ionization that produces plasma on sub pico- second time scales. As Keldysh[1] showed, the rate of tunnel ionization of atoms caused by the electric field $E$ of a laser with frequency $\omega$ is

$$\gamma = (\pi/2)^{1/2}(I_0/\hbar)(|E|/E_A)^{1/2}\exp(-E_A/|E|), \tag{1.1}$$

where $E_A = (4/3)(2m)^{1/2} I_0^{3/2}/e\hbar$ is the characteristic atomic field, $I_0$ is the ionization energy, and $\hbar$ is the reduced Planck's constant. This expression is written in the limit when the electron quiver energy $mv_0^2/2 > I_0$. Here $v_0 = e|E|/m\omega$ is the amplitude of the electron quiver velocity due to the laser. One may write $mv_0^2/2 = 1500 P_{16}\lambda_\mu^2$ in eV where $P_{16}$ is the laser intensity in units of $10^{16}$ W/cm$^2$ and $\lambda_\mu$ is the laser wavelength in microns. At a lower laser intensity, plasma is produced via impact ionization. Plasma produced locally, with size $r_0$, undergoes ambipolar diffusion on the time scale of $\tau_d = r_0/c_s$, where $c_s = \sqrt{T_e/m_i}$ is the ion sound speed and $m_i$ is the ion mass.

### 1.1.2 Impact ionization

Impact ionization is a process in which a laser, shining on a target, heats the seed electrons (produced via tunneling or some other process). Heating involves collisions between the electrons and atoms or ions, without which the electron current density is $\pi/2$ out of phase with the laser field and there is no time average heating. In a collision, the momentum of an electron is randomized but only a small fraction of the energy (of the order of twice the electron to atom mass ratio) is exchanged. A collision does two things – it provides an in-phase component to the current density with the electric field and raises the thermal energy of electrons. As the kinetic energy of the electrons exceeds the ionization energy of the atoms, the electrons ionize the atoms on colliding with them, at the rate

$$\partial n_e/\partial t = \alpha n_a n_e, \tag{1.2}$$

where $n_a$ is the number density of neutral atoms, $n_e$ is the number density of electrons, and $\alpha$ is the coefficient characterizing the ionizing collisions. The process grows exponentially in time.

The collision cross-section for impact ionization rises with the temperature $T_e$ (in energy units) as $\exp(-I_0/T_e)$, hence, higher the intensity of the laser, higher the heating rate and higher the rate of ionization. In atoms with many electrons, the ionization energy to reach a singly ionized state is the lowest and it increases with the state of ionization.

Not all collisions are ionizing – some are elastic collisions, and some raise the atoms to excited states (called Rydberg states whose orbital radii are proportional to the square of the number of the state). The atoms in excited states stay there for a significant time and have a much higher collision cross-section for ionizing collisions, They play an intermediary role in ionization when the electron energy is less than the ionization energy. In one collision, the atom goes to a Rydberg state and in the next collision it becomes ionized. Sharma et al.[2] have observed up to seven states of ionization of Xe using nano-second pulses of moderate energy, ~$10^9$ W/cm$^2$. Rajeev et al.[3] have employed high cross-section of Rydberg states to induce charge exchange collisions of ions with atoms for the production of high energy atoms.

## 1.2 Electromagnetic and Electrostatic Waves

An unmagnetized plasma supports three principal kinds of waves: electromagnetic wave with the dispersion relation $\omega = \left(\omega_p^2 + k^2c^2\right)^{1/2}$, a low-frequency ion-acoustic wave with $\omega = kc_s/\left(1 + k^2v_{\text{th}}^2/\omega_p^2\right)^{1/2}$, and a high frequency Langmuir wave with $\omega = \left(\omega_p^2 + 3k^2v_{\text{th}}^2/2\right)^{1/2}$, where $\omega$ is the angular wave frequency and $k$ is the wave number. The ion-acoustic and Langmuir waves are Landau damped when their phase speed $\omega/k$ is low enough to interact resonantly with the ions and electrons, respectively. The ion-acoustic wave is strongly Landau damped on ions unless the electron temperature is much higher than the ion temperature; the Langmuir wave becomes strongly Landau damped at short wavelengths due to the resonant interaction with electrons.

The presence of density gradient in the plasma brings about the novel phenomenon of mode conversion. It is especially important for an electromagnetic wave that is obliquely incident onto an inhomogeneous plasma with the electric field component along the density gradient. The wave then drives a density oscillation. At the critical density $n = n_c = m\varepsilon_0\omega^2/e^2$, the driven density oscillation is an electron plasma (Langmuir) wave. Thus, an electromagnetic wave energy can be directly converted to the Langmuir wave through "resonance absorption", and is eventually absorbed by the plasma. This is an example of how modes and resonant mode coupling are important for the absorption of electromagnetic waves in plasma.

## 1.3 Parametric Instabilities

For a large-amplitude electromagnetic wave, the nonlinear coupling with the electrostatic modes, caused by parametric instabilities becomes dominant.[2–12] These couplings can occur in the underdense region as well as near the critical layer. In an underdense plasma, an electromagnetic wave can decay into another electromagnetic wave and an electrostatic wave, that is, a Langmuir wave or the ion-acoustic wave, resulting in stimulated Raman and Brillouin scattering, respectively. Near the critical density the electromagnetic wave can decay into a Langmuir wave and an ion-acoustic wave. The energy of the electromagnetic wave is then diverted to the electrostatic modes, which in turn can heat the particles through Landau damping and collisions. The nonlinear interaction of electromagnetic waves with collective modes in the plasma, therefore, qualitatively changes the nature of its propagation, absorption, and scattering. Plasma inhomogeneity plays an important role in these parametric wave–wave interactions. It limits the region of resonant wave–wave coupling, and enhances the threshold for parametric instabilities.

There exist comprehensive texts treating many aspects of electromagnetic waves in plasmas.[6–8] This monograph aims at reviewing the physical processes of the linear and nonlinear collective interactions of electromagnetic waves with unmagnetized plasmas. We have several specific applications in mind, including laser-driven-fusion and advanced electron and proton accelerators.

## 1.4 Laser Driven Fusion

Laser-driven inertial confinement fusion (ICF) is a major scientific challenge. It aims at heating and compressing a deuterium–tritium (D–T) target using an intense laser light at temperatures above 10 keV and superhigh density. The D–T nuclei undergo an exothermic fusion reaction producing energy. The basic fusion reaction is as follows.

$$ {}_1D^2 + {}_1T^3 \rightarrow {}_2He^4 + {}_0 n^1 + 17.6 \text{ MeV}. \tag{1.3}$$

Deuterium is available in abundance in sea water, in the form of $D_2O$, and can be extracted at reasonable cost. Tritium can be obtained from lithium, available widely in the soil through neutron bombardment.

$$ {}_3Li^6 + {}_0 n^1 \rightarrow {}_2He^4 + {}_1 T^3. \tag{1.4}$$

The Lawson criterion for energy breakeven, that is, when the energy produced via thermonuclear fusion equals the energy employed in plasma production and heating, requires at 10 keV plasma temperature the product of density and plasma confinement time $n\tau > 10^{14}$ cm$^{-3}$ sec. For inertial confinement, $\tau$ is the ambipolar diffusion time $\tau_d = r_0/c_s$, required for the sound wave to traverse a microsphere of radius $r_0$ of a few hundred microns; thus, the densities required are about thousand times higher than a normal solid density. This is achieved by shock compression of spherical D–T targets using the laser itself or by laser induced X-rays. There are two approaches to ICF: direct-drive and indirect-drive.

### 1.4.1 Direct-drive ICF

In direct-drive ICF, the laser deposits its energy in the corona, that is, the underdense region and the critical layer via collisional and resonance absorption. The heat is transported to the ablation layer by thermal conduction where a shock wave is generated. This shock wave compresses the core. The target capsule (3–5 mm in radius) comprises D–T gas at the core, surrounded by cryogenic D–T ice (thickness 160–600 μm) inside a CH shell. Multi-beam laser (e.g., 192 beams in the National Ignition Facility (NIF) at Lawrence Livermore Laboratory) of 0.351 μm wavelength and nanosecond pulse length symmetrically irradiates the spherical capsule. Craxton et al.[13] have in an extensive review article divided the implosion into four phases (cf. Fig. 1.1). At first, the laser is absorbed by the target, leading to the formation of a hot plasma corona and the ablation of the target (at the ablation surface). Suitably tailored laser pulses produce a sequence of shock waves that propagate into the target. After the shock has reached the inner surface, a rarefaction wave moves outward toward the ablation surface, and the shell begins to accelerate inwards towards the target center. The laser intensity increases during the accelerating phase up to ~$10^{15}$ W/cm$^2$. After the main shock wave reflects from the target center and reaches the shell/D–T layer, a deceleration phase begins and the kinetic energy of the shell is converted into thermal energy, leading to that the D–T fuel is compressed and heated.

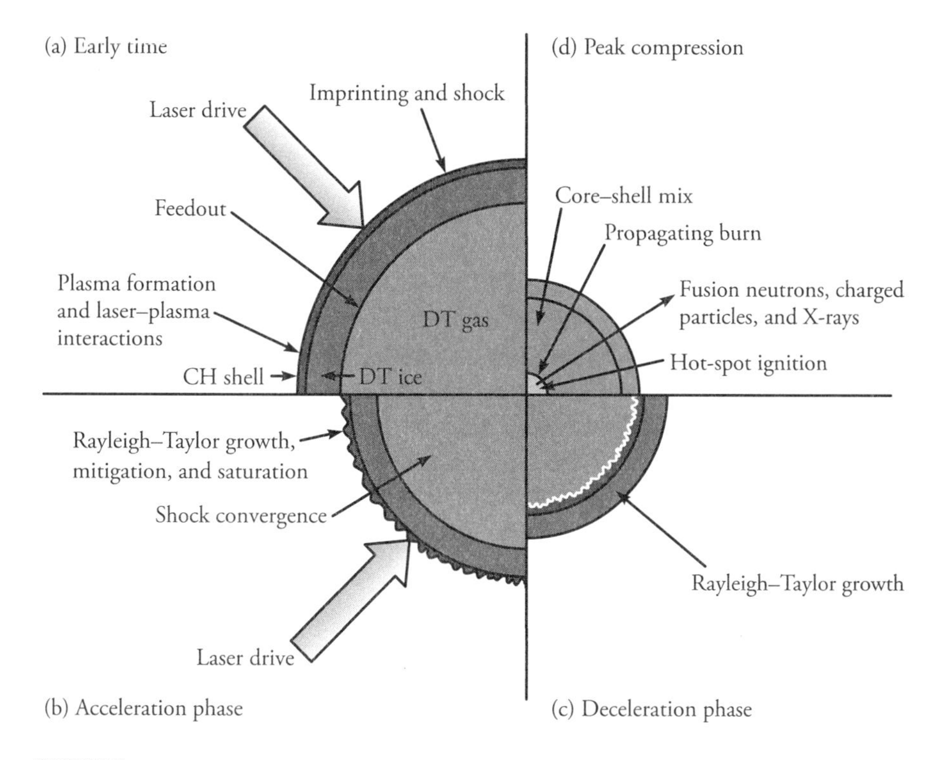


**Fig. 1.1** Different stages of implosion in direct-drive fusion as shown by Craxtom et al.[13]

During the deceleration phase, the Rayleigh–Taylor instability is a serious concern as it may destroy the symmetry and inhibit compression. Peak compression occurs in the final phase when fusion neutrons and X-rays are produced along with $\alpha$ particles. These particles deposit energy in the D–T fuel bringing more compressed fuel to the fusion temperature and leading to a propagating burn (ignition). A 10 keV temperature and a compressed mass density to the radius product of $\rho R = 300$ mg/cm$^2$ are required for ignition.

## 1.4.2 Indirect-drive ICF

In indirect drive ICF, the D–T filled capsule is placed at the center of a cylindrical enclosure (a hohlraum) coated on the interior with a high-Z material , for example, gold (see Fig. 1.2). The inside wall of the hohlraum is irradiated using multi-beam lasers. This produces X-rays which ablate the fuel filled capsule. The lasers are launched into outer cones and inner cones, through the entrances at the ends to maintain an X-ray flux symmetry at the poles and equator of the capsule.[14,15] Typical experiments at NIF utilize hohlraums filled with helium at densities in the range 0.96–1.6 mg/cm$^3$ to minimize the expansion of the interior high–Z gold wall, and help the laser reach the wall for the full pulse duration. The laser turns the gas inside the hohlraum to a low density plasma ($n/n_{cr} \leq 0.1$) with the electron temperature $T_e \sim 3$ keV.

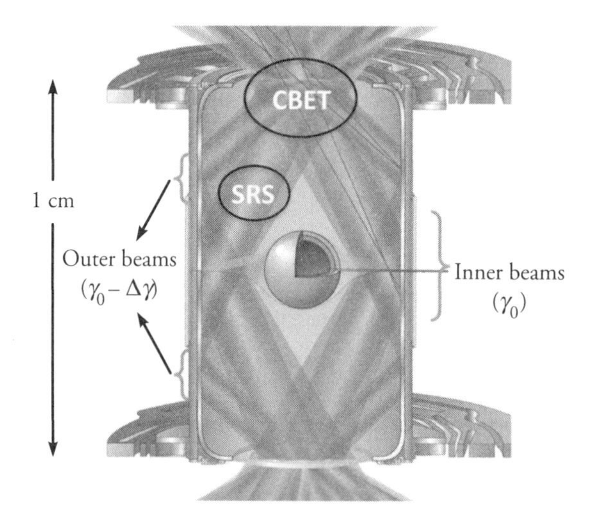


Fig. 1.2 Indirect-drive fusion: Hohlraum carrying a spherical D–T capsule in the center. Multi-laser beams enter through the end entrances, irradiate the gold-coated hohlraum wall as inner and outer beams and produce X-rays that cause implosion of the capsule. Cross-beam energy transfer (CBET) may take place at the entrance. (cf. Montgomery.[14])

The nonlinear interaction of intense laser light with the corona plasma is primarily due to parametric instabilities. Below the critical density, the laser excites large amplitude Langmuir and ion-acoustic waves, through the parametric Raman and Brillouin scattering instabilities.

These waves can produce high-energy electrons and ions, respectively, that can preheat the core and diminish the efficiency of shock compression.

## 1.5 Charged Particle Acceleration

The early work of Rosenbluth and Liu[16] on the beat wave excitation of high amplitude high phase velocity Langmuir waves by two collinear lasers in a plasma and that of Tajima and Dawson[17] on wake-field excitation by a short pulse laser (of pulse duration $\omega_p^{-1}$) initiated extensive studies on electron acceleration by large amplitude plasma waves. In these experiments, a gas jet target is impinged by a short pulse laser. The front of the pulse creates a plasma and the remainder of the pulse drives a wake-field plasma wave. The phase velocity of the plasma wave equals the group velocity of the laser. The plasma wave traps energetic electrons and accelerates them to GeV energy. At higher laser intensity, one obtains an electron evacuated ion bubble in the wake of the laser. In the moving frame of the bubble, electrons travel backward on the periphery of the bubble and surge to the stagnation point. The ion space charge field pulls these electrons into the bubble, accelerating them to GeV energies.

High power laser interaction with thin foil targets has opened up the possibility of proton acceleration via radiation pressure acceleration (RPA). A laser of intensity $I_L$ carries $I_L/c$ momentum per unit area per second. After being reflected from an overdense plasma foil, the laser exerts $2I_L/c$ radiation pressure or ponderomotive force on the plasma electrons. As the electrons move, the ion space charge left behind pulls them back, creating a double layer. The latter is accelerated by the radiation pressure as a whole, only limited by the Rayleigh–Taylor (RT) instability. Particle-in-cell (PIC) simulations with foils of two ion species,[18] reveal that a proton layer can detach from the heavier ions and the RT instability is mainly localized to the heavier ions. Experiments with diamond-like carbon (DLC) foils of 2–5 nm thickness, with adsorbed hydrogen on the rear, have resulted in proton energies of the order of 60 MeV. One considers the possibility of achieving quasi-mono-energy ion beams of 200 MeV energy in the near future. It would be a major breakthrough for cancer therapy.

If thicker foils are used, the electrons in the skin layer are pushed by the ponderomotive force to the rear where they create a high field sheath. The sheath accelerates adsorbed protons via the target normal sheath acceleration (TNSA) process.

## 1.6 Coherent X-rays

Laser produced plasmas are also useful sources of coherent radiation, for example, terahertz generation, harmonic generation, X-ray laser, and gamma ray generation. The mechanisms of X-ray laser and gamma ray generation are based on the same principles as a free electron laser (FEL). In an FEL,[18–23] a relativistic electron beam propagates through a periodic transverse magnetic field called a wiggler. The wiggler appears to the beam as an incoming electromagnetic wave whose stimulated Compton backscattering produces double Doppler-shifted coherent radiation. In this process, the wiggler and the radiation seed signal exert a ponderomotive force on the electrons causing space charge bunching and emission of coherent radiation. The wavelength of the radiation is proportional to the wiggler period and inversely proportional to the square of the energy of the beam. Currently, there is a growing need for employing lasers as wigglers and laser accelerated electrons as the beam that upconverts the laser to hard X-rays and gamma rays. The electrostatic field of the laser produced ion channel also gives rise to wiggle motion of the electrons and can lead to the realization of an X-ray laser.

## 1.7 Outline of the Book

We discuss these processes in subsequent chapters. In Chapter 2, we consider the properties of linear waves in a plasma. With Maxwell's equations and the equations of motion for the plasma as a starting point, we discuss the temporal and spatial dispersion and obtain expressions for the energy density and energy flow for electromagnetic and electrostatic waves in a dispersive medium. We employ Vlasov theory for electrostatic waves whereas for electromagnetic waves, a fluid description of the plasma is considered sufficient. We explain the phenomena of diffraction divergence, dispersion broadening, duct propagation, and anomalous resistivity. In Chapter 3, we study resonance absorption and Brunell absorption for electron heating. In Chapter 4, the laser coupling to surface plasmons is investigated and the phenomenon of surface enhanced Raman scattering is discussed. In Chapter 5 we study the electron response to a large amplitude electromagnetic wave and deduce an expression for the relativistic ponderomotive force for circular and linear polarizations of the laser. In Chapter 6, laser driven electron acceleration due to beat wave and wake-field mechanisms are studied. In Chapter 7, we discuss laser acceleration of quasi mono-energetic ions by radiation pressure acceleration (RPA) and target normal sheath acceleration (TNSA) mechanisms. In Chapter 8, we develop the paraxial ray theory of self-focusing of Gaussian laser beams in collisional and collisionless plasmas. For a plane uniform beam, we study the growth of the filamentation instability. Nonlinearities arise through relativistic mass variation of the electrons, the ponderomotive force, and differential Ohmic heating induced density redistribution. In Chapter 9, we study coherent radiation generation by free electron laser and ion channel X-ray lasers. Parametric instabilities in homogeneous plasma are explained in Chapter 10. We begin with the motion of a simple pendulum whose length is modulated by external means. The motion is governed by Mathieu's equation which we solve using perturbation theory. In the case of a parametric oscillator with two degrees of freedom, we solve two coupled mode equations. For plasmas, we discuss the physics of three- and four-wave parametric processes, deduce the coupled mode equations, and solve them to obtain the growth rate. Chapter 11 deals with parametric instabilities in inhomogeneous plasma. For backscattering processes, we employ the Wentzel–Kramers–Brillouin (WKB) theory that provides a convective amplification factor. For side scattering as well as for parametric instabilities near the quarter critical and critical densities, we use full wave theory to discuss convective and absolute parametric instabilities. In Chapter 12 we derive the nonlinear Schrödinger equation, and obtain steady state solitary solutions. In inhomogeneous plasma we study accelerating solitons, soliton excitation by lasers and the transition to chaos. In Chapter 13 we introduce particle-in-cell (PIC) and Vlasov simulations. In Chapter 14 we give an outline of high-field effects such as quantum electrodynamics (QED), and radiation reaction, and oscillating plasma mirrors.

## References

1. Keldysh, L. V. 1965. "Ionization in the Field of a Strong Electromagnetic Wave." *Sov. Phys. JETP* 20 (5): 1307–1314.
2. Sharma, P., R. K. Vatsa S. K. Kulshreshtha, J. Jha, D. Mathur, and M. Krishnamurthy. 2006. "Energy Pooling in Multiple Ionization and Coulomb Explosion of Clusters by Nanosecond-Long, Megawatt Laser Pulses" *The Journal of Chemical Physics* 125 (3): 034304.
3. Rajeev, R., T. Madhu Trivikram, K. P. M. Rishad, V. Narayanan, E. Krishnakumar, and M. Krishnamurthy. 2013. "A Compact Laser-driven Plasma Accelerator for Megaelectronvolt-energy Neutral Atoms" *Nature Physics* 9 (3): 185–90.
4. C. S. Liu, M. N. Rosenbluth, and R. B. White. 1974. "Raman and Brillouin Scattering of Electromagnetic Waves in Inhomogeneous Plasmas." *The Physics of Fluids* 17 (6): 1211–19.
5. Kruer, W. L. 1988. *The Physics of Laser Plasma Interactions*. Massachusetts: Addison-Wesley.

   Jaroszynski, Dino A., R. A. Bingham, and R. A Cairns (editors). 2017. *Laser-Plasma Interactions,* 1st edition, Scottish Graduate Series. CRC Press..
6. Liu, Chuan Sheng, and V. K. Tripathi. 1986. "Parametric Instabilities in a Magnetized Plasma." *Physics Reports* 130 (3): 143–216.
7. Stix, Thomas H. 1992. *Waves in plasmas*. New York: Springer
8. Bekefi, G. 1966. *Radiation Processes in Plasmas.* New York: Wiley.
9. Ginzburg, V. L. 1970. *The Propagation of Electromagnetic Waves in Plasma.* 2nd edition. Oxford: Pergamon.
10. Nishikawa, Kyoji. 1968. "Parametric Excitation of Coupled Waves. II. Parametric Plasmon-photon Interaction." *Journal of the Physical Society of Japan* 24 (5): 1152–1158.
11. Nishikawa, Kyoji and C. S. Liu. 1976. "General Formalism of Parametric Excitation." In *Advances in Plasma Physics vol.* 6, edited by A. Simon and W. B. Thompson. p. 3; C. S. Liu and P. K. Kaw, "Parametric Instabilities in Homogeneous Unmagnetized Plasmas" *ibid.*, p. 83; C S. Liu, "Parametric Instabilities in Inomogeneous Unmagnetized Plasma" *ibid.*, p. 121. New York: John Wiley.
12. Kadomtsev, Boris B., and V. I. Karpman. 1971. "Nonlinear Waves." *Soviet Physics Uspekhi* 14 (1): 40.
13. Craxton, R. S., K. S. Anderson, T. R. Boehly, V. N. Goncharov, D. R. Harding, J. P. Knauer, R. L. McCrory et al. 2015. "Direct-drive Inertial Confinement Fusion: A Review." *Physics of Plasmas* 22 (11): 110501.
14. Montgomery, David S. 2016. "Two Decades of Progress in Understanding and Control of Laser Plasma Instabilities in Indirect Drive Inertial Fusion." *Physics of Plasmas* 23 (5): 055601.

15. Lindl, John. 1995. "Development of the Indirect-drive Approach to Inertial Confinement Fusion and the Target Physics Basis for Ignition and Gain." *Physics of Plasmas* 2 (11): 3933–4024.
16. Rosenbluth, M. N., and C. S. Liu. 1972. "Excitation of Plasma Waves by Two Laser Beams." *Physical Review Letters* 29 (11): 701.
17. Tajima, Toshiki, and John M. Dawson. 1979. "Laser Electron Accelerator." *Physical Review Letters* 43 (4): 267.
18. Dahiya, Deepak, Ashok Kumar, and V. K. Tripathi. 2015. "Influence of Target Curvature on Ion Acceleration in Radiation Pressure Acceleration Regime." *Laser and Particle Beams* 33 (2): 143–149.
19. Marshall, T. C. 1985. *Free Electron Lasers.* New York: Macmillan.
20. Roberson, C. W. and P. Sprangle. 1989. "A Review of Free-Electron Lasers". *Physics of Fluids B: Plasma Physics* l (3): 10.1063.
21. Chen, Kuan-Ren, J. M. Dawson, A. T. Lin, and T. Katsouleas. 1991. "Unified Theory and Comparative Study of Cyclotron Masers, Ion-channel Lasers, and Free Electron Lasers." *Physics of Fluids B: Plasma Physics* 3 (5): 1270–1278.
22. Flyagin, V. A., A. V. Gaponov, I. Petelin, and V. K. Yulpatov. 1977. "The Gyrotron." *IEEE Transactions on Microwave Theory and Techniques* 25 (6): 514–521.
23. Walsh, J. E. 1980. "Cerenkov and Cerenkov–Raman Radiation Sources in Free-electron Generators of Coherent Radiation." *Phys. of Quant. Electron, vol.* 7, edited by S. F. Jacobs, H. S. Pilloff, M. Sargent III, M. O. Scully and R. Spitzer. Massachusetts: Addison-Wesley p. 255.
24. Sprangle, Phillip. 1976. "Excitation of Electromagnetic Waves from a Rotating Annular Relativistic E-beam." *Journal of Applied Physics* 47 (7): 2935–2940.

# 2 Linear Waves

## 2.1 Introduction

An unmagnetized plasma supports two kinds of waves, electromagnetic and electrostatic waves. The electromagnetic waves have oscillatory electric and magnetic fields associated with them; while the electrostatic waves have only an oscillatory electric field, expressible as the gradient of a scalar potential. At low amplitudes, the responses of electrons and ions to these waves are linear, that is, the density and velocity perturbations in the plasma vary linearly with the wave's electric field, and the wave velocity does not depend on the wave amplitude. In this chapter, we will develop an analytical framework to understand the response of charged particles to waves and study the properties of linear waves.

## 2.2 Maxwell's Equations

Maxwell's equations governing electromagnetic fields in a medium are as follows.[1]

$$
\begin{aligned}
&\nabla \cdot \vec{D} = \rho, \\
&\nabla \cdot \vec{B} = 0, \\
&\nabla \times \vec{E} = -\frac{\partial \vec{B}}{\partial t}, \\
&\nabla \times \vec{H} = \vec{J} + \frac{\partial \vec{D}}{\partial t}, \\
&\vec{D} = \varepsilon_0 \vec{E} + \vec{P}, \\
&\vec{B} = \mu_0 (\vec{H} + \vec{M}),
\end{aligned}
\tag{2.1}
$$

where $\varepsilon_0$ and $\mu_0$ are the free space electric permittivity and magnetic permeability, $\vec{E}$ and $\vec{B}$ are the electric and magnetic fields, $\vec{D}$ and $\vec{H}$ are the electric displacement and auxiliary magnetic field, $\vec{P}$ and $\vec{M}$ are polarization (defined as the electric dipole moment per unit volume caused by the displacement and orientation of the bound electrons of atoms and molecules) and magnetization (defined as the magnetic dipole moment per unit volume caused by the displacement and orientation of bound electrons of atoms and molecules, and $\rho$ and $\vec{J}$ are the charge and current densities due to free electrons and ions, related to each other through the continuity equation

$$\frac{\partial \rho}{\partial t} + \nabla \cdot \vec{J} = 0. \tag{2.2}$$

In a fully ionized plasma, $\vec{P} = \vec{M} = 0$. However, often quite commonly in plasma, $|\vec{P}| << \varepsilon_0 |\vec{E}|$ and $|\vec{M}| << |\vec{H}|$; hence, we can assume that $\vec{D} = \varepsilon_0 \vec{E}$ and $\vec{B} = \mu_0 \vec{H}$. The contributions of the plasma to the electric and magnetic fields appear through $\vec{J}$ and $\rho$, which are obtained by solving the kinetic or fluid equations, in terms of $\vec{E}$ and $\vec{B}$.

## 2.3 Kinetic Equation

The response of charged particles to space- and time-dependent fields depends on the particles' trajectories. Due to the large number of particle in a plasma, it is necessary to adopt a statistical description. A particle distribution function $f(t, \vec{r}, \vec{p})$ is defined as the density of particles of a given species in the six-dimensional phase space $(\vec{r}, \vec{p})$, where $\vec{r}$ is the position and $\vec{p}$ is the momentum. If collisions are ignored, $f$ must satisfy a continuity equation in the six-dimensional phase space,

$$\frac{\partial f}{\partial t} + \frac{\partial}{\partial \vec{r}} \cdot \left( \dot{\vec{r}} f \right) + \frac{\partial}{\partial \vec{p}} \cdot \left( \dot{\vec{p}} f \right) = 0, \tag{2.3}$$

where the over-dots denote total time derivative along trajectories in $(\vec{r}, \vec{p})$ space, given by $\dot{\vec{r}} = \vec{v}$, and $\dot{\vec{p}} = q\left(\vec{E} + \vec{v} \times \vec{B}\right)$ where $q$ is the particle electric charge. We here treat the non-relativistic limit where $\vec{p} = M\vec{v}$ with $M$ being the particle's mass and $\vec{v}$ the velocity. Treating $\vec{p}$ and $\vec{r}$ as independent variables, Eq. (2.3) becomes the Vlasov equation

$$\frac{\partial f}{\partial t} + \vec{v} \cdot \nabla f + q\left(\vec{E} + \vec{v} \times \vec{B}\right) \cdot \frac{\partial f}{\partial \vec{p}} = 0. \tag{2.4}$$

One may note from this equation that $df/dt = 0$, that is, on a phase space trajectory, following a particle, $f$ is a constant. In a non-relativistic plasma, one often defines $f(t, \vec{r}, \vec{v})$ as the particle density in $\vec{r}, \vec{v}$ space, giving the Vlasov equation

$$\frac{\partial f}{\partial t}+\vec{v}\cdot\nabla f+\frac{q}{M}(\vec{E}+\vec{v}\times\vec{B})\cdot\frac{\partial f}{\partial \vec{v}}=0$$

When electron–electron and electron–ion Coulomb collisions are present, the Vlasov equation must be modified to include these Coulomb collisions, leading to the Fokker–Planck equation:

$$\frac{\partial f}{\partial t}+\vec{v}\cdot\nabla f+\frac{q}{M}\left(\vec{E}+\vec{v}\times\vec{B}\right)\cdot\frac{\partial f}{\partial \vec{v}}=\left.\frac{\partial f}{\partial t}\right|_{\text{coll.}}, \tag{2.5}$$

where the right-hand side denotes the rate of change of the distribution function due to the collisions. In this book, we do not evaluate the collision term and refer the reader to standard text-books.[2–7]

We obtain the charge and current densities in terms of $f(t,\vec{r},\vec{v})$ as

$$\rho=\sum q\int f\, d^3\vec{v},$$

$$\vec{J}=\sum q\int \vec{v} f\, d^3\vec{v}, \tag{2.6}$$

where the summation extends over electron and ion species.

## 2.4 Fluid Equations

In many cases, it suffices to adopt a fluid description of the plasma. The equations governing the density, fluid velocity and temperature can be deduced from Eq. (2.5) by taking appropriate moments.

By definition, the density $n$ and average (or drift) velocity $\langle\vec{v}\rangle$ of a species of plasma are related to $f$ through the following relationships.

$$n=\int f\, d^3\vec{v},$$

$$\langle\vec{v}\rangle=\frac{1}{n}\int f\,\vec{v}\, d^3\vec{v}. \tag{2.7}$$

The velocity of the particles can be considered to be composed of two parts: a mean velocity $\langle\vec{v}\rangle$ and a random velocity $\vec{v}'=\vec{v}-\langle\vec{v}\rangle$. The average kinetic energy of random motion may be used to define the temperature $T$ via

$$\frac{3}{2}T=\frac{1}{2n}\int M\left(\vec{v}-\langle\vec{v}\rangle\right)^2 f d^3\vec{v}. \tag{2.8}$$

Hence the temperature is written in energy units. In SI units, $T$ should be replaced by $k_B T$ in Eq. (2.8), where $k_B$ is Boltzmann's constant. The fluid equations governing the density, drift velocity, and temperature can be obtained by multiplying Eq. (2.5) with 1, $M\vec{v}$, and $Mv^2/2$, respectively, and integrating over the velocity. Braginskii[8] derived the non-relativistic fluid equations that incorporate the effect of collisions, as

$$\frac{\partial n}{\partial t} + \nabla \cdot (n\vec{v}) = 0, \tag{2.9}$$

$$M\left(\frac{\partial \vec{v}}{\partial t} + \vec{v} \cdot \nabla \vec{v}\right) = q\left(\vec{E} + \vec{v} \times \vec{B}\right) - \frac{1}{n}\nabla(nT) - \vec{R}, \tag{2.10}$$

$$\frac{3}{2}\left(\frac{\partial T}{\partial t} + \vec{v} \cdot \nabla T\right) + \frac{3}{2}T\nabla \cdot \vec{v} = q\vec{E} \cdot \vec{v} + \frac{1}{n}\nabla \cdot (\chi \nabla T) - Q, \tag{2.11}$$

where $\langle \vec{v} \rangle$ has been written as $\vec{v}$ for the sake of brevity, $\vec{R}$ is the mean momentum lost per second by a particle in collisions with other species, $Q$ is the mean energy lost per second via elastic and inelastic collisions, and $x$ is the thermal conductivity. It is assumed that the velocity distribution function for the random velocity is isotropic.

Equation (2.9) is the continuity equation for the particle number density, ignoring ionization and recombination processes. Equation (2.10) is the momentum equation for the fluid element. It states that the rate of change of average particle momentum, following a fluid element, equals the force due to electromagnetic fields, pressure gradient, and collisional drag. In a fully ionized plasma, the collisional drag on electrons by ions is $\vec{R} = m\nu(\vec{v} - \vec{v}_i)$, where $m$ is the electron mass, $v$ is the electron–ion collision frequency, and $\vec{v}$ and $\vec{v}_i$ are the electron and ion drift velocities.

Equation (2.11) is the energy balance equation. Its first two terms sum up to $(3/2)dT/dt$, representing the rate of change of the particle's thermal energy following a fluid element, $-(3/2)T\nabla \cdot \vec{v}$ denotes the rate of heat generated via fluid compression, $q\vec{E} \cdot \vec{v}$ is the power gained from the electric field, and $-\frac{1}{n}\nabla \cdot (\chi \nabla T)$ is the power loss via thermal conduction.

In the following analysis, we designate electron mass, charge and temperature as $M = m$, $q = -e$, and $T = T_e$, respectively, and the electron density, drift velocity, and thermal velocity as $n$, $\vec{v}$, and $v_{\text{th}} = (T_e/m)^{1/2}$. For ions, the corresponding quantities are $m_i$, $q = Ze$, $T = T_i$, $n_i$, $\vec{v}_i$, and $v_{\text{thi}} = (T_i/m_i)^{1/2}$, where $Z$ is the charge state of the ions. When the dominant collisions of electrons are with ions, the expressions for $Q$ and $\chi$ for the electrons are

$$Q = \frac{2m}{m_i}\nu\frac{3}{2}(T_e - T_i), \tag{2.12}$$

$$\chi = \frac{n v_{\text{th}}^2}{\nu},$$

respectively, where

$$\nu = \nu_{ei} \equiv \left(\frac{2}{\pi}\right)^{1/2} \omega_p \frac{\ln(9N_D / Z)}{9N_D / Z}$$

$$= 2.9 \times 10^{-6} nZ \frac{In\ \Lambda}{T_e^{3/2}} \quad \left(T_e \text{ in eV, } n \text{ in cm}^{-3}\right)$$

is the electron–ion collision frequency,

$$N_D = \frac{4\pi}{3}\lambda_D^3\, n = \frac{4\pi}{3}\frac{v_{\text{th}}^3}{\omega_p^3} n = 1.7 \times 10^9 \left(\frac{T_e^3}{n}\right)^{1/2}$$

is the Plasma parameter describing the number of electrons in a Debye sphere, $\omega_p = (ne^2/m\varepsilon_0)^{1/2}$, is the electron plasma frequency, $\Lambda = 9N_D/Z$, and $\lambda_D = v_{\text{th}}/\omega_p$ is the Debye length. There are corresponding expressions $R_i$ $Q_i$ and $\chi_i$ in the ion fluid equations. Electron–ion collisions are primarily a momentum transfer process. If one passes an electromagnetic wave through a plasma or apply an AC electric field to it, the collisions give rise to a component of the electron velocity in phase with the electric field, leading to a time average power dissipation. The electron–ion collisions are sometimes termed inverse *Bremsstrahlung* as they are counter to *Bremsstrahlung* where electrons passing through the accelerating field of the scattering ions emit radiation. The Rutherford cross-section for electron–ion collisions goes as the inverse square of the electron energy; hence the collision frequency scales as $T_e^{-3/2}$ with the electron temperature $T_e$.

## 2.5 Plasma Response to AC Electric Field

In the limit of weak fields, the current density $\vec{J}$ is a linear function of $\vec{E}$. In a sinusoidal electric field, $\vec{E} = \vec{A}\exp(-i\omega t)$, the linearized momentum equation, Eq. (2.10) for the perturbed electron velocity in a plasma of equilibrium electron density $n_0$, is

$$m\frac{\partial \vec{v}_1}{\partial t} = -e\vec{E} - m\nu\vec{v}_1. \tag{2.13}$$

In the quasi-steady state, one may take the velocity $\vec{v}_1$ to have the same sinusoidal time dependence as the driver field, $\vec{v}_1 = \vec{a}\exp(-i\omega t)$. Substituting this in Eq. (2.13), that is, replacing $\partial/\partial t$ by $-i\omega$, we obtain

$$\vec{v}_1 = -\frac{e\vec{E}}{m(\nu - i\omega)} \tag{2.14}$$

and the current density

$$\vec{J} = -n_0 e \vec{v}_1 = \sigma \vec{E}, \tag{2.15}$$

where

$$\sigma = \frac{n_0 e^2}{m(\nu - i\omega)} \tag{2.16}$$

is the electrical conductivity. When the amplitude of the electric field is a slowly varying function of time (as compared to the phase), one may solve Eq. (2.13) iteratively, treating $a$ to be a slowly varying function of time. Equation (2.13), on substituting $\vec{v}_1 = \vec{a}\exp(-i\omega t)$, takes the form

$$\vec{a} = \frac{e\vec{A}}{m(\nu - i\omega)} - \frac{i}{\omega}\frac{\partial \vec{a}}{\partial t}.$$

Here we ignore $\partial \vec{a} / \partial t$ to obtain $\vec{a} = e\vec{A}/m(\nu - i\omega)$. This value of $\vec{a}$ is then substituted back in $\partial \vec{a} / \partial t$ to obtain

$$\vec{a} = -\frac{e}{m(\nu - i\omega)}\left(\vec{A} + \frac{i}{\omega}\frac{\partial \vec{A}}{\partial t}\right)$$

leading to the current density

$$\vec{J}(t) = \sigma \vec{E} + i\frac{\partial \sigma}{\partial \omega}\frac{\partial \vec{A}}{\partial t}e^{-i\omega t} \tag{2.17}$$

for $v < \omega$. One can recover Eq. (2.16) directly from Eq. (2.15) by writing and Taylor expanding $\sigma(\omega)$ as $\sigma(\omega + i\partial/\partial t) = \sigma(\omega) + i(\partial\sigma/\partial\omega)\partial/\partial t$ and letting $\partial/\partial t$ operate on the amplitude of the field.

In many situations, the conductivity may also have an explicit dependence on the wave number $k$ of the electric field due to thermal effects. In that case, if the amplitude has slow space–time dependence, for example, $\vec{E} = \vec{A}(z,t)e^{-i(\omega t - kz)}$, the current density (on Taylor expanding the conductivity $\sigma(\omega + i\partial/\partial t,\ k - i\partial/\partial z)$) can be written as

$$\vec{J}(t) = \sigma \vec{E} + i\left(\frac{\partial \sigma}{\partial \omega}\frac{\partial \vec{A}}{\partial t} - \frac{\partial \sigma}{\partial k}\frac{\partial \vec{A}}{\partial z}\right)e^{-i(\omega t - kz)}. \tag{2.18}$$

### 2.5.1 Plasma permittivity

For a monochromatic wave having the time dependence of the fields propotional to exp($-i\omega t$), the sum of current densities and displacement current density in Eq. (2.1), on replacing $\partial/\partial t$ by $-i\omega$, can be written as

$$\vec{J} + \frac{\partial \vec{D}}{\partial t} = \sigma \vec{E} - i\omega\varepsilon_0 \vec{E} = -i\omega\varepsilon_0 \varepsilon \vec{E}, \tag{2.19}$$

where the permittivity of the plasma is

$$\varepsilon = 1 + i\sigma / \omega\varepsilon_0 = 1 - \frac{\omega_p^2}{\omega^2 (1 + i\nu / \omega)}. \tag{2.20}$$

One may term $i\sigma/\omega\varepsilon_0$ as the plasma susceptibility $\chi$. From Eq. (2.2), one can write the perturbed charge density $\rho = \nabla \cdot \vec{J}/i\omega$. Thus, the first and fourth Maxwell's equations can be written as

$$\nabla \cdot (\varepsilon_0 \varepsilon \vec{E}) = 0,$$

$$\nabla \times \vec{H} = -\, i\omega\varepsilon_0 \varepsilon \vec{E}. \tag{2.21}$$

These equations resemble the ones for a dielectric medium of relative permittivity $\varepsilon$. In the discussion on electrostatic waves below, it will be useful to express the charge density perturbation in terms of the susceptibility,

$$\rho = \nabla \cdot \vec{J}/i\omega = -\varepsilon_0 \chi \nabla \cdot \vec{E}. \tag{2.22}$$

### 2.5.2 Wave equation

The wave equation governing the propagation of electromagnetic and electrostatic waves in plasma can be obtained by taking the curl of the third Maxwell's equation and using the fourth equation,

$$\nabla^2 \vec{E} - \nabla\left(\nabla \cdot \vec{E}\right) = \frac{1}{c^2 \varepsilon_0} \frac{\partial \vec{J}}{\partial t} + \frac{1}{c^2} \frac{\partial^2 \vec{E}}{\partial t^2}. \tag{2.23}$$

where $c = 1/\sqrt{\varepsilon_0 \mu_0}$ is the speed of light in vacuum. For monochromatic waves of frequency $\omega$, using Eq. (2.20), Eq. (2.23) can be written as

$$\nabla^2 \vec{E} - \nabla\left(\nabla \cdot \vec{E}\right) + \frac{\omega^2}{c^2} \varepsilon \vec{E} = 0. \tag{2.24}$$

For a plane wave, $\vec{E} = \vec{A}\exp[-i(\omega t - \vec{k}\cdot\vec{r})]$, the wave equation, on replacing $\nabla$ by $i\vec{k}$, gives

$$\left(k^2 - \frac{\omega^2\varepsilon}{c^2}\right)\vec{E} = \vec{k}\left(\vec{k}\cdot\vec{E}\right), \tag{2.25}$$

The scalar product of Eq. (2.25) with $\vec{k}$ gives

$$\varepsilon\vec{k}\cdot\vec{E} = 0. \tag{2.26}$$

There are two possibilities of satisfying Eq. (2.26): i) $\vec{k}\cdot\vec{E} = 0$ and ii) $\varepsilon = 0$, giving electromagnetic and electrostatic waves, respectively.

## 2.6 Electromagnetic Waves

For electromagnetic waves, with $\vec{k}\cdot\vec{E} = 0$, Eq. (2.25) gives the dispersion relation

$$k^2 = \frac{\omega^2}{c^2}\varepsilon = \frac{\omega^2}{c^2}\left[1 - \frac{\omega_p^2}{\omega^2 + \nu^2}(1 - i\nu/\omega)\right]. \tag{2.27}$$

The magnetic field of the wave (from the third Maxwell's equation) is

$$\vec{B} = \vec{k}\times\vec{E}/\omega. \tag{2.28}$$

One may notice the transverse nature of the electromagnetic waves as $\vec{E}$, $\vec{B}$ and $\vec{k}$ are mutually perpendicular to each other. In the absence of collisions, Eq. (2.27) can be written as

$$\omega^2 = \omega_p^2 + k^2c^2. \tag{2.29}$$

The wave number $k$ is real only when $\omega > \omega_p$ (cf. Fig. 2.1); it vanishes at the critical density

$$n_0 = n_{cr} = \frac{m\omega^2\varepsilon_0}{e^2}. \tag{2.30}$$

When $\omega < \omega_p$ (or $n_0 < n_{cr}$), $k$ is purely imaginary. Hence, $\vec{B}$ is $\pi/2$ out of phase with $\vec{E}$ (cf. Eq. (2.27)) and we shall see later that the time-averaged Poynting's vector $\vec{S}_{av} = 0$, that is, there is no average energy flux. The penetration depth of the field into the plasma is $c/\left(\omega_p^2 - \omega^2\right)^{1/2}$. When $\omega > \omega_p$, the refractive index of the plasma is

$$\eta = \left(1 - \omega_p^2/\omega^2\right)^{1/2} \tag{2.31}$$

and the product of phase and group velocities is $v_{ph}v_g = c^2$.

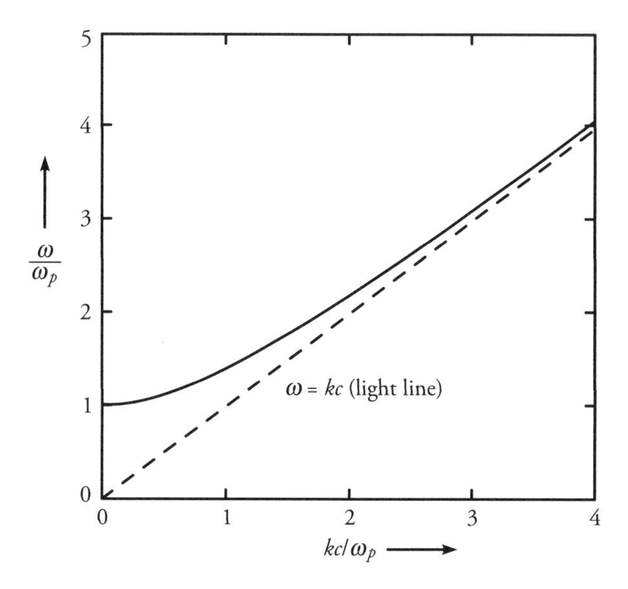


**Fig. 2.1** Electromagnetic wave dispersion relation.

The collisions introduce a component of the current density in phase with the electric field, causing power dissipation and the attenuation of waves. When $\omega > \omega_p >> \nu$, Eq. (2.27) can be simplified to obtain $k = k_r + ik_i$, with

$$k_r = \frac{\omega}{c}\left(1 - \frac{\omega_p^2}{\omega^2}\right)^{1/2},$$

and

$$k_i = \frac{\nu}{2c}\frac{\omega_p^2/\omega^2}{(1-\omega_p^2/\omega^2)^{1/2}}. \tag{2.32}$$

$k_i$ is known as the attenuation constant. The absorption efficiency of a high-frequency wave is a function of $n_0\nu/n_{cr}$. In an inhomogeneous laser-produced plasma, a laser of higher frequency penetrates deeper and to higher densities; and therefore it has a higher absorption efficiency. Furthermore, slower electrons, being more collisional, are heated preferentially via collisional absorption, unlike resonance absorption which produces high-energy electrons. Thus, a short wavelength laser is preferred for laser-driven fusion to avoid high-energy electrons. Experiments have reported absorption efficiency of $\geq 80\%$ for 0.35 μm light at $10^{15}$ W/cm$^2$ and moderate scale lengths.

## 2.7 Electrostatic Waves

For $\vec{k}\cdot\vec{E}\neq 0$, Eq. (2.26) gives

$$\varepsilon = 0 \tag{2.33}$$

which used in Eq. (2.25) yields $\vec{E}\parallel\vec{k}$, giving $\vec{B}=\vec{k}\times\vec{E}/\omega=0$ and $\rho=(\sigma/\omega)\vec{k}\cdot\vec{E}\neq 0$ (cf. Eq. (2.2)). This represents purely electrostatic (space charge) waves with $\vec{E}$ expressible in terms of a scalar potential $\phi$ as

$$\vec{E} = -\nabla\phi. \tag{2.34}$$

The electrostatic waves are characterized by the compression and rarefaction of charged particles. A key characteristic of electrostatic waves which follows from Eq. (2.34), is that the waves have longitudinal polarization where the electric field is parallel to the wave vector, $\vec{E}=-i\vec{k}\phi$.

### 2.7.1 Cold fluid approximation

If one uses Eq. (2.20) for the effective plasma permittivity, $\varepsilon$, Eq. (2.33) gives the frequency of these waves as

$$\omega \approx \omega_p - i\nu/2, \tag{2.35}$$

which is independent of $\vec{k}$ because of the cold fluid approximation. For a warm plasma, one must keep the pressure term in the momentum equation; this changes the picture dramatically.

### 2.7.2 Warm fluid

For an electrostatic wave with the potential

$$\phi = Ae^{-i(\omega t-\vec{k}\cdot\vec{x})}, \tag{2.36}$$

the momentum and continuity equations for the perturbed velocity and density of the electrons, on linearization and using $\partial/\partial t=-i\omega, \nabla = i\vec{k}$ and the adiabatic approximation (pressure divided by density cube $P/n^3$ = constant), yield

$$(\nu - i\omega)\vec{v}_1 = \frac{e}{m}i\vec{k}\phi - \frac{3v_{\text{th}}^2}{n_0}i\vec{k}n_1, \tag{2.37}$$

and

$$-i\omega n_1 + i\vec{k}\cdot(n_0\vec{v}_1) = 0, \tag{2.38}$$

respectively, giving

$$n_1 = -\frac{n_0 e k^2 \phi}{m(\omega^2 - 3k^2 v_{\text{th}}^2 + i\nu\omega)}. \tag{2.39}$$

Using Eq. (2.22), one may write $n_1$ in terms of the electron susceptibility $\chi_e$, as

$$n_1 = \varepsilon_0 \chi_e k^2 \phi / e, \tag{2.40}$$

with

$$\chi_e = -\frac{\omega_p^2}{\omega^2 - 3k^2 v_{\text{th}}^2 + i\nu\omega}. \tag{2.41}$$

Using the electron density perturbation in Poisson's equation,

$$\nabla^2 \phi = e(n - n_i)/\varepsilon_0 \tag{2.42}$$

and ignoring the ion motion due to their heavy mass, one obtains the dispersion relation for the Langmuir wave,

$$\varepsilon = 1 + \chi_e = 0,$$

or $$\omega^2 + i\nu\omega = \omega_p^2 + 3k^2 v_{\text{th}}^2. \tag{2.43}$$

If one neglects collisions (as $\nu << \omega_p$), one obtains the Bohm-Gross dispersion relation.[9]

$$\omega^2 = \omega_p^2 + 3k^2 v_{\text{th}}^2. \tag{2.44}$$

The frequency of the Langmuir wave is higher than the plasma frequency. The thermal effects provide a finite group velocity to the Langmuir wave.

The motions of the electrons and ions also give rise to a low-frequency electrostatic wave called the ion-acoustic wave, with the phase velocity of the ion-acoustic wave much smaller than the electron thermal velocity but greater than the ion thermal speed $v_{\text{ti}} = \sqrt{T_i/m_i}$. The electron density perturbation due to the ion-acoustic wave, from Eq. (2.39), for $\omega^2 << k^2 v_{\text{th}}^2$, is

$$n_1 = n_0 \frac{e\phi}{T_e} \tag{2.45}$$

For singly ionized ions, Eq. (2.39), on replacing the electron quantities by ion quantities and assuming the ion thermal velocity to be smaller than the phase velocity of the wave, gives

$$n_{1i} = -\varepsilon_0 \chi_i k^2 \phi / e = \frac{n_0 e k^2 \phi}{m_i \omega^2},$$

where

$$\chi_i = -\frac{\omega_{\mathrm{pi}}^2}{\omega^2}, \tag{2.46}$$

is the ion susceptibility. Using $n_1$ and $n_{1i}$ in Eq. (2.42), we obtain the dispersion relation,

$$\varepsilon = 1 + \chi_e + \chi_i = 1 + \frac{\omega_{\mathrm{pi}}^2}{k^2 c_s^2} - \frac{\omega_{\mathrm{pi}}^2}{\omega^2} = 0$$

or
$$\omega = \frac{k c_s}{\sqrt{1 + k^2 c_s^2 / \omega_{\mathrm{pi}}^2}}, \tag{2.47}$$

where $\omega_{\mathrm{pi}} = (n_0 e^2 / m_i \varepsilon_0)^{1/2}$ is the ion plasma frequency and $c_s = \sqrt{T_e / m_i}$ is the ion-acoustic speed. At small wave numbers, the frequency of the wave increases linearly with the wave number. At large wave numbers the ion-acoustic wave asymptotically approaches the ion plasma frequency with diminishing phase velocity.

## 2.8 Kinetic Theory of Electrostatic Waves

In a hot plasma, the electrons and ions move with large random velocities. To these particles, the frequency of a wave is Doppler shifted, $\omega - \vec{k} \cdot \vec{v}$, and is different for different particles. Particles with $\omega - \vec{k} \cdot \vec{v} \approx 0$ are in Cherenkov resonance with the electrostatic wave and can be efficiently accelerated or decelerated by the wave.

Let us consider an electrostatic wave $\vec{E} = -\hat{x} A \sin(\omega t - kx)$. In a frame moving with the phase velocity of the wave, we have $\vec{E} = \hat{x} A \sin kx$, the $x$-component of the electron velocity is $\Delta = v_x - \omega/k$. For near-resonant electrons, $\Delta << \omega/k$. We divide the electrons into two groups: (A) $\Delta > 0$ and (B) $\Delta < 0$. Initially, electrons of both groups are uniformly distributed along the

$x$-axis. Group A electrons ($v_x > \omega/k$) are accelerated in the accelerating zones (cf. Fig. 2.2) to move faster into the decelerating zones, and retarded in the decelerating zones to spend more time there. Thus, there is a bunching of Group A electrons in the retarding zones, transferring energy to the wave. The slower-moving electrons ($v_x < \omega/k$) tend to bunch in the accelerating zones, gaining energy from the wave. The relative population of electrons with $v_x > \omega/k$ and $v_x < \omega/k$, that is, the slope of the particle velocity distribution function at $v_x = \omega/k$ decides the net damping or growth of the wave.

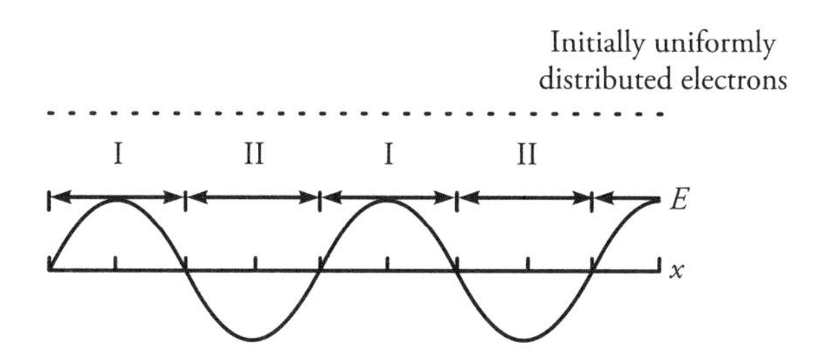


**Fig. 2.2** The electric field of an electrostatic wave in a frame moving with the phase velocity. Regions I refer to decelerating zones while II refer to accelerating zones.

To properly treat the induced emission and absorption of electrostatic waves due to Cherenkov resonance, we use the Vlasov equation.[4,6] Let the electrostatic potential of the wave be as given by Eq. (2.36). We expand the distribution function as $f = f_0 + f_1$, where $f_0$ is its equilibrium value and $f_1$ is a small-amplitude perturbation, to obtain the linearized Vlasov equation

$$\frac{\partial f_1}{\partial t} + \vec{v}.\nabla f_1 \equiv \frac{df_1}{dt} = -\frac{e}{m}\nabla\phi.\frac{\partial f_0}{\partial \vec{v}} \tag{2.48}$$

Solving for $f_1$ gives

$$f_1 = -\frac{e}{m}\int_{-\infty}^{t}\left(\nabla\phi\cdot\frac{\partial f_0}{\partial \vec{v}}\right)_{t'} dt'$$

$$= -\frac{e}{m}ik\frac{\partial f_0}{\partial v_x}A\int_{-\infty}^{t} e^{-i(\omega t' - kx(t'))}dt', \tag{2.49}$$

where the integration is over the unperturbed trajectory, $x(t') = x(t) + v_x(t' - t)$. The integrand depends on the space–time history of the particle giving rise to spatial and temporal dispersion. Equation (2.49) can be simplified to give

$$f_1 = -\frac{e\phi k \frac{\partial f_0}{\partial v_x}}{m(\omega - kv_x)}. \tag{2.50}$$

The density perturbation is obtained as

$$n_l = \int f_1 d^3v. \tag{2.51}$$

For resonant particles, the Doppler-shifted frequency $\omega - kv_x$ vanishes and $f_1$ attains a large value. One must be careful in carrying out the $v_x$ integration and properly account for the contribution from the singularity at $v_x = \omega/k$. Landau[10] treated the evolution of $f_1$ and $\phi$ as an initial value problem. He used a Laplace Fourier transform in time and carried out the $v_x$, integral by prescribing a contour that runs along the Re($v_x$) axis with a small semicircle around the pole assuming that $\omega$ has a small positive imaginary part, to write

$$\frac{1}{v_x - \omega / k} = P\left(\frac{1}{v_x - \omega / k}\right) + i\pi\delta(v_x - \omega / k), \tag{2.52}$$

where $P$ denotes the principal value. The singularity has great significance in physics as it represents the wave–particle interaction through the Cherenkov resonance. The Langmuir waves and the resonant electrons with velocity near the wave phase velocity can exchange energy, leading to the damping or growth of the waves, depending on the sign of the derivative of the distribution function at the resonant velocity. For a Maxwellian distribution function,

$$f_0 = n_0(\pi / 2)^{-3/2} v_{\text{th}}^{-3} \exp\left(-v^2 / 2v_{\text{th}}^2\right), \tag{2.53}$$

following Landau's prescription,[3,5] the perturbed electron density turns out to be

$$n_1 = \frac{\varepsilon_0 k^2}{e} \chi_e \phi, \tag{2.54}$$

where the electron susceptibility is

$$\chi_e = \frac{\omega_p^2}{k^2 v_{\text{th}}^2}\left[1 + \frac{\omega}{\sqrt{2} k v_{\text{th}}} Z\left(\frac{\omega}{\sqrt{2} k v_{\text{th}}}\right)\right], \tag{2.55}$$

where the plasma dispersion function[11] is

$$Z(\xi) = \frac{1}{\sqrt{\pi}} \int_{-\infty}^{\infty} \frac{e^{-x^2}}{x - \xi} dx, \quad \mathrm{Im}(\xi) > 0$$

An alternative form of the plasma dispersion function, valid for all $\xi$, is

$$Z(\xi) = e^{-\xi^2} \left( i\sqrt{\pi} - 2\int_0^{\xi} e^{x^2} dx \right).$$

The plasma dispersion function $Z(\xi)$ in the limiting cases of $\xi << 1$ and $\xi >> 1$ can be expressed as the Taylor expansion

$$Z(\xi) \approx -2\xi + .... + i\sqrt{\pi} e^{-\xi^2} \quad \text{for } \xi << 1,$$

and asymptotic expansion

$$Z(\xi) = -\frac{1}{\xi} - \frac{1}{2\xi^3} - \frac{3}{4\xi^5} - ... + i\sqrt{\pi} e^{-\xi^2} \quad \text{for } \xi >> 1. \tag{2.56}$$

Similarly, the ion response can be written as

$$n_{1i} = -\frac{\varepsilon_0 k^2}{Z_i e} \chi_i \phi,$$

where the ion susceptibility is

$$\chi_i = \frac{\omega_{\mathrm{pi}}^2}{k^2 v_{\mathrm{thi}}^2} \left[ 1 + \frac{\omega}{\sqrt{2} k v_{\mathrm{thi}}} Z\left( \frac{\omega}{\sqrt{2} k v_{\mathrm{thi}}} \right) \right]. \tag{2.57}$$

Using the expressions for $n_1$ and $n_{1i}$ in the Poisson's equation,

$$\nabla^2 \phi = e\,(n_1 - Z_i n_{1i}) / \varepsilon_0,$$

one obtains the dispersion relation for the space charge modes

$$\varepsilon = 1 + \chi_i + \chi_e = 0. \tag{2.58}$$

Equation (2.58) has two distinct roots.

### 2.8.1 Langmuir wave

The Langmuir wave is a high-frequency electrostatic wave with $\omega > kv_{th}, kv_{thi}$, that is, for which the distance traveled by electrons in a wave period is much shorter than the wavelength. In this limit, the ion susceptibility is $m/m_i$ times smaller than the electron susceptibility and Eq. (2.58), on writing $\omega = \omega_r - i\Gamma_d$ and using the expansion (2.56) for the electron susceptibility yields

$$\omega_r^2 = \omega_p^2 + 3k^2 v_{th}^2,$$

$$\Gamma_d = \sqrt{\pi}\frac{\omega_p^3}{2^{3/2}k^3 v_{th}^3}\omega_p \exp\left(\frac{-\omega_r^2}{2k^2 v_{th}^2}\right). \tag{2.59}$$

The frequency $\omega_r$ is greater than $\omega_p$ as in Eq. (2.44), and the factor 3 is inherent in the Vlasov theory. At short wavelengths as $k$ approaches $\omega_p/v_{th}$, the Langmuir wave is strongly Landau damped (cf. Fig. 2.3).

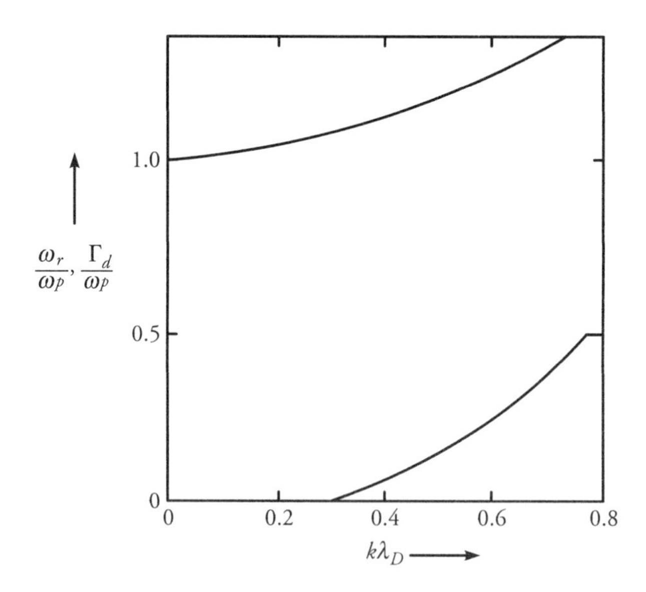


Fig. 2.3 Frequency (upper curve) and damping rate (lower curve) for Langmuir waves as a function of wave number, $\lambda_D = v_{th}/\omega_p$.

### 2.8.2 Ion-acoustic wave

In the limit $kv_{thi} << \omega << kv_{th}$, Eq. (2.55) with appropriate expansions yields the ion-acoustic wave: $\omega = \omega_r - i\Gamma_d$, with

$$\omega_r = \frac{kc_s}{\left(1 + k^2c_s^2 / \omega_{\text{pi}}^2\right)^{1/2}}$$

and

$$\Gamma_d = \sqrt{\frac{\pi}{8}\frac{\omega^3}{k^2c_s^2}}\left[\frac{\omega}{kv_{\text{th}}} + \frac{Z_iT_e}{T_i}\frac{\omega}{kv_{\text{thi}}}e^{-\omega^2/2k^2v_{\text{thi}}^2}\right], \tag{2.60}$$

where $c_s^2 = Z_iT_e / m_i$. To evaluate $\Gamma_d$, one can use $\omega \approx \omega_r$ in Eq. (2.60). In the expression for $\Gamma_d$ in Eq. (2.60), the first term in the bracket is due to resonant electrons while the second term is due to ions. The electron and ion contributions are comparable when $T_i/Z_iT_e \approx 0.1$ (see Ref. 12). The maximum value of $\omega_r$ is $\omega_{\text{pi}}$, the ion plasma frequency. However, at large values of $k$, when $k \to \omega/v_{\text{thi}}$, the ion Landau damping is severe. Weak damping of these waves is possible only when $Z_iT_e >> T_i$ and $k^2c_s^2 << \omega_{\text{pi}}^2$. It may be noted from Eq. (2.54) that for ion-acoustic waves, the electron response $n_1 \approx n_0e\phi/T_e$ follows the Boltzmann's law. Figure 2.4 shows the dispersion curve and the damping rate of an ion-acoustic wave. One may recall that the damping of waves is caused by the pole of $f_1$ that is, the particles having $\omega \approx \vec{k} \cdot \vec{v}$. When $\partial f_0/\partial v_x$ at $v_x = \omega/k$ is negative, that is, there are more particles traveling slower than the wave than those traveling faster than the wave, the wave becomes damped because the larger number of slower particles absorb energy from the wave and fewer faster particles give energy to the wave. The accuracy of the asymptotic formulas for electron and ion Landau damping are discussed in Ref. 12.

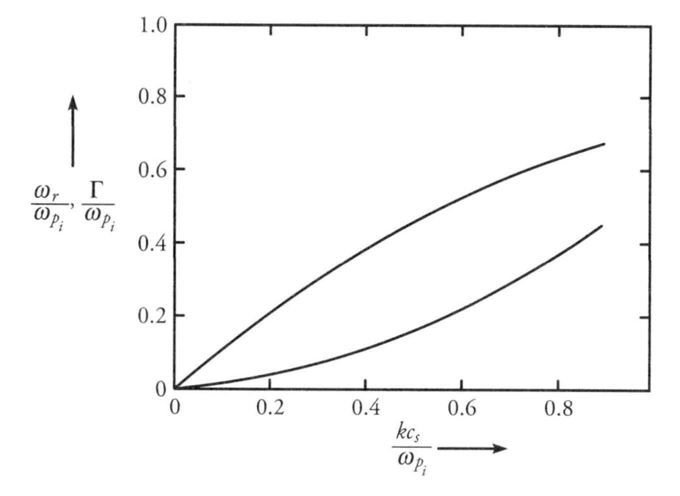


**Fig. 2.4** Frequency (upper curve) and damping rate (lower curve) of ion-acoustic waves as a function of wave number.

## 2.9 Energy Density and Energy Flow in Dispersive Media

Let us return to Eq. (2.1) to investigate the energy density of fields and the energy flow associated with a wave. Carrying out a scalar product between the fourth Maxwell's equation by $\vec{E}$ and using vector identity $\nabla.(\vec{E}\times\vec{H}) = \vec{H}.\nabla\times\vec{E} - \vec{E}.\nabla\times\vec{H}$, we obtain

$$\vec{J}\cdot\vec{E} = -\vec{E}\cdot\frac{\partial\vec{D}}{\partial t} - \vec{H}\cdot\frac{\partial\vec{B}}{\partial t} - \nabla\cdot(\vec{E}\times\vec{H}). \tag{2.61}$$

In a non-dispersive medium, where $\vec{J}$ and $\vec{D}$ have an instantaneous relationship with $\vec{E}$, as $\vec{B}$ has with $\vec{H}$, the energy densities of the electric and magnetic fields are given by $W_E = \vec{E}\cdot\vec{D}/2$ and $W_B = \vec{H}\cdot\vec{B}/2$. Equation (2.61) has a simple interpretation: The power dissipation per unit volume in the left-hand side of the equation equals the rate at which field energy decreases plus the power flux that enters the unit volume (the last term). The power flow density, called Poynting's vector, is

$$\vec{S} = \vec{E}\times\vec{H}. \tag{2.62}$$

For a plane electromagnetic wave, using Eq. (2.28) and that the average power flux

$$\vec{S}_{\text{av}} = \frac{1}{2}\text{Re}\left(\vec{E}\times\vec{H}^*\right) \tag{2.63}$$

where * denotes the complex conjugate and Re the real part of the quantity, the time-averaged Poynting's vector can be written as

$$\vec{S}_{\text{av}} = \frac{|\vec{E}|^2}{2\mu_0\omega}\text{Re}(\vec{k}). \tag{2.64}$$

The intensity of the wave is $|\vec{S}_{\text{av}}|$. In an over-dense plasma, when $\vec{k}$ is imaginary, there is no energy flux. For electron densities below the critical density, the electromagnetic waves carry power. The intensity is proportional to the refractive index and to the amplitude square of the wave. Thus for a given intensity, the amplitude of the laser is proportional to the inverse square root of the refractive index, that is, the laser amplitude goes up as the plasma density increases.

It is useful to express the electron oscillatory velocity in terms of average power flow density $S_{\text{av}}$:

$$\left|\frac{v_1}{c}\right| = \frac{e}{m\omega}\left(\frac{2\mu_0 S_{\text{av}}}{\eta c}\right)^{1/2} \approx 8\times10^{-3}\lambda_\mu\frac{S_{14}^{1/2}}{\eta^{1/2}}, \tag{2.65}$$

where $S_{14}$ is the $S_{av}$ expressed in units of $10^{14}$ W/cm², $\lambda_\mu$ is the free space wavelength in micrometers. Thus, for an Nd-glass laser with 1.06 μm wavelength and $3 \times 10^{16}$ W/cm² intensity in an underdense plasma $\eta \sim 1$, one has $v_1/c \sim 0.1$ and therefore, the relativistic effects become important.

For electrostatic waves, the magnetic field fluctuations and Poynting flux vanish in Eq. (2.61), and one needs to re-interpret $\vec{J}\cdot\vec{E}$. For sinusoidal fields, $\vec{J}$ is almost $\pi/2$ out of phase with respect to $\vec{E}$ and $\vec{J}\cdot\vec{E}$ does not represent power dissipation. It is instead the rate of energy stored in the oscillatory motion of electrons. To determine this quantity, let us consider a field with slowly varying amplitude. In this case, $\vec{J}$ is given by Eq. (2.17). In general, $\sigma = \sigma_r + i\sigma_i$; for high frequency fields, $\sigma_r << \sigma_i$. Using Eq. (2.17), $\vec{J}\cdot\vec{E}$ averaged over a time period $2\pi/\omega$ is,

$$(\vec{J}\cdot\vec{E})_{av} = \frac{1}{2}\mathrm{Re}\left(\vec{J}\cdot\vec{E}^*\right) = \frac{1}{2}\sigma_r A^2 - \frac{1}{4}\frac{\partial \sigma_i}{\partial \omega}\frac{\partial A^2}{\partial t}. \tag{2.66}$$

We can rewrite Eq. (2.61) (without magnetic field fluctuations) as

$$\frac{1}{2}\sigma_r E_0^2 = -\frac{\partial}{\partial t}\overline{W}_E \tag{2.67}$$

where

$$\overline{W}_E = \frac{\varepsilon_0}{2}A^2\left(1 - \frac{1}{\varepsilon_0}\frac{\partial \sigma_i}{\partial \omega}\right) = \frac{1}{4}\varepsilon_0 A^2 \frac{\partial(\omega\varepsilon_r)}{\partial \omega} \tag{2.68}$$

is the time-averaged energy density of the electric field plus the energy density of the particles' motion. $\varepsilon_r$ is the real part of $\varepsilon$. Here we have used Eq. (2.20) to define $\varepsilon$. One may note that $\overline{W}_E$ is the same as the electric field energy density in a dispersive dielectric of permittivity $\varepsilon_r$.

In electrostatic waves, spatial dispersion is important. Consider a simple case

$$\vec{E} = \hat{z}A(z,t)e^{-i(\omega t - kz)}$$

$$\vec{H} = 0. \tag{2.69}$$

In this case, $\vec{J}$ is given by Eq. (2.18) and Eq. (2.67) takes the form

$$\frac{1}{2}\sigma_r A^2 = -\frac{\partial \overline{W}_E}{\partial t} + \frac{\varepsilon_0}{4}\frac{\partial A^2}{\partial z}\omega\frac{\partial \varepsilon_r}{\partial k}. \tag{2.70}$$

As we have seen earlier, $\varepsilon_r(\omega, k) \approx 0$ for electrostatic waves; hence, one can write

$$\frac{\partial \varepsilon_r}{\partial k}\Delta k + \frac{\partial \varepsilon_r}{\partial \omega}\Delta\omega = 0.$$

or

$$\frac{\partial \varepsilon_r}{\partial k} = -\frac{\partial \varepsilon_r}{\partial \omega} v_g, \tag{2.71}$$

where $v_g = \Delta\omega/\Delta k$ is the group velocity of the wave. Then, the last term of Eq. (2.70), representing net power flux entering the unit volume, can be written as $-\partial \bar{S} / \partial z$, with

$$\bar{S} = \frac{\varepsilon_0}{4}\omega\frac{\partial \varepsilon_r}{\partial \omega}A^2 v_g, \tag{2.72}$$

as the power flow density. One may recognize that $\bar{S}$ is the product of the time-averaged energy density (of the field and the particles' quiver motion) and the group velocity. Energy transport by plasma waves can be important due to the long mean-free path of the plasma waves.[13]

## 2.10 Diffraction Divergence

Now let us consider the propagation of an electromagnetic wave of finite transverse extent through a plasma and study the effect of its diffraction. According to Huygen's principle, every point of a wave front is a source of secondary wavelets, and consequently, a wave front of finite extent, or nonuniform illumination, undergoes diffraction divergence. To have a quantitative estimate of this effect, we consider a cylindrically symmetric electromagnetic beam propagating along $z$ in an uniform plasma:

$$\vec{E} = \hat{x}A(r,z)e^{-i(\omega t - kz)}, \tag{2.73}$$

where $r = \sqrt{x^2 + y^2}$ and

$$k = \frac{\omega}{c}\left(1 - \frac{\omega_p^2}{\omega^2}\right)^{1/2}.$$

We assume a Gaussian shape of the beam at the waist,

$$A(r,0) = A_{00}e^{-r^2/2r_0^2},$$

with a slow amplitude variation along the $z$-axis, such that

$$\frac{\partial A}{\partial z} << kA. \tag{2.74}$$

Using Eq. (2.73) in the wave equation (Eq. (2.24) with $\nabla \cdot \vec{E} = 0$), one obtains, in the Wentzel–Kramers–Brillouin (WKB) approximation,

$$2ik\frac{\partial A}{\partial z} + \nabla_\perp^2 A = 0, \tag{2.75}$$

where in cylindrical coordinates $\nabla_\perp^2 = \partial^2 / \partial r^2 + (1/r)\partial / \partial r$.

Using the method proposed by Akhmanov et al.,[14] we introduce an eikonal representation $A = A_0(r,z)e^{iS(r,z)}$ and separate the real and imaginary parts of Eq. (2.75),

$$k\frac{\partial A_0^2}{\partial z} + \frac{\partial S}{\partial r}\frac{\partial A_0^2}{\partial r} + A_0^2 \nabla_\perp^2 S = 0, \tag{2.76a}$$

$$2k\frac{\partial S}{\partial z} + \left(\frac{\partial S}{\partial r}\right)^2 = \frac{1}{A_0}\nabla_\perp^2 A_0, \tag{2.76b}$$

In the paraxial ray approximation, we expand $S$ in powers of $r^2$ (odd power terms vanish),

$$S = S_0(z) + S_2(z)r^2/2. \tag{2.77}$$

Then Eq. (2.76a) takes the form

$$k\frac{\partial A_0^2}{\partial z} + S_2 r\frac{\partial A_0^2}{\partial r} + 2A_0^2 S_2 = 0. \tag{2.78}$$

Introducing a function $f(z)$ such that

$$S_2 = \frac{k}{f}\frac{df}{dz}, \tag{2.79}$$

Eq. (2.78) reduces to

$$\frac{\partial}{\partial z}\left(A_0^2 f^2\right) + \frac{r}{f}\frac{df}{dz}\frac{\partial}{\partial r}\left(A_0^2 f^2\right) = 0,$$

giving the general solution

$$A_0^2 f^2 = F(r/f).$$

where $F$ is an arbitrary function of $r/f(z)$.

At $z = 0$, $A_0^2$ is Gaussian (cf. Eq. 2.74); and hence, we have

$$F(r/f) = A_{00}^2 e^{-r^2/r_0^2 f^2},$$

and

$$A_0^2 = \frac{A_{00}^2}{f^2} e^{-r^2/r_0^2 f^2(z)} \tag{2.80}$$

which conserves the power flux, as $\int A_0^2 r dr$ is independent of $z$. Substituting for $A_0$ and $S$ in Eq. (2.77) and collecting the coefficients of various powers of $r^2$, we get

$$2k\frac{\partial S_0}{\partial z} = -\frac{2}{r_0^2 f^2},$$

$$\frac{d^2 f}{dz^2} = \frac{1}{R_d^2 f^3}, \tag{2.81}$$

where $R_d = kr_0^2$ is the characteristic length of the diffraction divergence. From the equation of the wavefront, $kz + S$ = constant, one obtains the radius of curvature of the wave front on the axis, $R = kS_2 = f/(df/dz)$. For an initially plane wave front ($f = 1$, $df/dz = 0$ at $z = 0$), Eq. (2.81) yields

$$f^2 = 1 + \frac{z^2}{R_d^2}. \tag{2.82}$$

The beam radius $r_0 f$ expands as the beam advances while the intensity falls (cf. Fig. 2.5). The diffraction length decreases with increasing plasma density; hence, the diffraction divergence is stronger in denser plasmas. The diffraction effect decreases at shorter wavelengths (larger $k$), making shorter wavelengths preferable for long distance point-to-point communication.

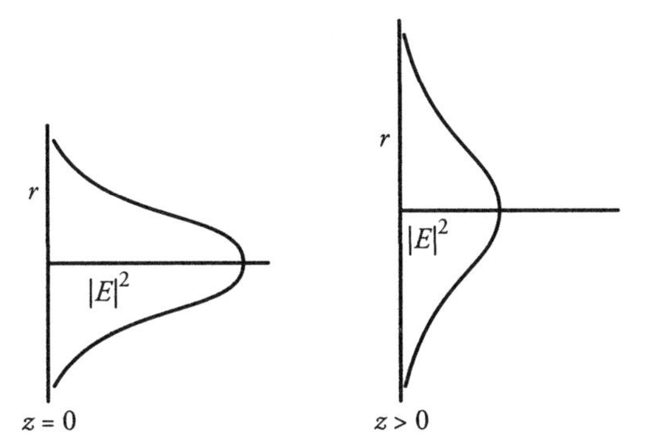


Fig. 2.5 Intensity profiles of an electromagnetic beam at $z = 0$ and $z > 0$. The width of the beam increases while the maximum intensity decreases due to diffraction effects.

## 2.11 Dispersion Broadening

An electromagnetic pulse is a superposition of many frequencies. Since plasma is a dispersive medium, different frequency components travel at different group velocities resulting in pulse distortion. Consider the propagation of an initially Gaussian (in time) electromagnetic pulse in a plasma

$$\vec{E} = \hat{x}A(z,t)e^{-i(\omega t - kz)} \tag{2.83}$$

With $A^2(0,t) = A_{00}^2 e^{-t^2/\tau^2}$. One may recall that $\partial\vec{J}/\partial t = n_0 e^2 \vec{E}/m$ (cf. the momentum equation), hence, on substituting Eq. (2.83) in the wave equation, Eq. (2.24), we obtain

$$2i\omega\left(\frac{\partial A}{\partial t} + v_g \frac{\partial A}{\partial z}\right) - \frac{\partial^2 A}{\partial t^2} + c^2 \frac{\partial^2 A}{\partial z^2} = 0, \tag{2.84}$$

where $v_g = c^2 k/\omega = c\left(1 - \omega_p^2/\omega^2\right)^{1/2}$. For $|\partial A/\partial t| << |\omega A|$, the last two terms in Eq. (2.84) are small and one gets $\partial A/\partial t = -v_g \partial A/\partial z$, that is, $A(z, t) = F(t - z/v_g)$. To obtain a solution to a higher order, we approximate $\partial^2 A/\partial z^2 = v_g^{-2}\partial^2 A/\partial t^2$ and introduce a new set of variables

$$z' = z,$$

$$t' = \left(t - \frac{z}{v_g}\right)\frac{\omega v_g}{\omega_p}.$$

Then, Eq. (2.84) takes the form

$$2ik\frac{\partial A}{\partial z'}+\frac{\partial^2 A}{\partial t'^2}=0 \tag{2.85}$$

which is the same as Eq. (2.75) with $\nabla_\perp^2$ replaced by $\partial^2/\partial t'^2$. Following the procedure outlined here, we obtain

$$E_0 = A_0 e^{iS},$$

$$A_0^2 = \frac{A_{00}^2}{f(z')}e^{-t'^2/\tau'^2 f^2(z')},$$

$$S=\phi(z')+\frac{1}{2}\beta(z')t'^2, \quad \beta=\frac{k}{f}\frac{df}{dz'},$$

$$\frac{d^2 f}{dz'^2}=\frac{1}{R_{\text{dis}}^2 f^3}, \tag{2.86}$$

where $\tau' = \tau v_g \omega/\omega_p$, $R_{\text{dis}} = k\tau'^2$. For $f(0) = 1$, $df/dz'|_{z'=0} = 0$, Eq. (2.86) yields

$$f^2 = 1+\frac{z^2}{R_{\text{dis}}^2}. \tag{2.87}$$

The pulse undergoes dispersion broadening over a scale length $R_{\text{dis}}$.

The dispersion broadening effect has an interesting application in determining the distance of pulsars, where higher frequencies propagate faster than lower frequencies, leading to a descending frequency with time in the observed pulse. Let two frequencies $\omega_1$, $\omega_2$, be generated simultaneously during an event in a pulsar. These two waves travel with group velocities $v_{g1} \approx c\left[1-\left(\omega_p^2/2\omega_1^2\right)\right]$ and $v_{g2} \approx c\left[1-\left(\omega_p^2/2\omega_2^2\right)\right]$, where $\omega_p$ ($<< \omega_1$, $\omega_2$) is the plasma frequency of the interstellar medium. The time delay $\Delta t$ between the two signals gives the distance $L$ of the pulsar from the Earth:

$$L \approx 2c\Delta t\frac{\omega_1^2}{\omega_p^2}\bigg/\left(1-\frac{\omega_1^2}{\omega_2^2}\right).$$

For $\omega_1 - \omega_2 = \Delta\omega$ with $\Delta\omega << \omega_1$, $\omega_2$ we have

$$L \approx -c\frac{\omega_1^3}{\omega_p^2}\frac{\Delta t}{\Delta\omega}.$$

## 2.12 Wave Propagation in Inhomogeneous Plasma

Let us now introduce a one-dimensional inhomogeneity in the plasma (say $\nabla n_0 \,||\, \hat{x}$) and consider the propagation of a monochromatic wave, with $\vec{E} \sim e^{-i\omega t}$, at an angle to the density gradient in the *x*–*z* plane (cf. Fig. 2.6). Now, $\varepsilon = 1 - \omega_p^2(x)/\omega^2$. One may note from the wave equation, Eq. (2.25), that the equations for $E_x$ and $E_z$ are coupled but independent of $E_y$. There are two independent modes of propagation: *S* polarized wave ($E_y \neq 0$, $E_x = E_z = 0$) and *P* polarized wave ($E_y = 0$).

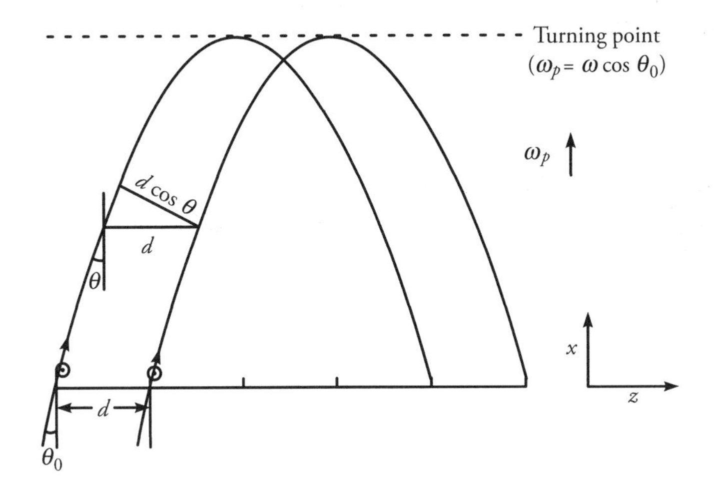


**Fig. 2.6** Power flux tube in an inhomogeneous plasma where the density increases with *x*, and $\theta_0$ is the angle of incidence at the entry point to the plasma.

### 2.12.1 S-polarization

$E_y \neq 0$; $E_x$, $E_z = 0$. In this case, the *y* component of Eq. (2.24) is

$$\frac{\partial^2 E_y}{\partial x^2} + \frac{\partial^2 E_y}{\partial z^2} + \frac{\omega^2}{c^2}\varepsilon(x)E_y = 0, \tag{2.88}$$

Expressing $E_y = f_1(x)\, f_2(z) e^{-i\omega t}$, Eq. (2.88) can be written as

$$\frac{1}{f_1}\frac{\partial^2 f_1}{\partial x^2} + \frac{\omega^2}{c^2}\varepsilon(x) = -\frac{1}{f_2}\frac{\partial^2 f_2}{\partial z^2} = k_z^2, \tag{2.89}$$

where $k_z^2$ is an arbitrary constant introduced because the left-hand side of Eq. (2.89) is independent of *z* and the right-hand side is independent of *x*. Hence, the left-hand and right-hand sides must be independent of *x* and *z*, that is, a constant. For $f_2$, we obtain

$$f_2 = e^{ik_z z}, \tag{2.90}$$

representing phase propagation along the $+z$ direction. One could also have written the $e^{-ik_z z}$ term; however, we suppress it. The equation for $f_1$ can be written as

$$\frac{\partial^2 f_1}{\partial x^2} + k_x^2(x) f_1 = 0 \tag{2.91}$$

where $k_x^2 = (\omega^2/c^2)\varepsilon - k_z^2$. As long as the $x$ variation of $k_x^2$ is slow, that is, $\partial k_x/\partial x << k_x^2$, we may write down a WKB solution

$$f_1 = Ae^{i\phi}, \tag{2.92}$$

where $\phi$ is a fast varying function of $x$ and $A$, a slowly varying function of $x$. By substituting $f_1$ in Eq. (2.91), neglecting $\partial^2 A/\partial x^2$, and equating the real and imaginary parts of the left-hand side to zero, to get

$$\phi = \int k_x dx, \qquad A = A_0/k_x^{1/2}, \tag{2.93}$$

Hence,

$$\vec{E} = \hat{y}\frac{A_0}{k_x^{1/2}} e^{-i\left(\omega t - k_z z - \int k_x dx\right)}. \tag{2.94}$$

One may interpret $k_x$ and $k_z$ as the $x$ and $z$ components of the wave vector. A very important characteristic of the wave propagation in an inhomogeneous plasma is that $k_z$ = constant, that is, the component of the propagation vector transverse to the direction of inhomogeneity is constant, and that the wave frequency $\omega$ = constant.

The ray trajectory is given by

$$\frac{dx}{dz} = \frac{v_{gx}}{v_{gz}}. \tag{2.95}$$

Since $\vec{v}_g = \partial\omega/\partial\vec{k} = c^2\vec{k}/\omega$, $v_{gx}/v_{gz} = k_x/k_z$, we have from Eq. (2.95),

$$z = k_z \int^x \frac{dx}{k_x(x)} + \text{const.} \tag{2.96}$$

Consider two rays, one passing through $x = 0$, $z = 0$, that is, having

$$z_1 = k_z \int_0^x \frac{dx}{k_x(x)} \tag{2.97}$$

and the other having $x = 0$, $z = d$,

$$z_2 = d + k_z \int_0^x \frac{dx}{k_x(x)}, \tag{2.98}$$

where subscripts 1 and 2 are used just to identify the two rays (cf. Fig. 2.6). At any value of $x$, the $z$ separation between the two rays is always $d$. The transverse separation between the two rays is $d\cos\theta = dk_x/k$, where $\theta$ is the angle a ray makes with the $x$-axis. The power flux carried by a tube bound between these rays and having a unit width in the $y$-direction is

$$P = \frac{1}{2}\left|\vec{E}^* \times \vec{H}\right| d\ \cos\theta = \frac{c^2 d\varepsilon_0}{2\omega} k_x \left|E_y\right|^2 = \text{constant}. \tag{2.99}$$

Thus the factor $k_x^{-1/2}$ in $E_y$ in Eq. (2.94) ascertains that the power flux in any tube is constant, as long as absorption is insignificant.

We notice that $k_x^2 = 0$ occurs when $\varepsilon = k_z^2 c^2/\omega^2$, that is,

$$\omega_p = \omega\cos\theta_0. \tag{2.100}$$

where $\theta_0$ is the incidence angle of the beam outside the plasma such that $k_z = (\omega/c)\sin\theta_0$.

Beyond this point $k_x^2$ is negative and the wave does not propagate. Equation (2.100) gives the turning point. However, near the turning point ($k_x \sim 0$), the amplitude would go to infinity according to Eq. (2.94), and the WKB approximation fails. If the density profile near the turning point is approximated to be linear

$$\omega_p^2 = \omega_{p0}^2\left(1 + \frac{x}{L_n}\right), \tag{2.101}$$

Eq. (2.98) reduces to the Airy equation

$$\frac{d^2 f_1}{d\xi^2} - \xi f_1 = 0, \tag{2.102}$$

giving the well-behaved solution

$$f_1 = a_1 A_i(\xi), \tag{2.103}$$

where

$$\xi = \frac{x - x_0}{\delta}, \quad x_0 = \left( \frac{\omega^2 \cos^2 \theta_0 - \omega_{p0}^2}{c^2} \right) \frac{c^2 L_n}{\omega_{p0}^2}$$

$$\delta = \left( c^2 L_n / \omega_{p0}^2 \right)^{1/3}.$$

The Airy function $A_i(\xi)$ is oscillatory for $\xi < 0$ and decays exponentially for $\xi > 0$. The scale length of the variation of $f_1$, that is, the effective wavelength of the wave, is $\delta$. The solution $f_1$ acquires large values (as compared to its values in the far underdense region) near the turning point.

### 2.12.2 P-polarization

$E_y = 0$; $E_x$, $E_z \neq 0$. For an electromagnetic wave polarized in the $x$–$z$ plane, $\nabla \cdot \vec{E} \neq 0$, we have from Eq. (2.24) (on taking its divergence) that $\nabla \cdot \vec{E} = -(1/\varepsilon)\vec{E} \cdot \nabla \varepsilon$. Substituting for $\nabla \cdot \vec{E}$ in Eq. (2.24), we obtain for the $x$-component

$$\nabla^2 E_x + \frac{\partial}{\partial x}\left( E_x \frac{\partial}{\partial x} \ln \varepsilon \right) + \frac{\omega^2}{c^2} \varepsilon E_x = 0. \tag{2.104}$$

Following the method of separation of variables as outlined earlier, we take the $z$ variation of $E_x$ as $e^{ik_z z}$ with $k_z$ = constant. Introducing a new function,

$$E_x = \frac{F(x)}{\varepsilon^{1/4}} e^{-i(\omega t - k_z z)}, \tag{2.105}$$

we obtain from Eq. (2.104),

$$\frac{d^2 F}{dx^2} + \left( \frac{\omega^2}{c^2} \varepsilon - k_z^2 - \frac{1d}{2dx}\left( \frac{1d\varepsilon}{\varepsilon dx} \right) - \frac{1}{4}\left( \frac{1d\varepsilon}{\varepsilon dx} \right)^2 \right) F = 0. \tag{2.106}$$

The turning point for the wave occurs at a density at which the big parenthesis of Eq. (2.106) vanishes. This does not happen at $\omega_p = \omega \cos\theta_0$ which is the turning point for the S-polarization. However, for long density scale lengths, the turning points of both polarizations are not very different. Away from the turning point the last two terms in the big parenthesis can be dropped, and then the WKB solution of Eq. (2.106) is the same as that of Eq. (2.91) for the S-polarized wave. However, near the turning point and the singular point $\varepsilon = 0$, Eq. (2.106) needs to be re-examined carefully for the inclusion of kinetic effects and the possibility of mode conversion to Langmuir waves. We defer this discussion to Chapter 3.

## 2.13 Duct Propagation

A depressed density plasma duct has a tendency to trap and guide electromagnetic radiation, as observed in several experiments. Consider, for example, a density profile

$$n_0 = n_{0I} \qquad \text{for } |x| < a$$

$$= n_{0II} \qquad \text{for } |x| > a \tag{2.107}$$

with $n_{0I} < n_{0II}$. The inner region has a higher refractive index than the outer one. When a ray propagating in the inner region reaches the duct boundary at a large angle of incidence, it is internally reflected back into and trapped inside the duct. Let us examine the propagation of a $\hat{y}$ polarized electromagnetic wave through the duct when $\partial/\partial y = 0$.

$$\vec{E} = \hat{y}A(x)e^{-i(\omega t - k_z z)}, \tag{2.108}$$

where $A(x)$ is governed by the wave equation, Eq. (2.91)

$$\frac{d^2 A}{dx^2} + k_x^2 A = 0, \tag{2.109}$$

$$k_x^2 = k_I^2 \equiv \frac{\omega^2 - \omega_{pI}^2}{c^2} - k_z^2 \qquad \text{for } |x| < a$$

$$= -\alpha_{II}^2 \equiv \frac{\omega^2 - \omega_{pII}^2}{c^2} - k_z^2 \qquad \text{for } |x| > a \tag{2.110}$$

with $\omega_{pI,II}^2 = n_{0I,II}e^2 / \varepsilon_0 m$. Waves are ducted when $k_I^2 > 0$ and $\alpha_{II}^2 > 0$, which we will assume here. When $\alpha_{II}^2 < 0$, the waves can escape the duct and propagate into the denser plasma. The symmetric solution of Eq. (2.109) satisfying the continuity of $E_y$ (the tangential electric field) at $x = \pm a$ and being well behaved at $|x| \to \infty$ can be written as

$$A = A_1 \cos k_1 x \qquad \text{for } |x| < a$$

$$= A_1 \cos k_1 a e^{-\alpha_{11}(|x|-a)} \qquad \text{for } |x| > a. \tag{2.111}$$

An integration of Eq. (2.109) across the discontinuity at $x = a$, from $a - \Delta$ to $a + \Delta$ with $\Delta \to 0$, yields $\partial A/\partial x$ which is continuous at $|x| = a$. This condition leads to the dispersion relation

$$k_I a \tan(k_I a) = \alpha_{II} a \equiv \left(\alpha^2 - k_I^2 a^2\right)^{1/2}, \qquad (2.112)$$

where $\alpha^2 = \left(\omega_{pII}^2 - \omega_{pI}^2\right)a^2/c^2$. For a given $\alpha$, Eq. (2.112) can be solved to obtain $k_I a$, which may have multiple values, corresponding to different modes of propagation. Equation (2.112) can be solved numerically or graphically. One may plot the left-hand and right-hand sides (LHS and RHS) as functions of $k_I a$ for a fixed $\alpha$. The points of intersection of the two curves give the roots $k_I a$ of Eq. (2.112). For $\alpha = 3$ only one root exists, that is, there is only one mode of propagation. For $\alpha = 5$ there exist two roots, that is, there are two modes of propagation, and so on. The root with the smallest $k_I a$ is called the fundamental mode. In terms of a known $k_I a$ and $\alpha$, the $\omega$ versus $k_z$ dispersion relation for a mode is obtained from the definition, Eq. (2.110), as

$$\omega^2 = \left(k_I^2 + k_z^2\right)c^2 + \omega_{pI}^2. \qquad (2.113)$$

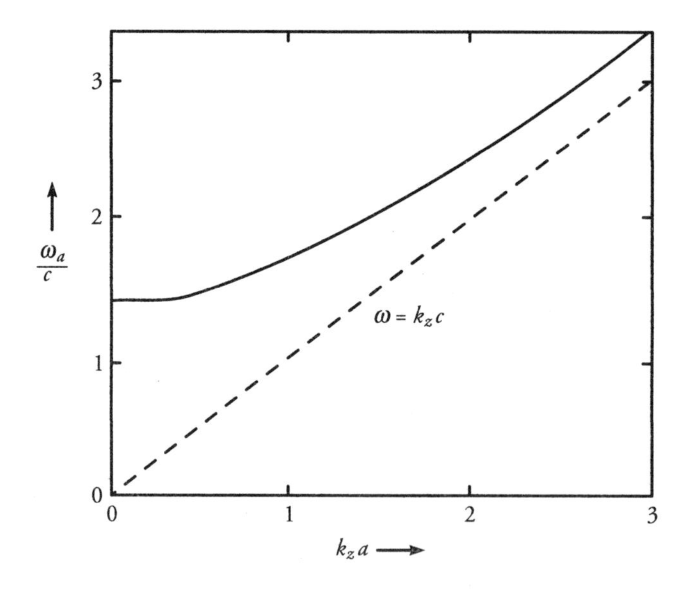


Fig. 2.7 Dispersion curve for the fundamental mode of duct propagation for $\alpha = 3$ and $\omega_{pI} a/c = 1$, giving $k_I a = 1.1$ and $k_{II} a = 2.51$.

Figure 2.7 shows the plot of the dispersion relation against the fundamental mode for $\alpha = 3$, $\omega_{pI} a/c = 1$. The mode structure is plotted in Fig. 2.8. The mode extends to the outer region but falls off rapidly away from the duct boundary. There is a lower frequency cutoff for ducted propagation. For higher modes, the cutoff frequency is higher. Asymptotically at large $k_z$, all modes approach $w = k_z c$.

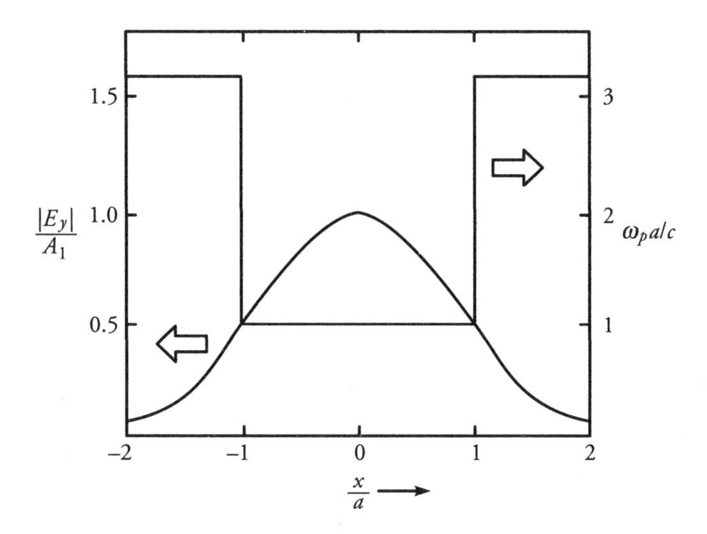


Fig. 2.8 Mode structure of the fundamental mode in a rectangular density duct.

The guided propagation of electromagnetic waves is important in high-power free electron lasers,[15,16] collective charged particle acceleration,[17] laser-driven fusion, and ionospheric propagation.[18,19]

In some applications, it is appropriate to model the density profile as parabolic,

$$n_0 = n_0^0\left(1 + x^2 / L^2\right). \tag{2.114}$$

In that case the equation governing $A$ takes the form

$$\frac{d^2A}{d\xi^2} + \left(\mu - \xi^2\right)A = 0, \tag{2.115}$$

giving mode structure and eigenvalues as

$$A = A_n H_n\left(\xi\right)e^{-\xi^2/2} \tag{2.116}$$

$$\mu \equiv \left(\frac{\omega^2 - \omega_{p0}^2}{c^2} - k_z^2\right)\frac{cL}{\omega_{p0}} = n + \frac{1}{2}; \quad n = 0, 1, \ldots \tag{2.117}$$

where $\xi = x(\omega_{p0}/cL)^{1/2}$, $\omega_{p0} = (n_0^0 e^2/\varepsilon_0 m)^{1/2}$, $H_n$ are Hermite polynomials and $A_n$ are constants. The fundamental ($n$ = 0) mode has a Gaussian distribution with half width $\delta \sim (cL/\omega_{p0})^{1/2}$ (cf. Fig. 2.9).

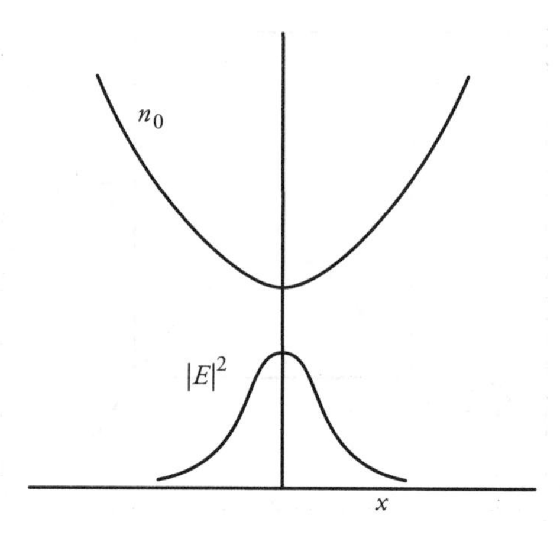


Fig. 2.9 Parabolic density profile and the mode structure of a trapped mode.

## 2.14 Anomalous Resistivity

The macroscopic behavior of a plasma is strongly influenced by the presence of density fluctuations. Let us apply a high-frequency electric field $\vec{E}_{\omega_0}$ to a plasma having a certain level of density fluctuations $n_k$. The field produces an oscillatory velocity $\vec{v}_{w_0}$ of the electrons, resulting in a current density $-n_k e\vec{v}_{\omega_0}$, driving a resonant/non-resonant mode at $\omega_0, \vec{k} \pm \vec{k}_0$. The Landau damping of the driven mode causes dissipation of the high frequency field, giving rise to anomalous resistivity.[20]

Consider the propagation of an electromagnetic wave $\vec{E}_0 = \hat{x}A_0 e^{-i(\omega_0 t - \vec{k}_0 . \vec{r})}$ in a plasma having static turbulence characterized by $\vec{k}$ spectrum of density fluctuations $n_k$. The electrons acquire an oscillatory velocity $\vec{v}_0 = e\vec{E}_0 / mi\omega_0$ that beats with $n_k$ to produce a nonlinear current and a nonlinear density perturbation

$$\vec{J}_{\omega_0}^{\mathrm{NL}} = -\frac{1}{2}\sum_k n_k e\vec{v}_0,$$

$$n_1^{\mathrm{NL}} = \sum_k \frac{n_k \vec{k}.\vec{v}_0}{2\omega_0}, \tag{2.118}$$

where we have used the continuity equation and the fact that $\vec{k}_0 \cdot \vec{E}_0 = 0$ for the electromagnetic wave. In the analysis that follows, we assume $k_0 << k$.

The nonlinear perturbations produce a field $\vec{E}$ that we approximate to be electrostatic $\vec{E} = -\nabla\phi = -i\sum_{\vec{k}} \vec{k}\phi_k$. Each $\phi_k$ produces a self-consistent linear density perturbation (cf. Eq. (2.54)):

$$n_{1k}^{L} = \frac{k^2 \varepsilon_0}{e} \chi_e\left(\omega_0, \vec{k}\right)\phi_k. \tag{2.119}$$

Using $n_{1k} = n_{1k}^{L} + n_{1k}^{\mathrm{NL}}$ in Poisson's equation, $-k^2\phi_k = en_{1k}/\varepsilon_0$, we obtain

$$\phi_k = -\frac{e}{\varepsilon_0 k^2} \frac{\vec{k}.\vec{v}_0 n_k}{2i\omega_0 \varepsilon\left(\omega_0, \vec{k}\right)}, \tag{2.120}$$

where $\varepsilon(\omega_0, \vec{k}) = 1 + \chi_e(\omega_0, \vec{k})$ is the plasma dielectric function. The electric potential $\phi_k$ also produces an oscillatory velocity

$$\vec{v}_{1k} = -\frac{e\vec{k}\phi_k}{m\omega_0}$$

which beats with $n_k$ to produce nonlinear current density at $(\omega_0, \vec{k}_0)$:

$$\vec{J}_{\omega_0}^{\mathrm{NL}} = -\frac{1}{2}\sum_k n_k^* e\vec{v}_{1k} = \frac{\omega_p^2}{4\omega_0^2}\sum_k \frac{\left|n_k\right|^2 e\vec{k}.\vec{v}_0}{n_0 \varepsilon\left(\omega_0, \vec{k}\right)k^2}\vec{k}, \tag{2.121}$$

where $n_0$ is the equilibrium plasma density. Thus, the total current density at $\omega_0, \vec{k}_0$ is

$$\vec{J}_{\omega_0} = -n_0 e\vec{v}_0 + \vec{J}_{\omega_0}^{\mathrm{NL}} = -\frac{n_0 e^2}{mi\omega_0}\left[\vec{E}_0 - \frac{1}{4\varepsilon_0 m\omega_0^2}\sum_k \frac{\left|n_k\right|^2 e^2 \vec{k}.\vec{E}_0}{n_0 \varepsilon\left(\omega_0, \vec{k}\right)k^2}\vec{k}\right]. \tag{2.122}$$

The coefficient of $\vec{E}_0$ in this expression is the conductivity tensor. For $\vec{k}$ parallel to $\vec{E}_0$, the anomalous conductivity is

$$\sigma = -\frac{n_0 e^2}{mi\omega_0}\left[1 - \frac{\omega_p^2}{4\omega_0^2}\sum_k \frac{\left|n_k\right|^2}{n_0^2 \varepsilon\left(\omega_0, \vec{k}\right)}\right]. \tag{2.123}$$

The imaginary part of $\varepsilon\left(\omega_0, \vec{k}\right)$ gives rise to *a* real part of the conductivity $\sigma$ and correspondingly of the resistivity $1/\sigma$, leading to the anomalous absorption of the laser. In cases when the turbulence spectrum is not discrete, the summation in Eq. (2.123) is replaced by an integration. Dawson and Oberman[21] studied the resistivity due to thermal fluctuations and found a considerable enhancement. At a fluctuation level higher than the thermal level, the resistivity is significantly large.

## 2.15 Discussion

The linear modes of plasma have distinct frequency regimes. Ion-acoustic waves exist with frequency less than the ion plasma frequency, and suffer Landau damping primarily due to resonant interactions with ions. At lower frequencies the ion-acoustic waves suffer negligible dispersion broadening in course of their propagation. However, as the frequency approaches the ion plasma frequency, the dispersion broadening becomes strong. In a later chapter, we shall see that the dispersion broadening can be offset by nonlinear effects and the waves can propagate as solitons.

Around the electron plasma frequency, one has Langmuir waves. They have no dispersion in a cold plasma, but in hot plasma they have strong dispersive effects due to the electron temperature. These waves suffer Landau damping due to resonant interactions with electrons when their frequency exceeds the plasma frequency by a few percent. In the vicinity of the plasma frequency, the Langmuir wave has a large phase velocity; it also has the potential to accelerate electrons to ultra-relativistic energies.

Electromagnetic waves in un-magnetized plasmas exist at frequencies higher than the electron plasma frequency. Their phase velocity is greater than the speed of light in vacuum, $c$; however, the group velocity is less than c. Waves of finite spot size suffer diffraction divergence over a characteristic Rayleigh diffraction length. Short pulses are susceptible to dispersion broadening. Nonlinear effects can effectively counter these effects.

## References


1. Jackson, J. D. 1999. *Classical Electrodynamics.* 3rd edition. NY: John Wiley and Sons.
2. Akhiezer, A. I. 1975 "Plasma Electrodynamics Vol. 1: Linear Theory - Vol. 2: Non-linear Theory and Fluctuations." *International Series of Monographs in Natural Philosophy.* Oxford: Pergamon Press.
3. Chen, F. F. 2016. *Introduction to Plasma Physics and Controlled Fusion.* 3rd edition. New York: Springer.
4. Stix, T. H. 1992. "Waves in Plasmas American Institute of Physics, New York." *Google Scholar*: 238–246.
5. Alexandrov, A. F., L. S. Bogdankevich and A. A. Rukhadze. 1984. *Principles of Plasma Electrodynamics,* Springer Verlag Series in Electrophysics, Vol. 9. Springer Verlag.
6. Goldston, R. J. and P. H. Rutherford. 1995. *Introduction to Plasma Physics.* Bristol and Philadelphia: Institute of Physics Publishing Ltd.
7. Ginzburg, V. L. 1970. *The Propagation of Electromagnetic Waves in Plasma.* 2nd edition. Oxford: Pergamon.
8. Braginskii, S. I. 1958. "Transport Phenomena in a Completely Ionized Two-temperature Plasma." *Sov. Phys. JETP* 6 (33): 358–69.
9. Bohm, D. and E. P. Gross. 1949. "Theory of Plasma Oscillations. A. Origin of Medium-Like Behavior." *Physical Review* 75 (12): 1851–64.

10. Landau, L. D. 1946. "On the Vibrations of the Electronic Plasma." *Journal of Physics (USSR)* 10 (1): 25–34
11. Fried, B. D. and S. D. Conte. 1961. *The Plasma Dispersion Function: The Hilbert Transform of the Gaussian.* New York and London: Academic Press.
12. McKinstrie, C. J., R. E. Giacone, and E. A. Startsev. 1999. "Accurate Formulas for the Landau Damping Rates of Electrostatic Waves." *Physics of Plasmas* 6 (2): 463–466.
13. Rosenbluth, M. N. and C. S. Liu. 1976. "Cross-field Energy Transport by Plasma Waves." *The Physics of Fluids* 19 (6): 815–18.
14. Akhmanov, Sergei Aleksandrovich, Anatolii Petrovich Sukhorukov, and R. V. Khokhlov. 1968. "Self-focusing and Diffraction of Light in a Nonlinear Medium." *Physics-Uspekhi* 10 (5): 609–36.
15. Tripathi, Vipin K., and Chuan Sheng Liu. 1990. "Plasma Effects in a Free Electron Laser." *IEEE Transactions on Plasma Science* 18 (3): 466–71.
16. Roberson, C. W., and P. Sprangle. 1089. "A Review of Free-electron Lasers." *The Physics of Fluids B: Plasma Physics* 1: 3
17. Esarey, E., A. Ting, and P. Sprangle. 1988. "Relativistic Focusing and Beat Wave Phase Velocity Control in the Plasma Beat Wave Accelerator." *Applied Physics Letters* 53 (14): 1266–1268.
18. Sodha, M. S., A. K. Ghatak, and V. K. Tripathi. 1976. "V Self Focusing of Laser Beams in Plasmas and Semiconductors." In *Progress in Optics* 13: 169–265. Elsevier.
19. Perkins, F.W., Oberman, C. and Valeo, E.J., 1974. "Parametric Instabilities and Ionospheric Modification." *Journal of Geophysical Research* 79(10): 1478–96.
20. Kruer, W. L., and J. M. Dawson. 1972. "Anomalous High-frequency Resistivity of a Plasma." *The Physics of Fluids* 15 (3): 446–53.
21. Dawson, John, and Carl Oberman. 1962. "High-frequency Conductivity and the Emission and Absorption Coefficients of a Fully Ionized Plasma." *The Physics of Fluids* 5 (5): 517–24.

# 3 Resonance Absorption and Brunel Absorption

## 3.1 Introduction

A P-polarized electromagnetic wave propagating through an inhomogeneous plasma, $n_0(x)$, at an angle to the density gradient with the electric vector polarized in the plane of incidence (the plane that contains the density gradient and the propagation vector, $x$–$z$ plane in Fig. 3.1) bends away from the density gradient as it penetrates deeper into higher density plasmas. The $z$ component of the wave vector, $k_z$ remains constant, while the $x$ component $k_x = \left((\omega^2 - \omega_p^2(x))/c^2 - k_z^2\right)^{1/2}$, where $\omega_p(x)$ is the plasma frequency and $\omega$ is the wave frequency, decreases. The wave suffers reflection from the layer (turning point $x = x_T$) where $k_x = 0$, that is, $\omega_p = \omega \cos\theta_0$ with $\theta_0$ being the angle of incidence at the entry into the plasma. The wave amplitude rises up to the vicinity of the turning point and decreases for $x > x_T$ where the wave is evanescent. The wave is accompanied by density oscillations as the electron oscillatory velocity $\vec{v}$ caused by the wave has a component along the density gradient and $\partial n/\partial t = -\nabla.(n_0\vec{v}) \neq 0$. If the wave field tunnels through the evanescent region to the critical layer $x = x_c$ (where the plasma frequency equals the wave frequency), the space charge oscillations are resonantly enhanced as the frequency of oscillations equals the natural frequency; results in the excitation of an electrostatic plasma wave. The diversion of energy from the electromagnetic wave caused by these enhanced oscillations is called resonant absorption.

For a linear density profile, $\omega_p^2 = \omega^2 x / L_n, x_T = L_n \cos^2\theta_0$ is where $n_0 = n_{cr}\cos^2\theta_0$ and $x_{cr} = L_n$ is where $n_0 = n_{cr}$ (cf. Fig. 3.1). The effective wavelength of the wave is $\lambda_{em} = (c^2 L_n/\omega^2)^{1/3}$. Significant tunneling of the electromagnetic fields through the evanescent region to the critical layer occurs when the width of this region is comparable to a wavelength, $x_{cr} - x_T = L_n \sin^2\theta_0 \simeq \lambda_{em}$. In this case, a large amplitude plasma wave is excited near the critical layer. As the plasma wave propagates toward lower densities, it acquires a larger wave number, and deposits its energy to the electrons via Landau damping. This leads to strong absorption of radiation and production of hot electrons, and which can account for nearly 60% absorption of obliquely incident radiation.[1–4] In a collisional plasma, heating also occurs via inverse Bremsstrahlung (collisions) near the critical layer.[1]

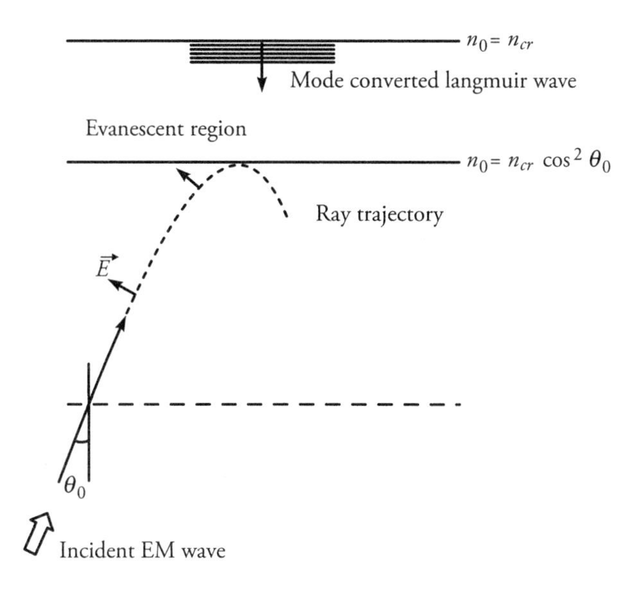


**Fig. 3.1** Schematic of the turning point and mode conversion layer.

In an overdense plasma with a sharp boundary. Brunel[5] pointed out another mechanism of absorption for a P-polarized wave (with a component of the electric field vector normal to the plasma surface). The wave induces an oscillatory surface charge. In half the wave period, this charge is negative; it is caused by the accumulation of electrons, and extends into vacuum. An electron released from the surface at time $t$ sees, besides the laser field, a static space charge field due to the electrons emitted earlier. As a result, when it returns back to the surface it arrives with finite energy, causing a drain on the wave. Brunel developed an elegant capacitor model for the mechanism and termed this collisionless absorption mechanism vacuum electron heating. Kumar and Tripathi[6] extended the theory to the case of vacuum heating by surface plasma waves.

In Section 3.2, we will provide a heuristic treatment of resonance absorption in a collisional plasma. In Section 3.3, we study linear mode conversion of an electromagnetic wave into a Langmuir wave including thermal effects, using the method suggested by Piliya.[2] In Section 3.4 we study Brunel's mechanism of heating. Finally, the results are summarized in Section 3.5.

## 3.2 Resonance Absorption: Heuristic Approach

We here study the tunneling of an electromagnetic wave beyond its turning point to reach the resonance at the critical density where it converts to an electrostatic wave. For the sake of simplicity, we consider a cold collisional plasma with a linear density profile whose plasma frequency profile can be written as

$$\omega_p^2 = \omega_{p0}^2 \, x / L_n \qquad \text{for } x > 0$$

$$= 0 \qquad\qquad \text{for } x < 0. \tag{3.1}$$

A P-polarized electromagnetic wave is incident on the plasma at $x = 0$ at an angle of incidence $\theta_0$ (cf. Fig. 3.1) with the electric and magnetic fields

$$\vec{E} = \vec{A}_0\, e^{-i\left(\omega t - k_z z - (\omega/c)\cos(\theta_0)\, x\right)},$$

$$\vec{H} = \hat{y}\, H_0\, e^{-i\left(\omega t - k_z z - (\omega/c)\cos(\theta_0)\, x\right)}, \tag{3.2}$$

where $\vec{A}_0 = A_0\left(\cos\theta_0\,\hat{z} - \sin\theta_0\,\hat{x}\right)$, $H_0 = -A_0/\mu_0 c$, and $k_z = (\omega/c)\sin\theta_0$. Inside the plasma, $\vec{H}$ and $\vec{E}$ can be written in the Wentzel–Kramers–Brillouin (WKB) approximation as

$$\vec{H} = \hat{y}\, F(x)\, e^{-i\left(\omega t - k_z z - \int k_x x\right)},$$

$$\vec{E} = -\frac{\vec{k}\times\vec{H}}{\omega\varepsilon_0\varepsilon}, \tag{3.3}$$

where $k_x = \left(\omega^2\varepsilon/c^2 - k_z^2\right)^{1/2}$, and $\varepsilon = 1 - \left(\omega_p^2/\omega^2\right)\left(1 - i\nu/\omega\right) = 1 - x/L_n - i\nu\, x/\omega L_n$, $\nu$ is the collision frequency and we have taken $\omega = \omega_{p0}$. The wave meets a turning point ($k_x = 0$) at $x_T = L_n\cos^2\theta_0$ beyond which its field is evanescent; the field acquires a large value at the critical layer $x = x_{cr} = L_n$, where plasma frequency equals the wave frequency. To assess the field amplitude at the critical layer we proceed as follows.

The time-averaged Poynting's vector of the wave is

$$\vec{S}_{av} = \frac{|H|^2}{2\omega\varepsilon_0\varepsilon}\vec{k}.$$

Its component in the direction of the density gradient (if collisional absorption is ignored), $S_{avx}$ remains constant; hence,

$$F^2 = H_0^2\,\varepsilon\,\omega / k_x c. \tag{3.4}$$

Near the turning point ($k_x \to 0$, $x \sim x_T = L_n\cos^2\theta_0$), the WKB approximation fails and the effective wavelength of the wave (as we shall see in the next section) is $\lambda_{em} = (c^2 L_n/\omega^2)^{1/3}$. In that region, one may take $\varepsilon\omega/k_x c \sim ck_x/\omega \sim (c/\omega\lambda_{em})$; hence,

$$F(x_T) \approx H_0\left(\frac{c}{\omega L_n}\right)^{1/6}. \tag{3.5}$$

From Eq. (3.3), one may write the longitudinal component of the electric field as follows.

$$E_x \approx \frac{k_z H_y}{\omega \varepsilon_0 \varepsilon} = \frac{\sin\theta_0}{\varepsilon_0 c \varepsilon} H_y. \tag{3.6}$$

The distance between the turning point $x_T$ and the critical layer $x_{cr}$ is $\Delta_x = L_n - L_n \cos^2\theta_0 = L_n \sin^2\theta_0$. The field tunneling factor can be taken as

$$T \approx e^{-\Delta x/\lambda_{em}} = e^{-\tau/2}$$

where

$$\tau = 2\left(\omega L_n/c\right)^{2/3} \sin^2\theta_0. \tag{3.7}$$

Hence, the electric field near the critical density layer is

$$E_x \approx \frac{T F(x_T)\sin\theta_0}{\varepsilon_0\ c\ \varepsilon}. \tag{3.8}$$

It is resonantly enhanced as the real part of $\varepsilon$ tends to zero and the imaginary part of $\varepsilon$ is small.

### 3.2.1 Absorption coefficient

In the turning point-critical layer region, $E_x$ is much larger than $E_z$. It imparts a oscillatory velocity to the electrons, $v_x = eE_x/[m\ i(\omega + i\nu)]$ and heats them at a time-averaged rate

$$R = \frac{1}{2}\ \mathrm{Re}\left[-eE_x^* v_x\right] = \frac{e^2 \left|E_x\right|^2 \nu}{2\, m\, \omega^2}, \tag{3.9}$$

where Re denotes the real part of the quantity. The power absorbed from the wave per unit area of the $y$–$z$ plane is

$$P_{abs} = \int R\ n_0\ dx$$

$$\approx \frac{\omega_{p0}^2 \sin^2\theta_0 F^2(x_T) T^2}{2\ \omega^2 c^2 \varepsilon_0} \int_0^{L_n} \frac{\nu\, dx}{\left(1 - \dfrac{x}{L_n}\right)^2 + \dfrac{\nu^2}{\omega^2}}$$

$$\approx \frac{\pi \sin^2 \theta_i \omega L_n A_0^2}{4 c^2 \mu_0} \left( \frac{c}{\omega L_n} \right)^{1/3} e^{-\tau}, \tag{3.10}$$

which is independent of collision frequency $\nu$. We define the absorption coefficient as

$$\eta_{\text{abs}} = \frac{P_{\text{abs}}}{s_{\text{avx}}} = \frac{2\mu_o c P_{\text{abs}}}{A_0^2} = \frac{\pi}{4} \tau e^{-\tau} \tag{3.11}$$

The absorption coefficient has a maximum value of $\eta_{\text{abs}} \simeq 0.3$ at $\tau = 1$. This estimate is not too far from a more rigorous treatment[1].

## 3.3 Laser Mode Conversion to Plasma Wave in a Warm Plasma

Now we consider a warm collisionless plasma with the electron temperature $T_e$. The density profile is linear (cf. Eq. 3.1). A P-polarized electromagnetic wave enters the plasma at $x = 0$ with the incident fields given by Eq. (3.2). Inside the plasma, the electric field is

$$\vec{E} = \vec{A}\,(x)\, e^{-i\,(\omega t - k_z z)}. \tag{3.12}$$

As it propagates to the turning point and its field tunnels to the critical layer one expects a Langmuir wave to be excited there, drawing energy from the electromagnetic wave. Around the critical layer, $\vec{E}$ is the sum of electromagnetic and Langmuir wave fields.

The wave field imparts an oscillatory velocity to the electrons, which on solving the linearized momentum equation,

$$m \frac{\partial \vec{v}}{\partial t} = -e\vec{E} - \frac{T_e}{n_0} \nabla n$$

turns out to be

$$\vec{v} = \frac{e\vec{E}}{mi\omega} + \frac{v_{th}^2}{i\omega n_0} \nabla n, \tag{3.13}$$

where $v_{\text{th}} = (T_e/m)^{1/2}$. Using Poisson's equation, $\nabla \cdot \vec{E} = -ne/\varepsilon_0$, one may substitute for the density perturbation $n$ in Eq. (3.13) and write the current density as

$$\vec{J} = -n_0 e\vec{v} = -\frac{n_0 e^2}{mi\omega} \vec{E} + \frac{\beta^2 c^2}{i\omega} \varepsilon_0 \nabla \left( \nabla \cdot \vec{E} \right), \tag{3.14}$$

where $\omega_p = (n_0\, e^2/m\, \varepsilon_0)^{1/2}$ and $\beta = v_{\text{th}}/c$.

### 3.3.1 Coupled mode equations for electromagnetic and electrostatic waves

For a P-polarized wave ($E_x, E_z, H_y \neq 0$), Maxwell's equations $\nabla \times \vec{E} = i\omega\mu_0\vec{H}$, $\nabla \times \vec{H} = \vec{J} - i\omega\varepsilon_0\vec{E}$ on replacing $\partial/\partial z$ by $ik_z$, give

$$ik_z\, E_x - \frac{\partial}{\partial x}\, E_z = i\omega\, \mu_0\, H_y,$$

$$-ik_z\, H_y = -\, i\omega\varepsilon_0 \left( \varepsilon E_x + \frac{\beta^2 c^2}{\omega^2} \left( \frac{\partial^2 E_x}{\partial x^2} + ik_z \frac{\partial E_z}{\partial x} \right) \right),$$

$$\frac{\partial}{\partial x} \left( \mu_0 H_y - \frac{k_z \beta^2}{\omega} E_x \right) = -\frac{i\omega}{c^2} \left( \varepsilon - \beta^2 \frac{k_z^2 c^2}{\omega^2} \right) E_z. \tag{3.15}$$

We define a new function

$$G(x) = \mu_0 H_y - \frac{k_z}{\omega} \beta^2 E_x \tag{3.16}$$

and write the above set as

$$ik_z E_x - \frac{\partial E_z}{\partial x} = i\omega \left( G + \frac{k_z}{\omega} \beta^2 E_x \right) \tag{3.17}$$

$$-ik_z G = -\frac{i\omega\varepsilon'}{c^2} E_x - \frac{i\beta^2}{\omega} \frac{\partial^2 E_x}{\partial x^2} + \frac{k_z \beta^2}{\omega} \frac{\partial E_z}{\partial x}, \tag{3.18}$$

$$\frac{\partial G}{\partial x} = -\frac{i\omega\varepsilon}{c^2}\, E_z, \tag{3.19}$$

where $\varepsilon' = \varepsilon - k_z^2\, c^2\, \beta^2 / \omega^2$, $\varepsilon = 1 - \omega_p^2(x)/\omega^2$. Using the value of $\partial E_z/\partial x$ from Eq. (3.17) in (3.18) and of $E_z$ from (3.19) in (3.17), we get

$$\frac{\partial^2 E_x}{\partial \xi^2} - \frac{c^2}{v_{\text{th}}^2} \xi E_x = \left( k_z \omega\, c^2 \lambda_{\text{em}}^2 / v_{\text{th}}^2 \right) G, \tag{3.20}$$

$$\frac{\partial^2 G}{\partial \xi^2} - \frac{1}{\xi} \frac{\partial G}{\partial \xi} - \left( \xi + k_z^2 \lambda_{\text{em}}^2 \right) G = -k_z^2 \lambda_{\text{em}}^2 \left( G + \frac{\xi}{\lambda_{\text{em}}^2 k_z \omega} E_x \right), \tag{3.21}$$

where $\xi = (x - L_n)/\lambda_{em}$, neglecting terms proportional to $k_z^2\beta^2$ and using $\beta << 1$, and where $\lambda_{em} = (c^2L_n/\omega^2)^{1/3}$ is the characteristic electromagnetic wavelength. We may notice two different scale lengths for $E_x$ and $G$, viz, $\xi \sim v_{th}/c$ and $\xi \sim 1$ respectively. Equation (3.20) represents the Langmuir wave with the electromagnetic wave as a driver, while Eq. (3.21) is the electromagnetic wave coupled to the Langmuir wave

## 3.3.2 Mode conversion

If $v_{th} \to 0$, Eq. (3.20) gives $E_x = -\lambda_{em}^2 k_z \omega G/\xi$ and the right-hand side of Eq. (3.21) identically vanishes. For $(c^2/v_{th}^2)\xi >> 1$ one may write down an asymptotic (WKB) solution to Eq. (3.20)

$$E_x = -\omega k_z \lambda_{em}^2 \frac{G(\xi)}{\xi} + a_1 \frac{v_{th}}{c}(-\xi)^{1/4} e^{i\frac{2c}{3v_{th}}(-\xi)^{3/2}}, \tag{3.22}$$

where the first term is the particular solution corresponding to the transverse wave and the last term, with a constant $a_1$, is the complementary solution representing a plasma wave propagating towards $-\hat{x}$. A plasma wave going along $+\hat{x}$ has been omitted as there is no source for these waves in the underdense region.

The WKB solution to Eq. (3.21), for $\xi >> 1$, can be written as

$$G = b_1\left(-\xi - k_z^2\lambda_{em}^2\right)^{-1/4} e^{-i\frac{2}{3}\left(-\xi - k_z^2\lambda_{em}^2\right)^{3/2}} + b_2\left(-\xi - k_z^2\lambda_{em}^2\right)^{-1/4} e^{i\frac{2}{3}\left(-\xi - k_z^2\lambda_{em}^2\right)^{3/2}}. \tag{3.23}$$

We have dropped the particular integral which is $\approx a_1\left(v_{th}^2/c^3\right)(-\xi)^{+3/4} i\left(k_z c/\omega\right)$. The two terms in Eq. (3.23) represent incident and reflected electromagnetic waves going along $+\hat{x}$ and $-\hat{x}$. To evaluate $b_2$ and $a_1$ in terms of $b_1$, the incident wave amplitude, we solve Eqs. (3.20) and (3.21) around $\xi \simeq 0$ and match the solutions to the WKB solutions for large $\xi$.

Since the scale length of $G$ is very long, we may replace $G(\xi)$ in Eq. (3.20) around $\xi << 1$ by $G(0)$. Then Eq. (3.20) yields

$$E_x = k_z\omega\left(\frac{c}{v_{th}}\right)^{2/3} \lambda_{em}^2 G(0) W\left(\frac{c^{2/3}}{v_{th}^{2/3}}\xi\right) \tag{3.24}$$

where $W(\eta)$ with $\eta = \xi c^{2/3}/v_{th}^{2/3}$ is a solution of the inhomogeneous Airy equation $W'' - \eta W = 1$,

$$W(\eta) = i\int_0^\infty e^{-i\left(\eta t + t^3/3\right)} dt \tag{3.25}$$

For $(-\eta) >> 1$,

$$W(\eta) \sim (-\eta)^{-1/4} \pi^{1/2} e^{-i\pi/4 + i2/3(\eta)^{3/2}} - \frac{1}{\eta}. \tag{3.26}$$

Matching solution (3.24) to (3.22) asymptotically we get

$$a_1 = k_z \omega \left( \frac{c}{v_{\text{th}}} \right)^{3/2} \lambda_{\text{em}}^2 \pi^{1/2} e^{-i\pi/4} G\ (0). \tag{3.27}$$

Figure 3.2 illustrates the mode structure of the field component $E_x$.

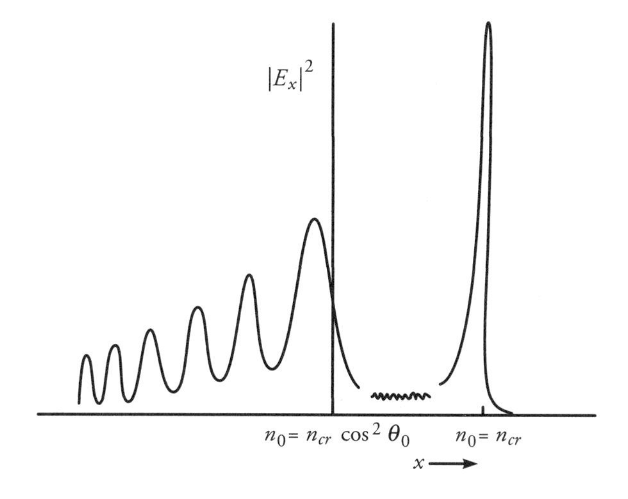


**Fig. 3.2** The mode structure of the field component $E_x$ near the turning point ($n_0 = n_{cr} \cos^2 \theta_0$) and the critical layer ($n_0 = n_{cr}$).

Regarding Eq. (3.21), one must notice that its right-hand side, for $\xi > v_{\text{th}}^2 / c^2$, where Eq. (3.22) could be used for $E_x$, is proportional to the plasma wave amplitude $a_1$ and goes as exp $[(ic/v_{\text{th}})\ (2/3)\ (-\xi)^{3/2}]$ which is too rapid for the transverse wave. Consequently, the contribution of the right-hand side, beyond $\xi > v_{\text{th}}^2 / c^2$, to $G$ is very small. Let $G_1(\xi)$ and $G_2(\xi)$ be the solutions of the homogeneous equation (3.21), i.e. when its right-hand side put to zero. Then the solution of Eq. (3.21) can be written as

$$G(\xi) = C_1 G_1(\xi) - qG_1(\xi) \int_{-\infty}^{\xi} \frac{G_2(\xi')}{\xi'} \left( G(\xi') + \frac{\xi' k_z}{q\omega} E_x(\xi') \right) d\xi'$$

$$-qG_2(\xi) \int_{\xi}^{\infty} \frac{G_1(\xi')}{\xi'} \left( G(\xi') + \xi' \frac{k_z}{\omega q} E_x(\xi') \right) d\xi', \tag{3.28}$$

where $q = k_z^2\ \lambda_{em}^2$ and one must remember that $G(\xi)$ is a function of $q$. Since the contribution to the integrals arises only from the vicinity of $\xi' \le v_{th}^2/c^2$, one may replace $G_1(\xi)$ and $G_2(\xi)$ inside them by $G_1(0)$, $G_2(0)$. Further, from Eqs. (3.20) and (3.24)

$$G(\xi') + \xi' \frac{k_z}{q\omega} E_x(\xi') = G(0) \frac{d^2W}{d\eta'^2}.$$

Hence

$$G(\xi) = C_1 G_1(\xi) - q_1 G_1(\xi) G_2(0) G(0) \int_{-\infty}^{\eta} \frac{1}{\eta'} \frac{d^2W}{d\eta'^2}\, d\eta'$$

$$-qG_2(\xi) G_1(0) G(0) \int_{\eta}^{\infty} \frac{1}{\eta'} \frac{d^2W}{d\eta'^2}\, d\eta' \tag{3.29}$$

and

$$G(0) = C_1 G_1(0) - qG_1(0) G_2(0) G(0) \int_{\infty}^{\infty} \frac{1}{\eta'} \frac{d^2W}{d\eta'^2}\, d\eta' \tag{3.30}$$

where $\eta = \xi\ c^{2/3}/v_{th}^{2/3}$ and the principal value of the integral is to be considered;

$$\int_{-\infty}^{\infty} \frac{1}{\eta'} \frac{d^2W}{d\eta'^2}\, d\eta' = i \int_0^{\infty} dt \int_{-\infty}^{\infty} \frac{t^2}{\eta'} \exp\left[-i\left(\eta' t + \frac{t^3}{3}\right)\right] d\eta'$$

$$= \int_0^{\infty} dt\, t^2\, e^{-i\frac{t^3}{3}} \int_{-\infty}^{\infty} \frac{\sin\alpha}{\alpha}\, d\alpha = -i\pi, \tag{3.31}$$

$$G(0) \ = \ C_1 \ \frac{G_1(0)}{1 - i\pi\ q\ G_1(0)\ G_2(0)}. \tag{3.32}$$

$G_1(0)$ and $G_2(0)$ are function of $q$

$$G_i(0) \ = \ A_i(q) \left[-\frac{2A_i(q)}{\pi\ A_j'(q)}\right]^{1/2}$$

$$G_2(0) = B_i(q)\left[-\frac{2B_i(q)}{\pi\, B_j'(q)}\right]^{1/2}$$

where $A_i$ and $B_i$ are the Airy functions and the prime denotes their derivatives with respect to $q$. For $(-\xi) > 1$, Eq. (3.29) gives

$$G_1(\xi) = C_1 G_1(\xi) + i\pi q\, G_1(0) G(0)\, G_2(\xi). \tag{3.33}$$

In this limit

$$G_1(\xi) = (-\xi - q)^{-1/4} \sin\left[\frac{3}{2}(-\xi - q)^{3/2} + \delta_0\right]$$

$$G_2(\xi) = (-\xi - q)^{-1/4} \cos\left[\frac{2}{3}(-\xi - q)^{3/2} + \delta_0\right];$$

hence, G($\xi$) can be cast in the form (3.23) giving

$$b_{1,2} = \frac{ic_1}{2}\left\{\frac{1 \pm \pi q\, G_1^2(0) - i\pi q\, G_1(0)\, G_2(0)}{1 - i\pi q\, G_1(0)\, G_2(0)}\right\} e^{\mp i\delta_0} \tag{3.34}$$

with $\delta_0$ as a real constant. The absorption coefficient can now be obtained as

$$A = 1 - \left|\frac{b_2}{b_1}\right|^2 \tag{3.35}$$

which depends only on a single parameter $q = k_z^2\, \lambda_{\text{em}}^2 = (\omega L_n / c)^{2/3} \sin^2\theta_0$, where it was used that $k_z = (\omega/c) \sin\theta_0$. The quantity $q$ represents the number of Airy wavelength between the turning point and the critical layer. Figure 3.3 demonstrates the variation of $A$ with $q$. A maximizes to ~ 0.5 at $q$ ~ 0.3. Such behavior can be understood as follows. For mode conversion one requires (i) a component of the wave field to be along the density gradient ($E_x \neq 0$) and (ii) the separation between the critical layer and the turning point to be small. At normal incidence $\theta_0 = 0$, $q = 0$, the electric vector $\vec{E} \parallel \hat{z}$. Hence, there are no density oscillations and no mode conversion. At large angles of incidence, the separation between the critical layer and turning point is too large, hence the electromagnetic wave field cannot tunnel into the critical layer and mode conversion cannot take place. The maximum mode conversion (hence maximum absorption) occurs for an optimum value of $q$ ~ 0.3.

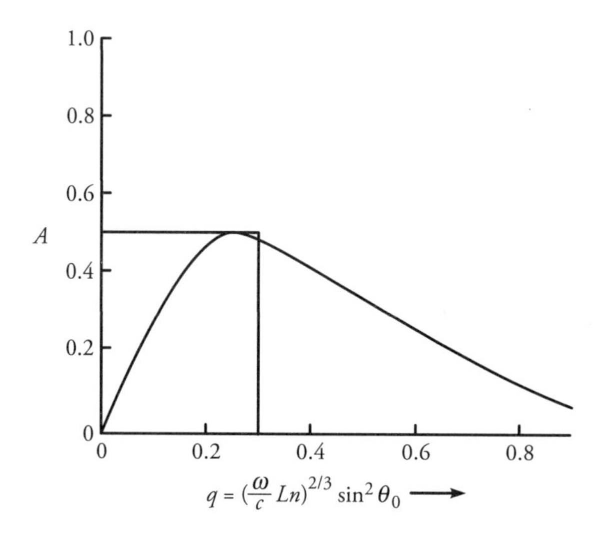


Fig. 3.3 The absorption coefficient as a function of the parameter $q$.

## 3.4 Brunel Absorption

Now, let us we consider a sharp boundary with an overdense plasma and study the collisionless absorption of an obliquely incident P-polarized laser. We employ Brunel's capacitor model to deduce the absorption coefficient.

Consider an overdense plasma–vacuum interface located at $x = 0$ with $x < 0$ representing the plasma of electron density $n_0^0$ and $x > 0$ the vacuum – (cf. Fig. 3.4)

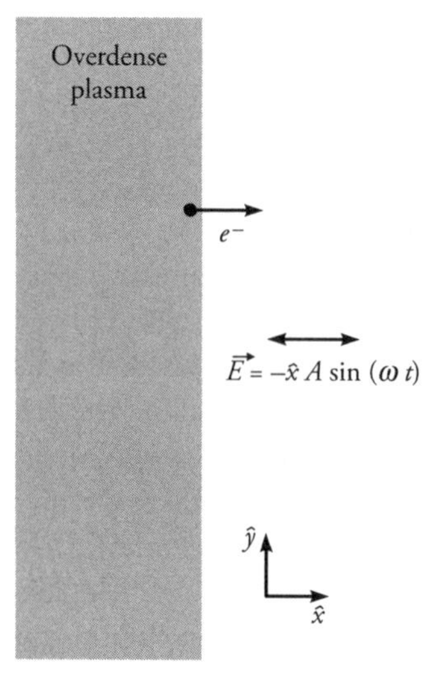


Fig. 3.4 Schematic of an overdense plasma–vacuum interface. In the vacuum region, an oscillatory electric field is applied.

We apply an AC field in the vacuum region

$$\vec{E} = -\hat{x} A \sin \omega t \tag{3.36}$$

with $\omega << \omega_p = \left(n_0^0 e^2 / m\varepsilon_0\right)^{1/2}$. At $t = 0$, the field is zero inside and outside the plasma. At $t > 0$, $E_x$ is $-ve$ in the vacuum. Since $E_x = 0$ always inside the plasma, there must exist a surface change density $\sigma$ on the surface at $x = 0$, such that the space charge field $E_s\,\hat{x}$ inside the plasma cancels the applied field

$$E_s = -\frac{\sigma}{\varepsilon_0} = -E_x = A \sin \omega t. \tag{3.37}$$

This surface charge is not localized on the surface but spreads into the vacuum ($x > 0$). Let the total number of electrons per unit area at $x > 0$ be $N$. Then

$$-Ne = \sigma = -A\,\varepsilon_0 \sin \omega t,$$

or

$$N = \left(A\varepsilon_0 / e\right) \sin \omega t. \tag{3.38}$$

For an electron released from the $x = 0$ surface at time $t_l$, the space charge ahead is $-eN(t_l)$ and the value of this space charge will remain (for this electron) same at later times. Thus, the net electric field seen by this electron (born at $t_l$) is

$$E_x + E_s\left(t_l\right) = -A \sin \omega t + A \sin \omega t_l. \tag{3.39}$$

The equation of motion for this electron,

$$\frac{dv_{xl}}{dt} = \frac{eA}{m}\left(\sin \omega t - \sin \omega t_l\right), \tag{3.40}$$

gives

$$v_{x1}\left(t\right) = v_{osc}\left(-\cos \omega t + \cos \omega t_l\right) - \omega v_{osc}\left(t - t_l\right) \sin \omega t_l, \tag{3.41}$$

$$x_l\left(t\right) = -\frac{v_{osc}}{\omega}\left(\sin \omega t - \sin \omega t_l\right) + v_{osc}\left(t - t_l\right) \cos \omega t_l$$

$$-\frac{v_{osc}}{\omega}\frac{\omega^2}{2}\left(t - t_l\right)^2 \sin \omega t_l, \tag{3.42}$$

where $v_{osc} = eA/m\omega$. We have used the boundary condition that at $t = t_1$, $x_1 = 0$, $v_{xl} = 0$.

The electrons released between $t_1$ and $t_1 + dt_1$ would, at time $t$, lie between $x_1$ and $x_1 - dx_1$,

$$-ndx_l = \frac{dN}{dt_l} dt_l,$$

or

$$n = -\frac{dN/dt_l}{dx_l/dt_l} \tag{3.43}$$

where $n$ is the electron density at $x_l(t)$. From Eqs. (3.38) and (3.43) we get

$$\frac{dN}{dt_l} = \left(A\varepsilon_0\, \omega/e\right)\cos\ \omega t_l,$$

$$\frac{dx_l}{dt_l} = -\frac{v_{\text{osc}}}{2}\omega^2\left(t - t_l\right)^2 \cos\omega t_l.$$

Thus,

$$n = \frac{A\varepsilon_0\omega}{e\left(v_{\text{osc}}/2\right)\ \omega^2\left(t - t_l\right)^2}$$

$$= \frac{2\ n_{\text{cr}}}{\omega^2\left(t - t_l\right)^2}, \tag{3.44}$$

where $n_{cr} = \omega^2\, m\, \varepsilon_0/e^2$ is the critical density.

Equation (3.42) gives $t_1$ as a function of $t$ and $x_1$, hence Eq. (3.44) gives $n$ as a function of $x$ ($=x_1$) at different $t$.

The electrons released from the plasma surface at $t_1$ would return back to plasma at time $t = t_l'$ when $x_1(t_l') = 0$, that is,

$$-\sin\left(\omega t_l'\right) + \sin\left(\omega t_l\right) + \omega\left(t_l' - t_l'\right)\cos\ \omega t_l - \frac{1}{2}\omega^2\left(t_l' - t_l\right)^2 \sin\left(\omega t_l\right) = 0. \tag{3.45}$$

Using this value of $t_l'$ one can obtain the velocity of the returning electrons at time $t_l'$ when they reach back the plasma, $v_{xl}\left(t_l'\right)$. The energy re-deposited by the electrons in the plasma in a half cycle is

$$W_{\text{abs}} = \int_0^{\pi/\omega} \frac{1}{2} m v_{\text{xl}}^2 (t_l') \, n \, v_{\text{xl}} (t_l') dt_l$$

Brunel defined a new quantity via

$$W_{\text{abs}} = \frac{1}{2} \eta \, n_0 \, m v_{\text{osc}}^2 = \frac{1}{2} \eta \frac{v_{\text{osc}}}{\omega} \varepsilon_0 A^2, \tag{3.46}$$

where $N_0 = A\varepsilon_0/e$ is the maximum number of electrons per unit area drawn from the surface. By evaluating $W_{\text{abs}}$ numerically, the quantity $\eta$ was found by Brunel to be $\approx 1.57$

The power absorbed per unit area per second is

$$I_{abs} = \frac{1}{2} \eta \frac{v_{osc}}{2\pi} \varepsilon_0 \, A^2 = \frac{\eta e}{4\pi} \frac{\varepsilon_0}{m\omega} A^3 .$$

One may note that the power absorption is not proportional to the square of the field amplitude but varies as its cube. Thus, Brunel absorption is a nonlinear affect.

## 3.5 Discussion

Resonance absorption constitutes one of the most important absorption mechanisms in laser produced plasmas. A pedagogic treatment of the resonance absorption mechanism in a cold, unmagnetized plasma is given by Frenning.[7] Resonance absorption has been studied numerically[8] as well as in the laboratory using microwaves[9] and using intense laser pulses interacting with solid metal targets.[10] In the presence of a self-generated magnetic field, transverse to the density gradient, an electromagnetic wave propagating in the X-mode possesses finite $E_x$ (the component of electric field along the density gradient) even at normal incidence and strong absorption may occur near the upper hybrid resonance layer.[4] Woods et al.[11] and Cairns and Lashmore Davies[12] have developed an elegant formalism of mode conversion of electromagnetic waves in an inhomogeneous magnetized plasma. One is also referred to the classical paper[13] and book by Stix,[14] as well as the brief review by Mjølhus.[15]

Brunel absorption, or vacuum heating, is a nonlinear mechanism, significant at mildly relativistic laser intensities. It can also cause significant absorption of large amplitude surface plasma waves. Signatures of vacuum heating have been observed in the interacting of intense laser pulses with solid targets.[16,17]

## References


1. Ginzburg, V. L. 1970. *The Propagation of Electromagnetic Waves in Plasma.* 2nd edition. Oxford: Pergamon.
2. Piliya, A. D. 1966. "Wave Conversion in an Inhomogeneous Plasma." *Soviet Physics Technical Physics-USSR* 11 (5): 609.
3. Kruer, W. L. 1988. *The Physics of Laser Plasma Interactions.* Massachusetts: Addison-Wesley.
4. Grebogi, C., C. S. Liu, and V. K. Tripathi. 1977. "Upper-Hybrid-Resonance Absorption of Laser Radiation in a Magnetized Plasma." *Physical Review Letters* 39 (6): 338.
5. Brunel, F. 1987. "Not-so-Resonant, Resonant Absorption." *Physical Review Letters* 59 (1): 52.
6. Kumar, Pawan, and V. K. Tripathi. 2011. "Vacuum Electron Heating by Surface Plasma Wave." *Applied Physics Letters* 99 (2): 021502.
7. G. Frenning. 1999. "Radio Wave Propagation and Linear Resonance Absorption in an Inhomogeneous, Cold, Unmagnetised Plasma." *IRF Scientific Report 258,* Swedish Institute of Space Physics.
8. Forslund, D. W., J. M. Kindel, Kenneth Lee, E. L. Lindman, and R. L. Morse. 1975. "Theory and Simulation of Resonant Absorption in a Hot Plasma." *Physical Review A* 11 (2) : 679.
9. Stenzel, R. L., A. Y. Wong, and H. C. Kim. 1974. "Conversion of Electromagnetic Waves to Electrostatic Waves in Inhomogeneous Plasmas." *Physical Review Letters* 32 (12): 654.
10. Meyerhofer, D. D., H. Chen, J. A. Delettrez, B. Soom, S. Uchida, and B. Yaakobi. 1993. "Resonance Absorption in High-intensity Contrast, Picosecond Laser–plasma Interactions." *Physics of Fluids B: Plasma Physics* 5 (7): 2584–88.
11. Woods, A. M., R. A. Cairns, and C. N. Lashmore-Davies. 1986. "A Full Wave Description of the Accessibility of the Lower-Hybrid Resonance to the Slow Wave in Tokamaks." *The Physics of Fluids* 29 (11): 3719–29.
12. Cairns, R. A., and C. N. Lashmore-Davies. 1983. "A Unified Theory of a Class of Mode Conversion Problems." *The Physics of Fluids* 26 (5): 1268–74.
13. Stix, Thomas H. 1965. "Radiation and Absorption via Mode Conversion in an Inhomogeneous Collision-free Plasma." *Physical Review Letters* 15 (23): 878.
14. Stix, Thomas H. 1992. *Waves in Plasmas.* New York: Springer Science & Business Media.
15. Mjølhus, E., 1990. "On Linear Conversion in a Magnetized Plasma." *Radio science, 25*(6): 1321–39.
16. Beg, F. N., A. R. Bell, A. E. Dangor, C. N. Danson, A. P. Fews, M. E. Glinsky, B. A. Hammel, P. Lee, P. A. Norreys, and Ma Tatarakis. 1997. "A Study of Picosecond Laser–solid Interactions up to $10^{19}$ $Wcm^{-2}$." *Physics of Plasmas* 4 (2): 447–57.
17. Wharton, K. B., S. P. Hatchett, S. C. Wilks, M. H. Key, J. D. Moody, V. Yanovsky, A. A. Offenberger, B. A. Hammel, M. D. Perry, and C. Joshi. 1998. "Experimental Measurements of Hot Electrons Generated by Ultraintense ($>10^{19}$ $W/cm^2$) Laser-Plasma Interactions on Solid-Density Targets." *Physical Review Letters* 81 (4): 822–25.

# Plasmonics

## Surface Plasma Waves and their Coupling to Lasers

## 4.1 Introduction

A sharp plasma–vacuum boundary or a conductor–dielectric interface supports surface plasma waves (SPWs) that propagate along the surface and whose electromagnetic fields fall away with the distance on both sides of the surface.[1–4] It is a popular means of medium wave (0.3–3 MHz) radio communication across the globe, where the Earth's surface plays the role of a conductor.

At optical frequencies, the SPWs excited by lasers can possess much larger amplitudes and shorter wavelengths than that of the laser, and hence, they are of vital importance in laser ablation of materials,[5–6] data storage devices, high sensitivity sensors[7–8] and meta-materials.[9–10] In plasmas, SPWs can drastically reduce the microwave reflectivity (in some cases[11] from 95% to 5%) of an overdense plasma. Significant improvement in target normal sheath acceleration (TNSA) of protons has been observed when the target has a surface grating[12,13] and the laser is coupled to SPWs. An important feature of an SPW is that its wave number and surface localization resonantly increase as the wave frequency approaches the surface plasmon resonance frequency $\omega = \omega_p/\sqrt{2}$, where $\omega_p$ is the plasma frequency of the plasma.

Thin conducting layers[14,15] support electrostatic plasmons with unique properties, one of which is that their frequency depends as the square root of the wave number in the long wavelength limit. This is in sharp contrast to the ordinary plasmons in three dimensions which oscillate with the plasma frequency. The two-dimensional plasmons have dispersion properties similar to water gravity waves[16] (another example of surface waves).

Graphene[17,18] is a novel two-dimensional mono-atomic layer material that has large in-plane conductivity and supports SPWs in the terahertz (THz) range.[19–22] The electrons in grapheme, called massless Dirac Fermions, have a linear energy–momentum relation ($\varepsilon = pv_F$, $v_F$ being the Fermi speed); as a result, all the electrons, irrespective of their energy, move with the same speed, $v_F = 10^8$ cm/s. The frequency of the plasmonic mode in grapheme scales as the square root of the wave number ($\omega \sim k^{1/2}$) and one-fourth power of the areal electron density. The phase velocity of plasmons is much lesser than the speed of light in vacuum; hence, they are primarily electrostatic modes. When a transverse magnetic field is applied, the electrons are localized in circular orbits and graphene, due to finite Larmor radius effects, supports a variety of magneto-plasmonic modes,[23] viz., upper hybrid and Bernstein-like modes near the multiples of the electron cyclotron frequency.

Rough metal surfaces or surfaces embedded with nanoparticles show yet another remarkable feature of surface plasmon resonance that enhances the Raman scattering of the laser by adsorbed molecules by five to six orders of magnitude, making surface enhanced Raman scattering (SERS) an extremely effective diagnostic tool.[24–27] The electron cloud inside a metallic nanoparticle has a natural frequency of oscillation against ions. For a spherical nanoparticle, this resonance frequency is $\omega_p/\sqrt{3}$. When the laser frequency equals the resonance frequency, the induced oscillatory space charge field inside the nanoparticle becomes orders of magnitude larger than the laser field, and this field causes enhanced Raman scattering from the adsorbed molecules.

In this chapter, we will study these processes in some detail. The mode structure and propagation characteristics of SPWs over metal or plasma surfaces will be studied in Section 4.2. In Section 4.3, we will discuss graphene plasmonics. We will study laser excitation of SPWs in the presence of a surface ripple in Section 4.4. In Section 4.5, we will study the interaction of laser light with nanoparticles and examine the Coulomb explosion of ions; while, in Section 4.6, we will discuss the surface enhanced Raman scattering of laser light from molecules adsorbed on nanoparticles.

## 4.2 Surface Plasma Wave

Consider a plasma–vacuum interface at $x = 0$ (cf. Fig. 4.1) with vacuum at $x > 0$ and the plasma at $x < 0$ characterized by an effective relative plasma permittivity at frequency $\omega$,

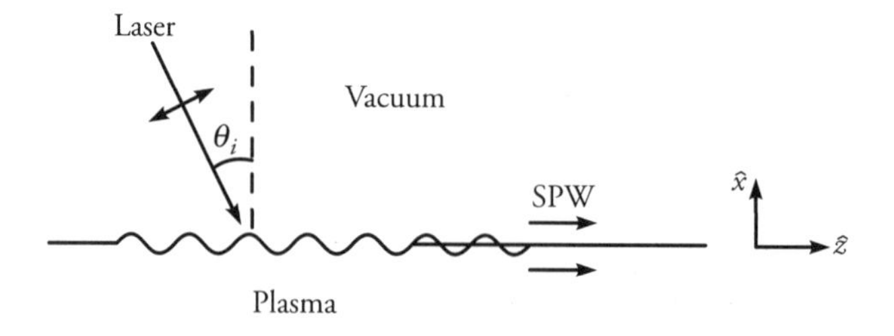


**Fig. 4.1** Schematic of surface plasma wave propagation over a plasma–vacuum interface. The SPW can be resonantly excited by a P-polarized laser that is obliquely incident at an angle $\theta_i$ on the rippled surface, $x = h\cos(qz)$.

$$\varepsilon_p = 1 - \frac{\omega_p^2}{\omega^2} \tag{4.1}$$

where $\omega_p = \left(n_0 e^2 / m\varepsilon_0\right)^{1/2}$ is the plasma frequency. A wave with frequency $\omega$ propagates along $\hat{z}$. We take the $t$ and $z$ variations of the fields as

$$\vec{E}, \vec{H} \sim e^{-i(\omega t - k_z z)}$$

with no variation along the $y$ direction. We look for a solution of the wave equation that falls away with the distance from the interface. Such solutions do not exist for $\vec{E} \parallel \hat{y}$, and hence, we consider $\vec{E} = \hat{x}E_x + \hat{z}E_z$. The wave equation governing $E_z$ is

$$\nabla^2 E_z + \frac{\omega^2}{c^2}\varepsilon_{\text{eff}} E_z = 0, \tag{4.2}$$

where $\varepsilon_{\text{eff}} = 1$ for $x > 0$ and $\varepsilon_{\text{eff}} = \varepsilon_p$ for $x < 0$. Replacing $\partial/\partial z$ by $ik_z$, Eq. (4.2) can be written as follows

$$\frac{\partial^2 E_z}{\partial x^2} - \alpha^2 E_z = 0,$$

$$\alpha^2 = \alpha_I^2 \equiv \left(k_z^2 - \frac{\omega^2}{c^2}\right)^{1/2} \qquad \text{for} \quad x > 0$$

$$= \alpha_{\text{II}}^2 \equiv \left(k_z^2 - \frac{\omega^2}{c^2}\varepsilon_p\right)^{1/2} \qquad \text{for} \quad x < 0$$

This gives us the following:

$$E_z = A(x)\, e^{-i(\omega t - k_z z)},$$

$$A(x) = A_0 e^{-\alpha_I x} \qquad \text{for} \quad x > 0$$

$$= A_0 e^{\alpha_{\text{II}} x} \qquad \text{for} \quad x < 0, \tag{4.3}$$

where we kept in view the fact that $E_z$ is continuous at $x = 0$ and $E_z$ must vanish as $|x| \to \infty$. Employing $\nabla \cdot \vec{E} = 0$ in both media, one obtains

$$E_x = \frac{ik_z}{\alpha_I} A_0 e^{-\alpha_I x} e^{-i(\omega t - k_z z)} \qquad \text{for} \quad x > 0$$

$$= -\frac{ik_z}{\alpha_{\text{II}}} A_0 e^{\alpha_{\text{II}} x} e^{-i(\omega t - k_z z)} \qquad \text{for} \quad x < 0. \tag{4.4}$$

If we need the normal component of the displacement vector to be continuous ($\varepsilon_0 \varepsilon_{\text{eff}} E_x$ = continuous) at $x = 0$, we obtain

$$\frac{\varepsilon_p}{\alpha_{\text{II}}} = -\frac{1}{\alpha_I},$$

which gives the following dispersion relation

$$k_z = \frac{\omega}{c}\left(\frac{\varepsilon_p}{1 + \varepsilon_p}\right)^{1/2}. \tag{4.5}$$

The spatial decrements

$$\alpha_I = \frac{\omega}{c}\left(\frac{-1}{1 + \varepsilon_p}\right)^{1/2}$$

and

$$\alpha_{\text{II}} = \frac{\omega}{c}\left(\frac{-\varepsilon_p^2}{1 + \varepsilon_p}\right)^{1/2}.$$

are real only when $1 + \varepsilon_p < 0$ or $\omega < \omega_p/\sqrt{2}$. Figure 4.2 shows the dispersion relation ($\omega$ vs $k_z$) of the SPW. For $\omega << \omega_p$, the wave number $k_z \approx \omega/c$, that is, the wave propagates at a speed close to $c$. As $\omega$ increases $k_z$ increases more rapidly and so do $\alpha_{\text{I}}$ and $\alpha_{\text{II}}$, giving a smaller phase velocity and a stronger localization of the fields. At $\omega \to \omega_p / \sqrt{2}$, the wave number $k_z$ becomes resonantly large, and the phase and group velocities of the wave become vanishingly small. Since the phase velocity of the wave is less than $c$ and it possesses a longitudinal electric field component $E_z$, it can be excited by an electron beam propagating along the surface, via Cherenkov interaction.

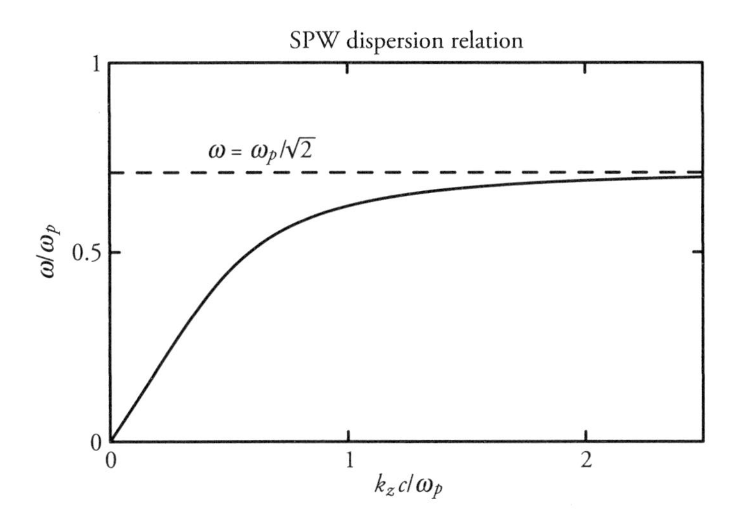


Fig. 4.2 The dispersion relation for the surface plasma wave. The frequency approaches SPW resonance, $\omega \to \omega_p / \sqrt{2}$, at $k_z \to \infty$.

## 4.3 Graphene Plasmons

Consider a graphene film mounted on a dielectric of relative permittivity $\varepsilon_g$ (cf. Fig. 4.3). Graphene is characterized by free electrons with the areal density $N_0^0$, the energy–momentum relation $\varepsilon = v_F p$, the velocity $\vec{v} = \partial\varepsilon / \partial\vec{p} = v_F \vec{p} / p$, and a two-dimensional (2D) equilibrium Fermi–Dirca distribution function

$$f_0^0 = \frac{1/(2\pi^2\hbar^2)}{e^{(\varepsilon-\varepsilon_F)/T} + 1}, \tag{4.6}$$

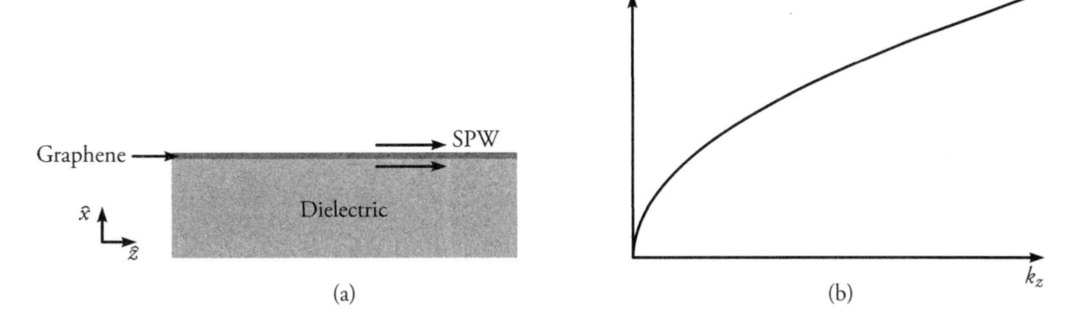


Fig. 4.3 (a) Schematic of SPW propagation over a graphene loaded dielectric; (b) Dispersion relation of SPWs.

where $\varepsilon_F$ is the Fermi energy, $v_F$ is the Fermi velocity (~ $10^8$ cm/s), and $T$ is the temperature in energy units, and $\hbar$ is the reduced Planck's constant. For $\varepsilon_F / T >> 1$, $N_0^0 = \varepsilon_F^2 / 2\pi\hbar^2 v_F^2$.

We can perturb the equilibrium by a space charge mode having the potential

$$\phi = Ae^{-k_z x} e^{-i(\omega t - k_z z)}, \; x > 0$$

$$= Ae^{k_z x} e^{-i(\omega t - k_z z)}, \; x < 0. \quad (4.7)$$

This equation is in compliance with Poisson's equation $\nabla^2 \phi = 0$ valid outside the graphene layer and the continuity of $\phi$ at $x = 0$. To incorporate the effect of the graphene layer, we can write the two-dimensional Vlasov equation for the free electrons,

$$\frac{\partial f}{\partial t} + \vec{v}.\nabla f - e(\vec{E} + \vec{v} \times \vec{B}).\frac{\partial f}{\partial \vec{p}} = 0, \quad (4.8)$$

where $\vec{E}$, and $\vec{B}$ are the electric and magnetic fields. In the presence of the space charge mode we write $f = f_0^0 + f_\omega^L$ and linearize the Vlasov equation,

$$\frac{\partial f_\omega^L}{\partial t} + \vec{v}.\nabla f_\omega^L = -e\nabla\phi.\frac{\partial f_0^0}{\partial \vec{p}}$$

to obtain

$$f_\omega^L = \frac{ek_z\phi}{\omega - k_z v_z} \frac{p_z}{p} \frac{\partial f_0^0}{\partial p}. \quad (4.9)$$

The corresponding areal electron density perturbation $N_\omega^L$, in the limit, $\omega > k_z v_F, E_F / T >> 1$ is

$$N_\omega^L = \int\int f_\omega^L dp_y dp_z = \chi_e k_z \varepsilon_0 \phi / e, \quad (4.10)$$

where

$$\chi_e = -\frac{N_0^0 e^2 k_z}{m^* \varepsilon_0 \omega^2} \quad (4.11)$$

is the electron susceptibility and $m^* = \varepsilon_F / v_F^2$ is the effective electron mass. We can deduce the same expression for density perturbation and $\chi_e$ by treating the electrons as a two-dimensional fluid with effective mass $m^*$ and using the fluid equations.

The jump condition on the normal component of displacement vector at $x = 0$ ,

$$\left.\frac{\partial\phi}{\partial x}\right]_{0^+} - \varepsilon_g \left.\frac{\partial\phi}{\partial x}\right]_{0^-} = \frac{e}{\varepsilon_0} N_\omega^L, \tag{4.12}$$

on using Eqs. (4.7) and (4.10) gives the dispersion relation $1 + \varepsilon_g + \chi_e = 0$, or

$$\omega^2 = \frac{N_0^0 e^2 k_z}{m^* \varepsilon_0 (1 + \varepsilon_g)}. \tag{4.13}$$

For a grapheme layer between two dielectrics of relative permittivities $\varepsilon_{g1}$ and $\varepsilon_{g2}$, the quantity $1 + \varepsilon_g$ should be replaced by $\varepsilon_{g1} + \varepsilon_{g2}$. The frequency of the plasmons scales as $k_z^{1/2}, N_0^{01/4}$ and falls in the terahertz range when the wavelength is in the micron range.

A background magnetic field perpendicular to the conducting layer gives rise to electron Bernstein-like magnetoplasmonic modes, as outlined in Appendix 4.1.

### 4.3.1 Electromagnetic plasmonic mode

To examine the electromagnetic character of the plasmonic mode, we can write the electric field of the mode as given by Eqs. (4.3) and (4.4) with the same $\alpha_I$ and $\alpha_{II}$ but with $\varepsilon_p$ replaced by $\varepsilon_g$. Using the jump condition on the normal component of the displacement vector,

$$E_x\big|_{0^+} - \varepsilon_g E_x\big|_{0^-} = -N_\omega^L e / \varepsilon_0, \tag{4.14}$$

we obtain the dispersion relation,

$$\frac{\varepsilon_g}{\alpha_{II}} + \frac{1}{\alpha_I} = \frac{N_0^0 e^2}{m^* \varepsilon_0 \omega^2}. \tag{4.15}$$

For $k_z >> \omega / c$, this dispersion relation reduces to Eq. (4.13), the electrostatic plasmon dispersion relation.

## 4.4 Surface Wave Coupling to Laser

In this section, we will study the excitation of a surface plasma wave by a laser. A basic problem arises due the wave number mismatch between the laser and the SPW. The wave number of the former is smaller than that of the latter for the same frequency. To compensate for it, we allow the metal/plasma to have a surface ripple (cf. Liu et al.[13]),

$$x = h \cos (qz) \tag{4.16}$$

with $q$ as the wave number and $h$ as the depth. A surface ripple is difficult to construct in a gaseous plasma; whereas it is straightforward to construct a surface grating on a metal. In a metal, the plasma permittivity can be written as $\varepsilon_p = \varepsilon_L - \omega_p^2 / \omega^2$, where $\varepsilon_L$ is the lattice dielectric constant. We may consider the ripple region ($-h < x < h$) as a density ripple with electron density

$$n = \frac{n_0^0}{2} + n_q, \quad n_q = \frac{n_0^0}{2} e^{iqz} . \tag{4.17}$$

In the case of a plasma, one can mount a metallic grating of width a fraction of a wavelength above the overdense plasma.[11]

A P-polarized laser is obliquely incident on the metal at an angle of incidence $\theta_i$ (cf. Fig. 4.1) with electric field

$$\vec{E}_i = A_{00}\left(\hat{z} - \tan\theta_i \hat{x}\right)\exp\left[-i\omega\left(t - \frac{z}{c}\sin\theta_i - \frac{x}{c}\cos\theta_i\right)\right] \tag{4.18}$$

In order to match the phases, the sum of the ripple wave number and the component of the laser wave vector along the $z$-axis must equal the SPW wave number,

$$q + \frac{\omega}{c}\sin\theta_i = k_z = \frac{\omega}{c}\left(\frac{\varepsilon_p}{1+\varepsilon_p}\right)^{1/2} . \tag{4.19}$$

Assuming that $\omega h/2\pi c << 1$, we write the transmitted laser field inside the plasma as

$$\vec{E}_T = TA_0\left(\hat{z} - i\frac{\omega}{c\alpha}\sin\theta_i \hat{x}\right)e^{\alpha x}\exp\left[-i\omega\left(t - \frac{z}{c}\sin\theta_i\right)\right] \tag{4.20}$$

where $\alpha \equiv \left(\frac{\omega^2}{c^2}\sin^2\theta_i - \frac{\omega^2}{c^2}\varepsilon_p\right)^{1/2}$, $T = \dfrac{2}{1 - \varepsilon_p \cos\theta_i / (i\alpha c / \omega)}$ .

The laser field imparts an oscillatory velocity to the electrons, $\vec{v}_\omega = e\vec{E}_T / (im\omega)$ that beats with the density ripple to produce a nonlinear current density

$$\vec{J}^{\mathrm{NL}} = -\frac{1}{2} n_q e \vec{v}_\omega , \tag{4.21}$$

which drives the surface plasma wave.

The electric and magnetic fields of the surface wave are governed by Maxwell's equations

$$\nabla \times \vec{E}_S = i\omega\, \mu_0\, \vec{H}_S, \tag{4.22}$$

$$\nabla \times \vec{H}_S = \vec{J}^{\mathrm{NL}} - i\omega\varepsilon_0\varepsilon_{\mathrm{eff}}\vec{E}_S, \tag{4.23}$$

which we will solve iteratively. First, we neglect $\vec{J}^{\mathrm{NL}}$ and obtain the SPW solution

$$\vec{E}_S,\, \vec{H}_S = \left[\vec{F}_1(x)\,,\, \vec{F}_2(x)\right] e^{-i\,(\omega\, t - k_z z)} \tag{4.24}$$

$$\vec{F}_1(x) = \begin{cases} \left(\hat{z} + i\dfrac{k_z}{\alpha_I}\hat{x}\right) e^{-\alpha_I x} & \text{for } x > 0 \\ \left(\hat{z} - i\dfrac{k_z}{\alpha_{\mathrm{II}}}\hat{x}\right) e^{\alpha_{\mathrm{II}} x} & \text{for } x < 0 \end{cases}$$

$$\vec{F}_2(x) = \begin{cases} \left(\dfrac{i\omega}{\mu_0 c^2 \alpha_I}\hat{y}\right) e^{-\alpha_I x} & \text{for } x > 0 \\ \left(\dfrac{-i\omega\varepsilon_p}{\mu_0 c^2 \alpha_{\mathrm{II}}}\hat{y}\right) e^{\alpha_{\mathrm{II}} x} & \text{for } x < 0 \end{cases} \tag{4.25}$$

where we have employed Eq. (4.22) to write $\vec{F}_2$.

For finite $\vec{J}^{\mathrm{NL}}$, we assume that the mode structure of the SPW remains unchanged, but that the SPW amplitude becomes a function of $z$,

$$\vec{E}_S = A_1(z)\, \vec{F}_1(x)\exp\left[-i(\omega t - k_z z)\right]$$

$$\vec{H}_S = A_2(z)\vec{F}_2(x)\exp\left[-i\,(\omega t - k_z z)\right] \tag{4.26}$$

Substituting these in Eqs. (4.22) and (4.23) and using the properties of $\vec{F}_1$ and $\vec{F}_2$, we obtain

$$\frac{\partial A_1}{\partial z} F_{1x} = -i\omega\, \mu_0 \left(A_2 - A_1\right) F_{2y} \tag{4.27}$$

$$\frac{\partial A_2}{\partial z}\left(\hat{z} \times \vec{F}_2\right) = -i\omega\, \varepsilon_0 \varepsilon_p \left(A_1 - A_2\right)\vec{F}_1 + \vec{J}_A^{\text{NL}} \tag{4.28}$$

$$\vec{J}_A^{\text{NL}} \equiv -\frac{n_0^0}{4}\frac{e^2 T A_0}{im\omega}\left(\hat{z} - \frac{i\omega \sin\theta_i}{c\alpha}\hat{x}\right) \tag{4.29}$$

Multiplying Eq. (4.17) by $F_{2y}^*$ and taking the scalar product between Eq. (4.28) and $\vec{F}_1^*$, and integrating over $\xi$ from $-\infty$ to $+\infty$, we obtain

$$\frac{\partial A_1}{\partial z} = -i\omega\mu_0\left(A_2 - A_1\right)\frac{I_2}{I_1}, \tag{4.30}$$

$$\frac{\partial A_2}{\partial z} = -i\omega\varepsilon_0\left(A_1 - A_2\right)\frac{I_3}{I_1} + h\left(\vec{J}_{\text{A}}^{\text{NL}} \cdot \vec{F}_1^*\right) / I_1, \tag{4.31}$$

where

$$I_1 = \int_{-\infty}^{+\infty} F_{1x}F_{2y}^* dx = \frac{k_x \omega\left(\alpha_{\text{I}}^2 - \alpha_{\text{II}}^2\right)}{2\mu_0 c^2 \alpha_{\text{I}}^3},$$

$$I_2 = \int_{-\infty}^{+\infty} F_{2y}F_{2y}^* dx = \frac{\omega^2\left(\alpha_{\text{I}} + \alpha_{\text{II}}\right)}{2\mu_0^2 c^4 \alpha_{\text{I}}^3 \alpha_{\text{II}}},$$

$$I_3 = \int_{-\infty}^{+\infty} \vec{F}_1 \cdot \vec{F}_1^* dx = \frac{k_z^2\left(\alpha_{\text{II}}^2 - \alpha_I^2\right)}{2\alpha_I^3 \alpha_{\text{II}}^2}. \tag{4.32}$$

To the zeroth order $A_1$ and $A_2$ are constant, hence $\partial A_1/\partial z = \partial A_2/\partial z = 0$, i.e., $A_1 - A_2 = 0$. To the next order $A_1 - A_2$ is nonzero but small. Hence $\partial A_2/\partial z$ in Eq. (4.31) can be taken $\approx \partial A_1/\partial z$. Eventually one obtains

$$\frac{\partial A_1}{\partial z} = R, \quad R \equiv \frac{+h\left(\vec{J}_A^{\text{NL}} \cdot \vec{F}_1^*\right)_{x=0}}{I_1\left(1 + \dfrac{\varepsilon_0}{\mu_0}\dfrac{I_3}{I_2}\right)} \tag{4.33}$$

Giving the SPW to laser amplitude ratio

$$\left|\frac{A_1}{A_L}\right| = \frac{\alpha_I^3\, \alpha_{\text{II}}^2\, \omega_p^2\, |T|\, \left(1 + \omega k_z \sin\theta_i / \alpha_I \alpha_{\text{II}} c\right)}{4\omega^2\, k_z\, \left(\alpha_{\text{II}}^2 - \alpha_I^2\right)}\, hz\cos\theta_i \tag{4.34}$$

where $A_L = A_0 \sec\theta_i$, and $z$ is of the order of the spot size $r_0$ of the laser. The SPW amplitude scales linearly with the depth of the ripple and spot size of the laser. It rises rapidly as $\left(\omega^2 - \omega_R^2\right)^{-2}$ when the frequency approaches the surface plasmon resonance $\omega_R = \omega_p/(\varepsilon_L + 1)^{1/2}$. Figure 4.4 shows the variation of $(A_1/A_L)$ as a function of $\omega/\omega_R$ for typical parameters $\varepsilon_L = 4, \omega b/c = 0.3, \omega r_0/c = 30, q_i = 45°$. For $\omega < 0.6\omega_R$, the amplitude ratio is less than 1. As $w$ increases and approaches $\omega_R$, the amplitude ratio rises rapidly to values far greater than 1. At resonance, the SPW amplitude is limited only by collisional effects. Near the surface plasmon resonance however, the requisite value of the ripple wave number becomes too large to be practical. Power conversion efficiency of ~ 50% is achievable for reasonable parameters.

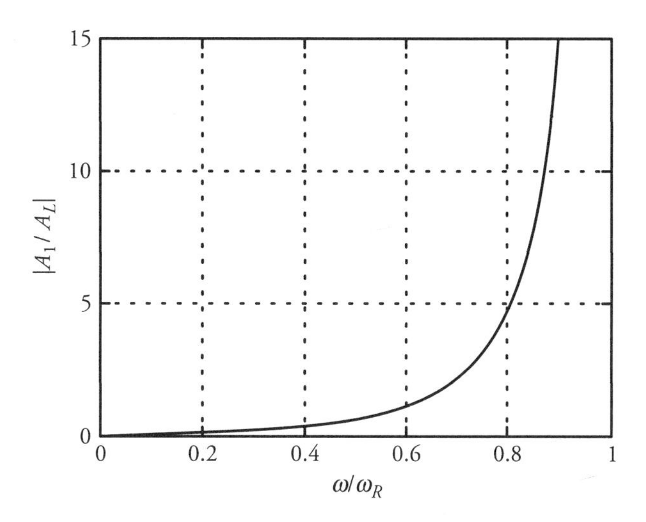


**Fig. 4.4** Normalized surface plasma wave amplitude $A_1$ as a function of the normalized frequency; $A_L$ is the normalized laser amplitude and $\omega_R = \omega_p/\left(\varepsilon_L + 1\right)^{1/2}$.

## 4.5 Surface Enhanced Raman Scattering

In this section, we will discuss the coupling of a laser to nanoparticles, and the resulting surface enhanced Raman scattering (SERS).

### 4.5.1 Response of a nanoparticle to a laser field

Let us begin with the linear response of nanoparticles to a laser field. Consider a nanoparticle of radius $r_p$, free electron density $n_{op}$, electron effective mass m', and lattice relative permittivity $\varepsilon_L'$ (cf. Fig. 4.5). It is subjected to a wave electric field

$$\vec{E} = \vec{A}(\vec{r})e^{i\omega t},$$

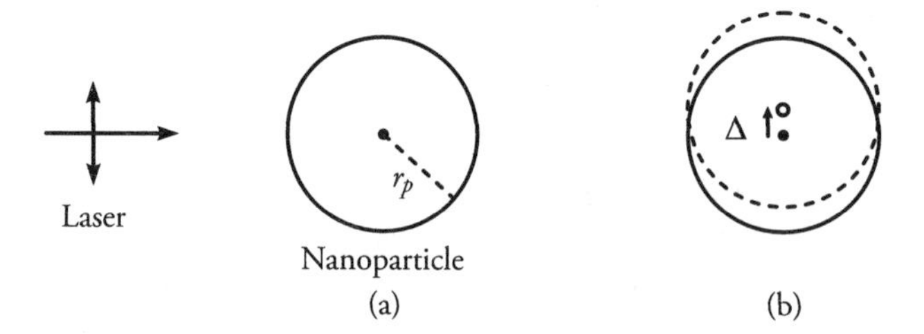


**Fig. 4.5** (a) Schematic of a nanoparticle of radius $r_p$ irradiated by a laser. (b) Overlap region of the electron sphere (dotted), displaced by a distance Δ, and the positively charged sphere (solid).

where the scale-length of the space variation of $\vec{A}$ is much longer than $r_p$. The field causes a displacement of the free electrons. As the electron sphere is displaced by a distance $\vec{\Delta}$ with respect to the ion sphere, a space charge field is created. If one ignores the lattice polarizability (i.e., $\varepsilon_L' = 1$), the space charge field at distance $\vec{r}$ from the center of the nanoparticle in the overlap region of the free electron sphere and ion sphere is $n_{0p}e\vec{r}/3\varepsilon_0$ due to the positive charge sphere and $-n_{0p}e(\vec{r}-\vec{\Delta})/3\varepsilon_0$ due to the electron sphere, that is $\vec{E}_s = \left(n_{0p}e/3\varepsilon_0\right)\vec{\Delta}$. When one includes the polarizability of the lattice, one obtains

$$\vec{E}_s = \left(n_{0p}e / \varepsilon_0(\varepsilon_L' + 2)\right)\vec{\Delta} \tag{4.35}$$

in the overlap region of the electron and ion spheres. This field exerts a restoration force $-e\vec{E}_s = -\omega_{\mathrm{pp}}^2\, m'\,\vec{\Delta}/(\varepsilon_L' + 2)$ $(\omega_{\mathrm{pp}}^2 = n_{\mathrm{op}}e^2/m'\varepsilon_0)$ on each electron. The equation of motion for the electrons

$$\frac{d^2\vec{\Delta}}{dt^2} + \frac{\omega_{pp}^2}{(\varepsilon_L' + 2)}\vec{\Delta} = \frac{e\vec{E}}{m'}$$

thus gives

$$\vec{\Delta} = \frac{e\vec{E}}{m'\left(\omega^2 - \omega_{\mathrm{pp}}^2/(\varepsilon_L' + 2)\right)}. \tag{4.36}$$

A resonance occurs at the surface plasmon resonance frequency $\omega = \omega_{\mathrm{SPR}} = \omega_{\mathrm{pp}}/(\varepsilon_L' + 2)^{1/2}$. The induced space charge field inside the particle is

$$\vec{E}_s = \beta\vec{E};\ \beta = \frac{\omega_{\mathrm{pp}}^2/(\varepsilon_L' + 2)}{\omega^2 - \omega_{\mathrm{pp}}^2/(\varepsilon_L' + 2)}. \tag{4.37}$$

which acquires a much larger value than the wave field $\vec{E}$ as $\omega$ approaches the resonance frequency. For non-spherical particles, $\beta$ is a tensor, and the resonance is shifted to $\omega = \omega_\alpha$, where $\omega_\alpha$ depends on the shape of the particle as well as on the free electron density. It is this field, $\vec{E}_s$, that polarizes the molecule adsorbed on the nanoparticle and causes SERS.

### 4.5.2 SERS of a surface plasma wave

Consider a conductor–vacuum interface embedded with nanoparticles of radius $r_p$ and an areal density of $N$ (cf. Fig. 4.6). The conductor is characterized by a free electron density $n_0$, electron effective mass m and lattice permittivity $\varepsilon_L$. Corresponding quantities for the nanoparticle are $n_{0p}, m'$, and $\varepsilon'_L$. Molecules are adsorbed on the surface of the nanoparticles. The SPW propagates over the conductor surface along $\hat{z}$ with fields given by Eqs. (4.3)–(4.4).

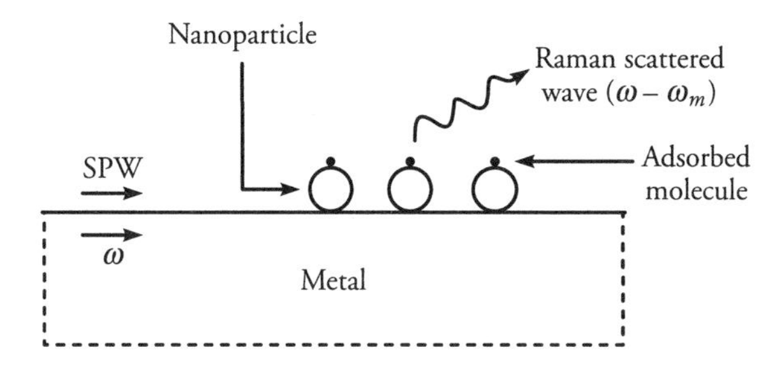


Fig. 4.6 Surface wave propagation over a metal–vacuum interface embedded with nanoparticles having adsorbed molecules. The Raman scattered wave, shifted by a frequency $\omega_m$, propagates in free space.

The SPW induces a space charge field inside the particles. The induced field polarizes the adsorbed molecules with the dipole moment

$$\vec{p}_{\text{mol}} = \alpha_R \beta \vec{E}, \tag{4.38}$$

where $\alpha_R$ is the effective Raman polarizability of the molecules, which gets modulated by the vibrational motion of the atoms. Thus, $\vec{p}_{\text{mol}}$ contains a Stokes frequency $\omega_s = \omega - \omega_m$, where $\omega_m$ is the vibrational frequency of the molecule. The dipole moment $\vec{p}_{\text{mol}}(\omega_R)$ produces a Raman-shifted electric field in the nanoparticle,

$$\vec{E}_1(\omega_s) = 2\,\vec{p}_{\text{mol}}(\omega_s) / 4\pi\varepsilon_0 r_p^3. \tag{4.39}$$

giving rise to the displacement and oscillatory velocity of the electrons

$$\vec{\Delta}_s = \frac{e\vec{E}_1}{m'\left(\omega_s^2 - \omega_{\text{pp}}^2 / (\varepsilon'_L + 2)\right)}$$

and

$$\vec{v}_s = -i\omega_s \vec{\Delta}_s . \tag{4.40}$$

Thus, the Raman shifted current density due to the nanoparticles is

$$\vec{J}_s = -N\delta(x)e\left(\frac{4\pi}{3}\eta_{op}v_p^3\right)\vec{v}_s$$

$$= -\frac{2}{3i\omega_s}\omega_{pp}^2 N\delta(x)\alpha_R\beta\beta_R\vec{E}, \tag{4.41}$$

where $\beta_R = \omega_{pp}^2 / \left((\varepsilon_L' + 2)\omega_s^2 - \omega_{pp}^2\right) \approx \beta$ as $\omega_m << \omega$. It is enhanced by a factor $\beta^2$ due to the surface plasmon resonance. This current density produces a Raman-shifted scattered waves. For constant $N$, the current density $\vec{J}_R$ varies proportional to exp $[-i\ (\omega_R t - k_z z)]$; hence, it would excite only a surface plasma wave. In order to have a Raman shifted scattered wave escaping into free space, we allow $N$ to have a surface ripple $N = N_0\ (1 + \mu \cos qz)$ and retain only the component of the Raman shifted current density that varies proportional to exp $[-i\ (\omega_R t - k_{Rz} z)]$, where $k_{Rz} = k_z - q < \omega/c$,

$$\vec{J}_s' = -\frac{\mu\omega_{pp}^2}{3i\omega_s} N\delta(x)\alpha_R\beta^2\vec{E}. \tag{4.42}$$

The wave equation governing the Raman shifted field $\vec{E}_s$ is

$$\nabla^2\vec{E}_s - \nabla\left(\nabla\cdot\vec{E}_s\right) + \frac{\omega_R^2}{c^2}\varepsilon_s'\vec{E}_s = -\frac{i\ \omega_s}{c^2\ \varepsilon_0}\vec{J}_s', \tag{4.43}$$

where $\varepsilon_s' = 1$ for $x > 0$, $\varepsilon_R' = \varepsilon_s = \varepsilon_L - \omega_p^2 / \omega_s^2$ for $x < 0$. The current density $\vec{J}_s'$ is finite only in the layer containing nanoparticles. Outside the layer, the solution of Eq. (4.33) can be written as

$$\vec{E}_s = A_{s1}\left(\hat{z} - \frac{k_{sz}}{k_{sx}}\hat{x}\right)e^{ik_{sx}x}e^{-i\ (\omega_s t\ -\ k_{sz}z)} \quad \text{for } x > 0$$

$$= A_{s1}\left(\hat{z} - \frac{ik_{sz}}{\alpha'}\hat{x}\right)e^{\alpha' x}e^{-i\ (\omega_s t\ -\ k_{sz}z)} \quad \text{for } x < 0, \tag{4.44}$$

where $\alpha' = \left(k_{sz}^2 - \omega_s^2\varepsilon_s / c^2\right)^{1/2}$, $k_{sx} = \left(\omega_s^2 / c^2 - k_{sz}^2\right)^{1/2}$.

$E_z$ is continuous across the boundary. The other boundary condition can be deduced by integrating the $z$ component of Eq. (4.33) across the interface from $0^-$ to $0^+$,

$$\left[\frac{\partial E_{sz}}{\partial x} - ik_{sz}E_{sx}\right]_{0^-}^{0^+} = -\frac{\mu}{c^2 \in_0}\frac{\omega_{pp}^2 N}{3}\alpha_R \beta^2 E_z . \tag{4.45}$$

$$\left|\frac{A_{s1}}{A_1}\right| = \frac{\alpha' \mu\, \omega_{pp}^2\, N\alpha_R\beta^2}{\varepsilon_0\omega_s^2\left|\varepsilon_s + i\alpha' / k_{sx}\right|} . \tag{4.46}$$

The intensity of the Raman shifted signal is

$$I_s = A_{s1}^2 / \left(2\mu_0 c \cos^2\theta_s\right),$$

where $\theta_s = \tan^{-1} k_{sz} / k_{sx}$ is the angle the Raman scattered wave makes with the surface normal.

For fullerene molecules adsorbed on silver nanoparticles attached to a silver base, $\alpha_R/4\pi\varepsilon_0 = 7.7 \times 10^{-29}$ m$^3$, $\omega_m = 8.48 \times 10^{12}$ rad/s, $\varepsilon_L = 4$, $n_0 \simeq 5 \times 10^{28}$ m$^{-3}$. For $\mu \sim 0.3$, $N = 10^{14}$ m$^{-2}$, and $\beta \sim 10^2$, one has $|A_{s1} / A_1| \sim 10^{-2}$.

## 4.6 Discussion

Surface plasma waves (SPWs) significantly influence laser–surface interactions.[28–31] They give rise to nonlinear effects at a much reduced laser intensity when the conditions for mode conversion are satisfied. SPW-based generators of terahertz radiation, employing optical rectification of lasers, are the focus of extensive research nowadays. SPWs have also shown potential for target normal sheath acceleration of protons.

### Appendix 4.1

# Magnetoplasmons in Graphene

Following the analysis formulated by Liu and Tripathi,[23] we consider the geometry shown in Fig. 4.7 with graphene mounted on a dielectric of relative permittivity $\varepsilon_g$. We apply a transverse static magnetic field $B_s\hat{x}$ to the graphene layer. Next, we perturb the equilibrium by a plasmonic mode having a potential given by Eq. (4.7). We can then write the distribution function as $f = f_0^0 + f_\omega^L$ and linearize the Vlasov equation (Eq. 4.8),

$$\frac{\partial f_\omega^L}{\partial t} + \vec{v}\cdot\nabla f_\omega^L - e\vec{v}\times\vec{B}_s\cdot\frac{\partial f_\omega^L}{\partial \vec{p}} \equiv \frac{df_\omega^L}{dt} = -e\nabla\phi\cdot\frac{\partial f_0^0}{\partial \vec{p}}$$

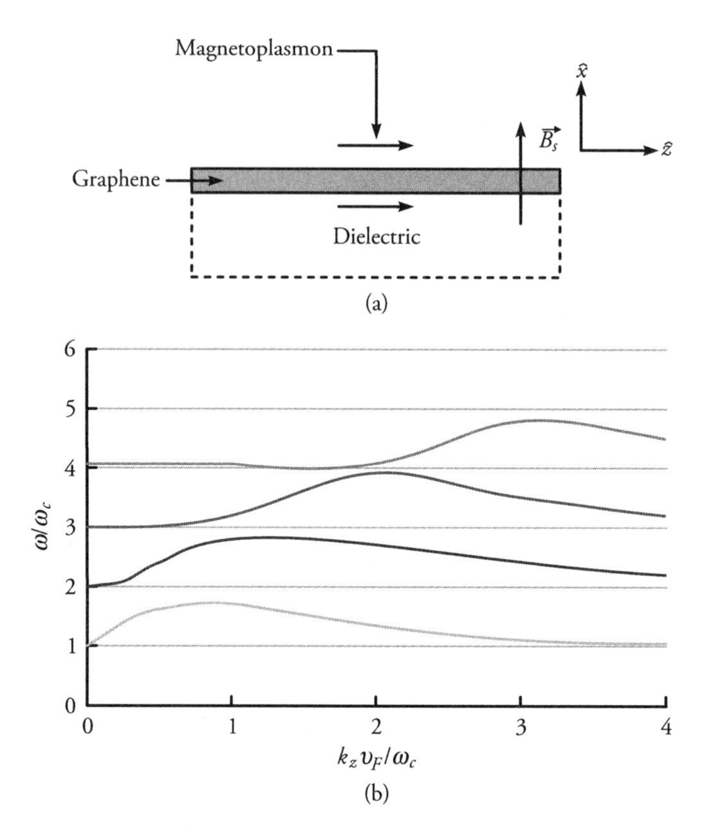


**Fig. 4.7** (a) Schematic of magnetoplasmon propagation on a graphene loaded dielectric in the presence of a transverse magnetic field. (b) Dispersion relation of plasmonic modes originating at integer multiples of cyclotron frequency for $G = N_0^0 e^2 / m^* \varepsilon_0 v_F \omega_c = 22$ corresponding to $N_0^0 = 10^{11}\,\text{cm}^{-2}$, $B_s = 1\,\text{Tesla}$, and $\varepsilon_g = 3$.

to obtain

$$f_\omega^L = -ieA \frac{1}{p} \frac{\partial f_0^0}{\partial p} \int_{-\infty}^{t} \vec{k} \cdot \vec{p}' e^{-i(\omega t' - k_z z')} dt' , \tag{A4.1}$$

where $z'$ and $\vec{p}'$ refer to the position and momentum of an electron at time $t'$ and the integration is to be carried out over unperturbed trajectory of the electrons in the static magnetic field, governed by the equation of motion,

$$\frac{d\vec{p}'}{dt'} = -e\vec{v}' \times \vec{B}_s = -\frac{ev_F}{p} \vec{p}' \times \vec{B}_s . \tag{A4.2}$$

Under the conditions that at $t' = t$, the electron has the position and momentum $y' = y$, $z' = z$, $p_z' = p_z = p\cos\theta, p_y' = p_y = -p\sin\theta$ (where $\theta$ is the gyrophase). Equation (A4.2) gives

$$p_z' = p\cos(\omega_c(t'-t)+\theta),\ p_y' = -p\sin(\omega_c(t'-t)+\theta),$$

$$z' = z + \frac{v_F}{\omega_c}[\sin(\omega_c(t'-t)+\theta) - \sin\theta],$$

$$y' = y + \frac{v_F}{\omega_c}[\cos(\omega_c(t'-t)+\theta) - \cos\theta], \quad \text{(A4.3)}$$

where $\omega_c = eB_s v_F/p$. Using Eq. (A4.3) and employing the Bessel function identity, $e^{i\alpha\sin\theta} = \sum_l J_l(\alpha)e^{il\theta}$, we obtain from Eq. (A4.1),

$$f_\omega^L = \frac{e\phi}{v_F}\frac{\partial f_0^0}{\partial p}\sum_l\sum_{l'}\frac{l\omega_c}{\omega - l\omega_c}J_l\left(\frac{k_z v_F}{\omega_c}\right)J_{l'}\left(\frac{k_z v_F}{\omega_c}\right)e^{i(l-l')\theta}. \quad \text{(A4.4)}$$

The areal density perturbation turns out to be

$$N_\omega^L = \int_0^\infty\int_0^{2\pi} f_\omega^L d\theta\, p dp = \frac{\chi_e\varepsilon_0}{e}k_z\phi,$$

where

$$\chi_e = -\frac{2N_0^0 e^2}{k_z m^* \varepsilon_0 v_F^2}\sum_l \frac{l\omega_c J_l^2\left(\frac{k_z v_F}{\omega_c}\right)}{\omega - l\omega_c}, \quad \text{(A4.5)}$$

is the electron susceptibility and $\omega_c = eB_s/m^*$ is the electron cyclotron frequency with effective mass $m^* = \varepsilon_F/v_F^2$, and we have assumed $\omega - l\omega_c > k_z v_F, E_F/T >> 1$. Using $N_\omega^L$ in the jump condition Eq. (4.12), we obtain the dispersion relation for the magnetoplasmons $1+\varepsilon_g+\chi_e = 0$, or

$$1+\varepsilon_g = \frac{2N_0^0 e^2}{m^*\varepsilon_0 k_z v_F^2}\sum_l \frac{l\omega_c J_l^2\left(\frac{k_z v_F}{\omega_c}\right)}{\omega - l\omega_c}. \quad \text{(A4.6)}$$

For $k_z v_F/\omega_c << 1$ (the small Larmor radius limit) only the l = 1, –1 terms are important. Equation (A4.6) gives

$$\omega^2 = \omega_c^2 + \frac{N_0^0 e^2 k_z}{m^*\varepsilon_0(1+\varepsilon_g)}. \quad \text{(A4.7)}$$

This is a kind of upper hybrid wave in a plasma[1]. For $\omega_c = 0$, Eq. (A4.6) reduces to the unmagnetized plasmon dispersion relation Eq. (4.13). With finite Larmor radius effects, Eq. (A4.6) offers many roots, each corresponding to a Bernstein magnetoplasmon mode with frequencies between the harmonics of the electron cyclotron frequency. At $k_z v_F / \omega_c \geq 1$, the $l^{th}$ harmonic Bernstein mode has

$$\omega \approx l\omega_c \left[ 1 + \frac{2N_0^0 e^2 J_l^2\left(\dfrac{k_z v_F}{\omega_c}\right)}{m^* \varepsilon_0 k_z v_F^2 \left(1+\varepsilon_g\right)} \right] \tag{A4.8}$$

In Fig. 4.7, we have plotted the dispersion relation for the Bernstein plasmonic modes. The modes are localized between cyclotron harmonics and possess large wave numbers. They offer the possibility of building tunable plasmonic devices with the magnetic field as the tuning parameter.

## References

1. Liu, C. S., and V. K. Tripathi. 2007. *Electromagnetic Theory for Telecommunications*. Cambridge India.
2. Raether, H. 1988. *Surface Plasmons on Smooth and Rough Surfaces and on Gratings*. Springer Tracts in Mod. Physics vol. 111. New York: Springer
3. Jordan, E. C., and K. G. Balmain. 1968. *Electromagnetic Waves and Radiating Systems*. Englewood Cliffs, New Jersey: Prentice Hall..
4. Smolyaninov, Igor I., David L. Mazzoni, and Christopher C. Davis. 1996. "Imaging of Surface Plasmon Scattering by Lithographically Created Individual Surface Defects." *Physical Review Letters* 77 (18): 3877.
5. Hutter, Eliza and Janos H. Fendler. 2004. "Exploitation of Localized Surface Plasmon Resonance." *Advanced Materials* 16 (19): 1685–706.
6. Garner, Quincy, and Pal Molian. 2013. "Formation of Gold Microparticles by Ablation with Surface Plasmons." *Nanomaterials* 3 (4): 592–605.
7. Cahill, Colin P., Kyle S. Johnston, and Sinclair S. Yee. 1997. "A Surface Plasmon Resonance Sensor Probe Based on Retro-reflection." *Sensors and Actuators B: Chemical* 45 (2): 161–6.
8. Clerc, D., and W. Lukosz. 1997. "Real-time Analysis of Avidin Adsorption with an Integrated-optical Output Grating Coupler: Adsorption Kinetics and Optical Anisotropy of Adsorbed Monomolecular Layers." *Biosensors and Bioelectronics* 12 (3): 185–94.
9. Pendry, J. B. 2004. "A Chiral Route to Negative Refraction." *Science* 306 (5700): 1353–5.
10. Fang, Nicholas, Hyesog Lee, Cheng Sun, and Xiang Zhang. 2005. "Sub-diffraction-limited Optical Imaging with a Silver Superlens." *Science* 308 (5721): 534–7.

11. Liu, C. S., V. K. Tripathi, and R. Annou. 2008. "Resonant Reduction in Microwave Reflectivity from an Overdense Plasma with the Employment of a Parallel Metal Grating." *Physics of Plasmas* 15 (6): 062103.
12. Ceccotti, T., V. Floquet, A. Sgattoni, A. Bigongiari, O. Klimo, M. Raynaud, C. Riconda et al. 2013. "Evidence of Resonant Surface-wave Excitation in the Relativistic Regime through Measurements of Proton Acceleration from Grating Targets." *Physical Review Letters* 111 (18): 185001.
13. Liu, C. S., V. K. Tripathi, Xi Shao, and T. C. Liu. 2015. "Nonlinear Surface Plasma Wave Induced Target Normal Sheath Acceleration of Protons." *Physics of Plasmas* 22 (2): 023105.
14. Stern, Frank. 1967 "Polarizability of a Two-dimensional Electron Gas." *Physical Review Letters* 18 (14): 546.
15. Grimes, C. C., and Gregory Adams. 1976 "Observation of Two-Dimensional Plasmons and Electron-ripplon Scattering in a Sheet of Electrons on Liquid Helium." *Physical Review Letters* 36 (3): 145.
16. Eliasson, Bengt, and Chuan Sheng Liu. 2016 "Nonlinear Plasmonics in a Two-dimensional Plasma Layer." *New Journal of Physics* 18 (5): 053007.
17. Novoselov, Kostya S., Andre K. Geim, Sergei V. Morozov, D. A. Jiang, Y. Zhang, Sergey V. Dubonos, Irina V. Grigorieva, and Alexandr A. Firsov. 2004 "Electric Field Effect in Atomically Thin Carbon Films." *Science* 306 (5696): 666–69.
18. Novoselov, Kostya S., Andre K. Geim, SVb Morozov, Da Jiang, MIc Katsnelson, IVa Grigorieva, SVb Dubonos, and and AA Firsov. 2005 "Two-dimensional Gas of Massless Dirac Fermions in Graphene." *Nature* 438 (7065): 197.
19. Grigorenko, A. N., Marco Polini, and K. S. Novoselov. 2012 "Graphene Plasmonics." *Nature Photonics* 6 (11): 749.
20. Wunsch, B., T. Stauber, F. Sols, and F. Guinea. 2006 "Dynamical Polarization of Graphene at Finite Doping." *New Journal of Physics* 8 (12): 318.
21. Hwang, E. H., and S. Das Sarma. 2007 "Dielectric Function, Screening, and Plasmons in Two-Dimensional Graphene." *Physical Review B* 75 (20): 205418.
22. Eliasson, Bengt, and Chuan Sheng Liu. 2018 "Semiclassical Fluid Model of Nonlinear Plasmons in Doped Graphene." Physics of Plasmas 25 (1): 012105.
23. Liu, Chuan Sheng, and Vipin K. Tripathi. 2017. "Kinetic Magnetoplasmons in Graphene and their Excitation by Laser." *Journal of Nanophotonics* 11 (3): 036015.
24. Moskovits, Martin. 1985. "Surface-enhanced Spectroscopy." *Reviews of Modern Physics* 57 (3): 783.
25. Vo-Dinh, Tuan, Leonardo R. Allain, and David L. Stokes. 2002 "Cancer Gene Detection using Surface-enhanced Raman Scattering (SERS)." *Journal of Raman Spectroscopy* 33 (7): 511–6.
26. Stokes, David L., and Tuan Vo-Dinh. 2000 "Development of an Integrated Single-fiber SERS Sensor." *Sensors and Actuators B: Chemical* 69(1–2): 28–36.
27. Liu, C. S., Gagan Kumar, and V. K. Tripathi. 2006. "Laser Mode Conversion into a Surface Plasma Wave in a Metal Coated Optical Fiber." *Journal of Applied Physics* 100 (1): 013304.
28. Mayergoyz, Isaak D. 2013 *Plasmon Resonances in Nanoparticles.* World Scientific Series in Nanoscience and Nanotechnology 6. Singapore: World Scientific Publishing Co.

29. Sarid, Dror, and William A. Challener. 2010. *Modern Introduction to Surface Plasmons: Theory, Mathematica Modeling, and Applications*. Cambridge University Press.
30. Pelton, Matthew, and Garnett W. Bryant. 2013. *Introduction to Metal-Nanoparticle Plasmonics*. Vol. 5. John Wiley & Sons.
31. Kumar, Pawan. 2015. "Multi-megagauss Magnetic Field Generation by Amplitude Modulated Surface Plasma Wave over a Rippled Metal Surface." *Optics Letters* 40 (2): 190–2.

# Motion in a Strong Electromagnetic Wave

## Ponderomotive Force and Self-Generated Magnetic Field

## 5.1 Introduction

An electromagnetic wave acts on the electrons through the electromagnetic force $-e(\vec{E}+\vec{v}\times\vec{B})$, where the Lorentz force is inherently nonlinear as it involves the product of the velocity (the response) and the wave field. At low amplitudes of the wave, one may treat the wave fields and particle response linearly as first order perturbed quantities and ignore the higher order term, which is the product of the perturbed quantities. In this way, one obtains the linear response, where the current density $\vec{J}$ is proportional to $\vec{E}$.

At high amplitudes, as the electron quiver velocity approaches the speed of light ($v \to c$), the nonlinear terms become significant, and a relativistic effect arises through the velocity dependence of the mass. This changes the motion significantly.[1–5] Further, when the wave amplitude has a slow space–time dependence, the electrons experience a quasi-static force known as the ponderomotive force. This force modifies the equilibrium force balance with far-reaching consequences.

In this chapter, we will study the electron response to a large amplitude electromagnetic wave in a collisionless cold plasma. In Section 5.2, we will present an exact solution of the single electron motion in a monochromatic electromagnetic wave of constant amplitude with either circular or linear polarization. In Section 5.3, we will allow the wave amplitude to have slow variations in space and time, and deduce the ponderomotive force in the non-relativistic case. The relativistic ponderomotive force is derived in Section 5.4. In Section 5.5, we will study the one-dimensional nonlinear propagation of an electromagnetic wave in uniform underdense plasma, dominated by the relativistic mass nonlinearity. In overdense plasma, we consider the reflection problem where ponderomotive force and mass nonlinearities are prevalent. We demonstrate the equivalence of the ponderomotive force and the radiation pressure in

Section 5.6. In Section 5.7, we will study the self-generated magnetic field due to a circularly polarized laser.

## 5.2 Relativistic Electron Motion in a Plane Wave

### 5.2.1 Circular polarization

Consider a right-hand circularly polarized electromagnetic wave,

$$\vec{E} = \left(\hat{x} + i\hat{y}\right) A e^{-i(\omega t - kz)} \tag{5.1a}$$

$$\vec{B} = \vec{k} \times \vec{E} / \omega = -i\left(k / \omega\right)\vec{E}, \tag{5.1b}$$

where $\vec{k} = k\hat{z}$ was used in Eq. (5.1b). The motion of a single electron in this field is governed by the relativistic equation of motion,

$$\frac{d\vec{p}}{dt} = -e(\vec{E} + \vec{v} \times \vec{B}), \tag{5.2}$$

where $\vec{p} = m\gamma\vec{v}$ is the momentum, $\gamma = (1 + p^2/m^2c^2)^{1/2}$ is the Lorentz factor, and $-e$ and $m$ are the electron charge and rest mass, respectively. Since $E_y = i\,E_x$ and $B_y = i\,B_x$, one expects $v_y = iv_x$; hence, $\left(\vec{v} \times \vec{B}\right)_z = 0$ and $p_z$ is a constant of motion. We take $p_z = 0$. Then, Eq. (5.2) gives

$$\vec{p} = \frac{e\vec{E}}{i\omega}, \quad \vec{v} = \frac{e\vec{E}}{im\gamma\omega}, \tag{5.3}$$

where $\gamma = (1 + a^2)^{1/2}$ is the relativistic gamma factor with $a = eA/m\omega c$ being the normalized amplitude. The relativistic effects are important when $a$ is of the order of or greater than unity. At 1 μm laser wave length, $a = 1$ at the laser intensity $I_L \approx 10^{18}$ W/cm$^2$.

### 5.2.2 Linear polarization

For a plane polarized electromagnetic wave,

$$\vec{E} = \hat{x} A e^{-i(\omega t - kz)} \tag{5.4a}$$

$$\vec{B} = \hat{y} \frac{Ak}{\omega} e^{-i(\omega t - kz)}, \tag{5.4b}$$

The equations for energy and momentum balance (in component form) for an electron are

$$mc^2 \frac{d\gamma}{dt} = -eE_x v_x \tag{5.5a}$$

$$\frac{dp_z}{dt} = -ev_x B_y = -eE_x v_x k / \omega, \tag{5.5b}$$

$$\frac{dp_x}{dt} = -eE_x + ev_z B_y = -eE_x (\omega - kv_z) / \omega = \frac{d}{dt}\left(\frac{eE_x}{i\omega}\right), \tag{5.5c}$$

$$\frac{dp_y}{dt} = 0. \tag{5.5d}$$

The first two equations combine to give a constant of motion,

$$\gamma - \frac{p_z}{mc}\frac{\omega}{kc} = c_1, \tag{5.6}$$

while the third and fourth equations give

$$p_x - \frac{eE_x}{i\omega} = c_2,\ p_y = c_3. \tag{5.7}$$

Further, we have

$$\gamma^2 = 1 + P_x^2 + P_y^2 + P_z^2, \tag{5.8}$$

where $P_x = p_x/mc$, $P_y = p_y/mc$, and $P_z = p_z/mc$. We choose the constants of motion such that the oscillation center of the electron in a stationary plasma is at rest, that is, $<p_x> = 0$, $<p_y> = 0$, $<p_z> = 0$, where <> denotes the time average over a laser period. Thus,

$$c_1 = <\gamma>,\ \ c_2 = 0,\ \ c_3 = 0. \tag{5.9}$$

We may explicitly write $p_x$ in normalized units as

$$P_x = -a \sin(\omega t - kz),$$

where $a = eA/m\omega c$. Equations (5.6) and (5.8) yield

$$\gamma^2\left(1 - \frac{1}{\beta^2}\right) = 1 + \frac{\langle\gamma\rangle^2}{\beta^2} - 2\frac{\gamma\langle\gamma\rangle}{\beta^2} + P_x^2, \tag{5.10}$$

where $\beta = \omega/kc$. In an underdense plasma, $\beta^2 - 1 << 1$ and the time average of Eq. (5.10) yields

$$< \gamma > = \left(1 + a^2/2\right)^{1/2}. \tag{5.11}$$

Thus, for an electron in a plane polarized laser field,

$$v_x = \frac{eE_x}{mi\omega\gamma}, \quad \gamma = \left(1 + a^2/2\right)^{1/2}, \tag{5.12}$$

where the time average on $\gamma$ is implied. One may note from Eqs. (5.6) and (5.8) that $p_z$ is also finite. For the special case of $\omega = kc$, Eq. (5.12) becomes

$$v_z = \frac{a^2 c}{4(1 + a^2/2)} \cos\left[2\left(\omega t - \omega z/c\right)\right].$$

As the electron's position $x$ oscillates with frequency $\omega$ and $z$ with $2\omega$, the resulting electron trajectory in the $x$-$z$ plane has the shape of a figure '8'.

## 5.3 Non-Relativistic Ponderomotive Force

Now we will allow the electromagnetic wave amplitude to have a slow dependence on space. As an example, a laser of finite spot size propagating along $z$ has a dependence of the wave amplitude $\vec{A}$ on the cylindrical coordinate $r$, and due to the diffraction divergence and nonlinear convergence, a dependence on $z$,

$$\vec{E} = \vec{A}\,(\vec{r})\, e^{-i(\omega t - kz)}. \tag{5.13}$$

Using Faraday's law $\nabla \times \vec{E} = -\partial \vec{B} / \partial t$, the wave magnetic field can be written as

$$\vec{B} = \frac{1}{\omega}\left(\vec{k} \times \vec{A} - i\nabla \times \vec{A}\right) e^{-i\,(\omega t - kz)} \tag{5.14}$$

The electron fluid response to the wave is governed by the momentum equation,

$$m\frac{\partial \vec{v}}{\partial t} + m(\vec{v} \cdot \nabla)\vec{v} = -e(\vec{E} + \vec{v} \times \vec{B}). \tag{5.15}$$

The second terms on both sides of Eq. (5.15) are nonlinear. We solve Eq. (5.15) iteratively. First neglecting the nonlinear terms and replacing $\partial/\partial t$ by $-i\omega$, we get

$$\vec{v} = \frac{e\vec{E}}{mi\omega}. \tag{5.16}$$

To the next order, we retain the nonlinear terms which put together are called the ponderomotive force,

$$\vec{F}_p = -m(\vec{v}\cdot\nabla)\vec{v} - e\vec{v}\times\vec{B} \tag{5.17}$$

One may remember that $\vec{E}$, $\vec{B}$ and $\vec{v}$ are all real, given by the real parts of their complex representations. The products in Eq. (5.17) are evaluated by taking only the real parts of $\vec{v}$ and $\vec{B}$, resulting in

$$\vec{F}_p = \vec{F}_{p0} + \vec{F}_{p2\omega}, \tag{5.18}$$

where

$$\vec{F}_{p0} = \frac{1}{2}\mathrm{Re}\left[-m(\vec{v}\cdot\nabla)\vec{v}^* - e\vec{v}\times\vec{B}^*\right]$$

is the static ponderomotive force, and

$$\vec{F}_{p2\omega} = \frac{1}{2}\mathrm{Re}\left[-m(\vec{v}\cdot\nabla)\vec{v} - e\vec{v}\times\vec{B}\right]$$

is the second harmonic ponderomotive force. Using $\vec{v}$ and $\vec{B}$ from Eqs. (5.16) and (5.14), and employing the identities

$$\mathrm{Re}\left(\vec{A}\cdot\vec{B}^*\right) = \left(\vec{A}\cdot\vec{B}^* + \vec{A}^*\cdot\vec{B}\right)/2$$

and

$$\nabla(\vec{A}\cdot\vec{B}) \equiv (\vec{A}\cdot\nabla)\vec{B} + (\vec{B}\cdot\nabla)\vec{A} + \vec{A}\times(\nabla\times\vec{B}) + \vec{B}\times(\nabla\times\vec{A}), \tag{5.19}$$

we obtain[1]

$$\vec{F}_{p0} = -\frac{e^2}{4m\omega^2}\left[(\vec{A}\cdot\nabla)\vec{A}^* + (\vec{A}^*\cdot\nabla)\vec{A} + \vec{A}\times(\nabla\times\vec{A}^*) + \vec{A}^*\times(\nabla\times\vec{A})\right]$$

$$= -\frac{e^2}{4m\omega^2}\nabla\left(\vec{A}.\vec{A}^*\right) = -\frac{mc^2}{4}\nabla a^2, \tag{5.20}$$

where $a = e|E|/(m\omega c)$ is the normalized amplitude of the non-relativistic electron quiver velocity. Similarly, one obtains

$$\vec{F}_{p2\omega} = -\mathrm{Re}\left(\frac{e^2 E^2}{2m\omega^2} i\vec{k}\right). \tag{5.21}$$

For the sake of brevity, we may suppress the subscript 0 from $\vec{F}_{p0}$ and express it as $\vec{F}_p = e\nabla\phi_p$ with ponderomotive potential $\phi_p = -m|v|^2/4e$.

The static ponderomotive force pushes the electrons away from regions of high laser fields. This gives rise to space charge electric fields that pulls the ions in the direction of electrons. The changes in electron density cause self channeling and self focusing of electromagnetic waves. The second harmonic ponderomotive force causes the electron density to oscillate, which by the beating with the oscillatory electron velocity $\vec{v}$ gives rise to a third harmonic current density, thus leading to third harmonic radiation. The oscillatory ponderomotive force sometimes leads to a rapid heating of the electrons. In a circularly polarized wave, $E_y = iE_x$, $\vec{F}_{p2\omega}$ vanishes as the magnitude of the electric field does not change with time (only its direction rotates); hence, there is no harmonic generation.

### 5.3.1 Response to a pulse

Now we will allow the electromagnetic wave amplitude to have a slow time dependence as well,

$$\vec{E} = \vec{A}(\vec{r},t)e^{-i(\omega t - kz)}. \tag{5.22}$$

The wave magnetic field can be deduced iteratively by writing $\vec{B} = \vec{b}(\vec{r},t)\ e^{-i(\omega t - kz)}$ in Faraday's law, $\nabla \times \vec{E} = -\partial\vec{B}/\partial t$, giving

$$i\omega\vec{b} - \frac{\partial\vec{b}}{\partial t} = i\vec{k} \times \vec{A} + \nabla \times \vec{A}.$$

First, we will ignore $\partial\vec{b}/\partial t$, to get $\vec{b} = \vec{k} \times \vec{A}/\omega - (i/\omega)\nabla \times \vec{A}$. Next, we use this $\vec{b}$ in $\partial\vec{b}/\partial t$ in the aforementioned equation. We will ignore the double derivatives of $\vec{A}$ in time and space to obtain

$$\vec{B} = \left[\frac{\vec{k} \times \vec{A}}{\omega} - \frac{i}{\omega}\nabla \times \vec{A} - \frac{i\vec{k}}{\omega^2} \times \frac{\partial\vec{A}}{\partial t}\right] e^{-i(\omega t - kz)}. \tag{5.23}$$

One can obtain $\vec{B}$ immediately from $\vec{B} = \vec{k} \times \vec{E}/\omega$ by replacing $\vec{k}$ with $\vec{k} - i\nabla$ and $\omega$ with $\omega + i\partial/\partial t$, and $\omega^{-1}$ with $(\omega + i\,\partial/\partial t)^{-1} = \omega^{-1} - i\omega^{-2}\partial/\partial t$. We also operate $\nabla$ and $\partial/\partial t$ on $\vec{A}$.

Using this procedure the solution of the linearized equation of motion can be deduced from Eq. (5.16) as

$$\vec{v} = \frac{e}{mi\omega}\left(\vec{A} - \frac{i}{\omega}\frac{\partial \vec{A}}{\partial t}\right) e^{-i(\omega t - kz)}. \tag{5.24}$$

Using $\vec{v}$ and $\vec{B}$ in Eq. (5.17), we obtain

$$\vec{F}_{p0} = -\frac{e^2}{4m\omega^2}\nabla\left(\vec{E}.\vec{E}^*\right) \tag{5.25}$$

which is the same as Eq. (5.19). Here $\vec{F}_{p0}$ is a function of time on the slow time-scale of the amplitude variation.

## 5.4 Relativistic Ponderomotive Force

In the relativistic case, the time-averaged relativistic ponderomotive force due to a wave of space-dependent amplitude (cf., Eqs. 5.13 and 5.14), can be defined as

$$\vec{F}_p = -\frac{m}{4}(\vec{v}.\nabla\gamma\vec{v}^* + \vec{v}^*.\nabla\gamma v) - \frac{e}{4}(\vec{v}\times\vec{B}^* + \vec{v}^*\times\vec{B}). \tag{5.26}$$

Using $\vec{v} = e\vec{E}/(mi\omega\gamma)$, $\vec{B}$ from Eq. (5.14) and the identity given by Eq. (5.19) one obtains

$$\vec{F}_p = -\frac{e^2}{4m\omega^2\gamma}\nabla\left(\vec{A}\cdot\vec{A}^*\right) \tag{5.27}$$

For circular polarization, replacing $\vec{A}$ with $(\hat{x} + i\hat{y})A$, Eq. (5.27) can be written as

$$\vec{F}_p = e\nabla\phi_p \tag{5.28}$$

with the ponderomotive potential as

$$\phi_p = -mc^2(\gamma - 1)/e, \tag{5.29}$$

where $\gamma = (1 + a^2)^{1/2}$, $a = eA/m\omega c$. We have used the condition that as the laser amplitude tends to zero, ponderomotive potential must vanish.

For linear polarization, replacing $\vec{A}$ with $\hat{x}A$ in Eq. (5.27), we can obtain the same expressions for the ponderomotive force and potential but with $\gamma = (1 + a^2/2)^{1/2}$ and $a = eA/m\omega c$.

At low intensities ($a<<1$), the ponderomotive force scales as $\sim\nabla|E|^2$; whereas at high intensities ($a >> 1$), it scales as $\sim\nabla|E|$. A more detailed discussion of the ponderomotiuve force is given by Bauer.[2]

## 5.5 Nonlinear Wave Propagation in One Dimension

### 5.5.1 Underdense plasma

A continuous wave (CW) laser in a homogeneous underdense collisionless plasma propagates with a constant amplitude. It exerts no quasi-static ponderomotive force on the electrons, and hence, the electron density remains unmodified. However, the electron mass is modified by the relativistic gamma factor; and the modified effective plasma permittivity becomes

$$\varepsilon = 1 - \frac{\omega_p^2}{\omega^2 \gamma}. \tag{5.30}$$

The propagation constant can be written as

$$k = \frac{\omega}{c}\left(1 - \frac{\omega_p^2}{\omega^2 \gamma}\right)^{1/2}. \tag{5.31}$$

The plasma frequency is effectively reduced by the relativistic gamma factor In 3-D, this will lead to self-focusing of the laser beam. The ponderomotive force pushes away the electrons to reduce the electron density, further reducing the second term in Eq. (5.30), and leading to enhanced self-focusing. Thus, this process continues until all the electrons are depleted.[22] This will be discussed in Chapter 9.

### 5.5.2 Overdense plasma

In overdense plasma, an electromagnetic wave is evanescent and its amplitude decreases with distance. The wave exerts a ponderomotive force on the electrons and modifies the electron density. Such a situation arises in real life when a laser impinges from vacuum on to an overdense plasma formed from ionization of a solid foil.

Consider an overdense plasma half space ($z > 0$) of number density $n_0$ (cf. Fig. 5.1). A circularly polarized laser normally impinges on the plasma with the electric and magnetic fields,

$$\vec{E}_i = ic\vec{B}_i = A_0(\hat{x} + i\hat{y})e^{-i(\omega t - \omega z/c)} \tag{5.32}$$

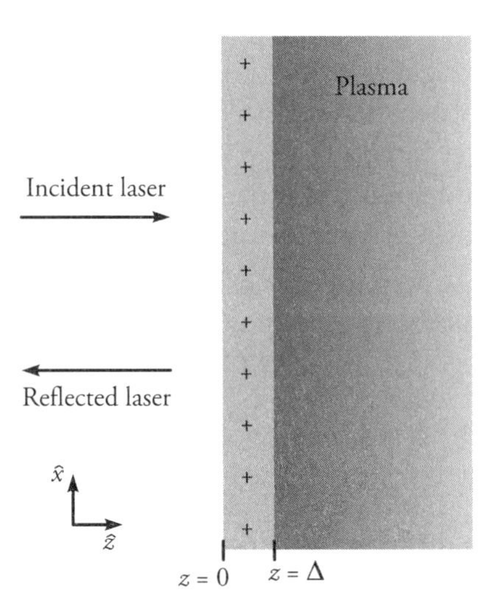


**Fig. 5.1** A plasma half space irradiated by a laser. The electrons are pushed by the ponderomotive force to a distance where the ponderomotive force is balanced by the space charge field of the ions left behind.

Following Tripathi et al.[3], the fields of the reflected and transmitted waves can be written as

$$\vec{E}_R = -ic\vec{B}_R = (\hat{x} + i\hat{y})RA_0 e^{-i(\omega t + \omega z/c)},$$

$$\vec{E}_T = (\hat{x} + i\hat{y})A_T(z)e^{-i\omega t},$$

$$\vec{B}_T = -(\hat{x} + i\hat{y})\frac{1}{\omega}\frac{\partial A_T}{\partial z}e^{-i\omega t}, \tag{5.33}$$

where $R$ is the amplitude reflection coefficient. We choose $A_T$ to be real and allow $A_0$ to be complex to account for the phase change on transmission. The transmitted wave exerts a ponderomotive force on the electrons, $\vec{F}_p = \hat{z}e\partial\phi_p / \partial z$, where $\phi_p = -(mc^2/e)(\gamma - 1)$, $\gamma = (1 + a_T^2)^{1/2}$ and $a_T = eA_T/m\omega c$, This force displaces the electrons, leading to the creation of a space charge field $\vec{E}_s = -\nabla\phi_s$. In the quasi-steady state, the net force on the electrons is zero, $\vec{F}_p - e\vec{E}_s = 0$, that is, $\phi_s = -\phi_p$. Using this in Poisson's equation, $\nabla^2\phi = e(n_e - n_0)/\varepsilon_0$, one obtains the modified electron density,

$$\frac{n_e}{n_0} = 1 + \frac{c^2}{\omega_p^2}\frac{\partial^2}{\partial z^2}(1 + a_T^2)^{1/2}, \tag{5.34}$$

Equation (5.35) is valid as long as $-\partial^2\left(1+a_T^2\right)^{1/2}/\partial z^2 < c^2/\omega_p^2$. In the opposite limit $n_e$ = 0, that is, the region is completely depleted of electrons only ions are left.[4,5] Physically, one expects that the ponderomotive force due to an intense laser would push the electrons up to a point say $z = \Delta$ where the ponderomotive force equals the net backward space charge force ($\phi_s = -\phi_p$). Beyond that point, the electron density is given by Eq. (5.34).

The effective relative plasma permittivity, including the electron density modification and relativistic mass variation, can be written as

$$\varepsilon = 1 - \frac{\omega_p^2 n_e}{\omega^2 n_0^0}\left(1+a_T^2\right)^{-1/2}, \tag{5.35}$$

The wave equation governing the normalized laser amplitude $a_T$ is

$$\frac{d^2 a_T}{dz^2} + \frac{\omega^2}{c^2}\varepsilon a_T = 0. \tag{5.36}$$

Using Eqs. (5.34) and (5.35), the wave equation can be written as

$$\frac{d^2 a_T}{dz^2} - \left[\frac{\omega_p^2}{c^2}\left(1+a_T^2\right)^{1/2} + \frac{1}{1+a_T^2}\left(\frac{da_T}{dz}\right)^2 - \frac{\omega^2}{c^2}\left(1+a_T^2\right)\right]a_T = 0. \tag{5.37}$$

One can solve it Eq. (5.37) numerically by integrating it backwards, starting with a large $z$, where $a_T^2 << 1$ and $a_T$ is proportional to $a_1 e^{-\alpha z}$, with $\alpha = (\omega/c)\left(\omega_p^2/\omega^2 - 1\right)^{1/2}$. At $z = \Delta$, the continuity of the tangential components of the electric and magnetic fields gives

$$\left(e^{i\omega\Delta/c} + \mathrm{R}e^{-i\omega\Delta/c}\right)a_0 = a_T(\Delta) \tag{5.38}$$

$$\left(e^{i\omega\Delta/c} - \mathrm{R}e^{-i\omega\Delta/c}\right)a_0 = \frac{c}{i\omega}\frac{da_T}{dz}\bigg|_\Delta, \tag{5.39}$$

where $a_0 = eA_0/m\omega x$. Adding these equations, we get

$$a_T(\Delta) - i\frac{c}{\omega}\frac{da_T}{dz}\bigg|_\Delta = 2a_0 e^{i\omega\Delta/c}. \tag{5.40}$$

Writing $a_0 e^{i\omega\Delta/c} = a_{00}e^{i\psi_0}$ and separating the real and imaginary parts we get

$$a_T(\Delta) = 2\, a_{00}\cos\psi_0 \tag{5.41}$$

and

$$\left.\frac{da_T}{dz}\right|_{\Delta} = -\frac{2\omega}{c} a_{00} \sin\psi_0 \tag{5.42}$$

giving

$$a_T^2(\Delta) + \frac{c^2}{\omega^2}\left(\frac{da_T}{dz}\right)_{\Delta}^2 = 4a_{00}^2. \tag{5.43}$$

Equation (5.43) relates the interior field amplitude to the incident laser amplitude. Figure 5.2 shows the variation of the normalized transmitted laser amplitude with the normalized distance for $\omega_p^2/\omega^2 = 10$ and $a_{00}$ = 3, 5, 7. The boundary of the electron layer is pushed by the distance a $\Delta$ from $z$ = 0. Beyond $\Delta$, the field falls off rapidly with distance; at higher $A_0$, the field falls off more rapidly. $\Delta$ is linked to the normalized laser amplitude through the balance of the ponderomotive force at the plasma boundary to the electrostatic force on the electrons by the ion space charge located between 0 and $\Delta$. The space charge electric field at $z = \Delta$ is $E_z = n_0 e\Delta/\varepsilon_0$; hence,

$$\Delta = -\left[\frac{c^2}{\omega_p^2}\frac{a_T}{\left(1 + a_T^2\right)^{1/2}}\frac{\partial a_T}{\partial z}\right]_{z=\Delta} \tag{5.44}$$

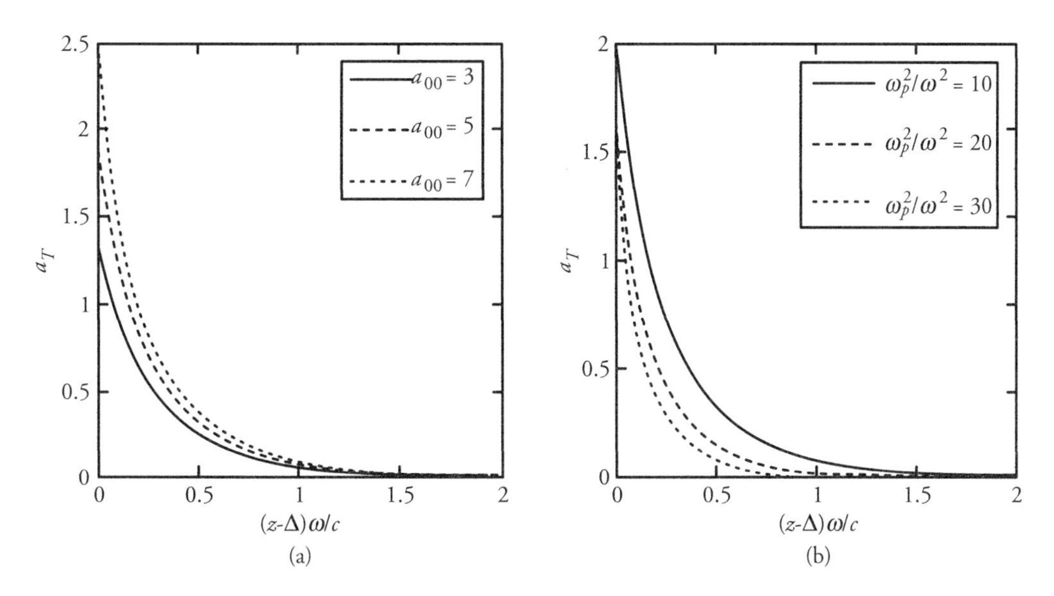


**Fig. 5.2** Distribution of the normalized transmitted field amplitude versus $z$. (a) $a_{00}$ = 3, 5, 7, and $\omega_p^2/\omega^2 = 10$, and (b) $\omega_p^2/\omega^2 = 10, 20, 30$ and $a_{00}$ = 5. (From Tripathi and Liu.[19])

In Chapter 7 we discuss this issue in greater detail in the context of radiation pressure acceleration of ions.

## 5.6 Ponderomotive Force and Radiation Pressure

We can relate the ponderomotive force acting on individual particles to the radiation pressure acting on the plasma. The force per unit area of the plasma is

$$\vec{F} = \hat{z}\int_0^l F_p n_e dz = -\hat{z}\, mc^2 \int_0^l \frac{\partial}{\partial z}\left(1 + a_T^2\right)^{1/2} n_e dz. \quad (5.45)$$

To evaluate this integral, we use Eq. (5.37). We multiply Eq. (5.37) by $(2da_T/dz)dz$, and integrate over $z$ from $\Delta$ to $\infty$,

$$\int_\Delta^l \frac{\partial}{\partial z}\left(1 + a_T^2\right)^{1/2} n_e dz = \frac{1}{2}\int_\Delta^l \frac{n_e}{\left(1 + a_T^2\right)^{1/2}} \frac{d\, a_T^2}{dz} dz$$

$$= -\frac{n_0^0}{2}\frac{c^2}{\omega_p^2}\left[\left(\frac{da_T}{dz}\right)_\Delta^2 + \frac{\omega^2}{c^2} a_T^2(\Delta)\right]$$

$$= -2n_0 a_{00}^2 \omega^2 / \omega_p^2, \quad (5.46)$$

where we have employed Eqs. (5.34) and (5.43). Using Eq. (5.46) in Eq. (5.45), we obtain

$$\vec{F} = 2mc^2 n_0 a_{00}^2 \frac{\omega^2}{\omega_p^2}\hat{z} = \frac{2I_0}{c}\hat{z}, \quad (5.47)$$

where $I_0$ is the incident laser intensity and $2I_0/c$ is the laser radiation pressure. Thus, the ponderomotive force per unit area equals the radiation pressure. This has a simple physical interpretation. A photon of energy $\hbar\omega$ carries the momentum $(\hbar\omega / c)\hat{z}$. Thus, a laser of intensity $I_0$ carries $(I_0 / c)\hat{z}$ momentum per unit area per second. On reflection, the momentum flow per unit area per second becomes $-(I_0 / c)\hat{z}$. Hence, the net momentum imparted to the plasma boundary per unit area per second is $(2I_0/c)\hat{z}$.

## 5.7 Self-Generated Magnetic Field due to a Circularly Polarized Laser

Self-generated magnetic fields, ranging from a few mega-gauss to hundreds of mega-gauss have been observed in laser produced plasmas.[6–9] One particular mechanism of magnetic field generation is the inverse Faraday effect[10–15].

Here we will present a formalism for the self-generated magnetic field due to a circularly polarized laser. We will consider the plasma electrons executing circular motions under the laser field (perpendicular to the direction of laser propagation) to be magnetic dipoles. When the laser spot size is finite or the transverse plasma density profile is nonuniform, the magnetization due to these dipoles gives a finite bound volume current or a bound surface current that produces quasi-static magnetic field.

Consider a plasma with the electron density $n_0$. A right-handed circularly polarized laser propagates through the underdense plasma with the electric field given by Eq. (5.1a). The laser induces an oscillatory electron velocity, given by Eq. (5.3). The electrons rotate in circular orbits in the $x$–$y$ plane with radius $R = v/\omega$. Each rotating electron can be viewed as a magnetic dipole having the dipole moment

$$\vec{m}_b = -\frac{e\omega}{2\pi}\pi\left(\frac{v}{\omega}\right)^2 \hat{z} = -\frac{ec^2a^2}{2\omega\gamma^2}\hat{z}, \tag{5.48}$$

giving the magnetization as

$$\vec{M} = n_0\vec{m}_b = -\frac{n_0ec^2a^2}{2\omega\gamma^2}\hat{z}.$$

The equivalent bound DC current is[16]

$$\vec{J}_b = \nabla\times\vec{M} = \hat{\varphi}\frac{ec^2}{2\omega}\frac{\partial}{\partial r}\left(\frac{n_0a^2}{1+a^2}\right). \tag{5.50}$$

This current is finite when the electron density or the laser amplitude is a function of $r$. In a long plasma slab with the density independent of $z$, one may employ Ampere's law (with Amperean loop of unit length along $z$, lying in the $x$–$z$ plane and extending from $x = r$ to $x = \infty$) to obtain the induced static magnetic field

$$\vec{B}_s(r) = \hat{z}\frac{ec^2\mu_0}{2\omega}\int_r^\infty \frac{\partial}{\partial r}\left(\frac{n_0a^2}{1+a^2}\right)dr = -\hat{z}\frac{e}{2\omega\varepsilon_0}\frac{n_0a}{1+a^2},$$

or

$$\frac{\omega_c}{\omega} = \frac{eB_s}{m\omega} = \frac{\omega_p^2}{2\omega^2}\frac{a^2}{1+a^2}. \tag{5.51}$$

This relation for magnetic field is the same as that obtained by Abdullaev and Frolov[7] and Bychenkov et al.[8] in the limit of laser spot size $r_0 < c/\omega_p$ and $a^2 < 1$ in an underdense plasma. One may recognize that the magnetic field due to a right-hand circularly polarized laser is directed opposite to the direction of laser propagation. The reason is that the gyrating electrons produce a magnetic dipole moment directed backward.

In the case of a Gaussian laser beam,

$$E_0^2 = A_0^2 e^{-r^2/r_0^2}, \tag{5.52}$$

the electrons experience a radial ponderomotive force $\vec{F}_p = e\nabla\phi_p$ with $\phi_p$ given by Eq. (5.30). The electron number density can be deduced as earlier,

$$n_e = n_0\left(1+\frac{c^2}{\omega_p^2}\nabla^2\sqrt{1+a^2}\right)$$

$$= n_0\left[1-\frac{c^2}{\omega_p^2 r_0^2}\frac{2a^2}{1+a^2}\left(\frac{1+a^2}{1+a^2/2}-\frac{r^2}{r_0^2}\right)\right]. \tag{5.53}$$

This expression is valid as long as the magnitude of the second term in the square parenthesis is less than one (as $n_e$ cannot be negative). For $r_0 \le c/\omega_p$, a situation arises at large laser amplitude when the second term in square parenthesis exceeds 1. In that case, an electron hole (a completely electron-evacuated ion channel) is created for $r < r_c$ where $r_c$ is given by

$$\frac{r_c^2}{r_0^2} = \frac{1+a^2}{1+a^2/2} - \frac{r_0^2\omega_p^2}{c^2}\frac{1+a^2}{2a^2}. \tag{5.54}$$

Using Eq. (5.53) in Eq. (5.51), we can obtain the normalized magnetic field

$$\frac{\omega_c}{\omega} = 0 \quad \text{for } r < r_c$$

$$= \frac{\omega_p^2}{2\omega^2}\frac{a^2}{1+a^2}\left[1-\frac{c^2}{r_0^2\omega_p^2}\frac{2a^2}{1+a^2}\left(\frac{1+a^2}{1+a^2/2}-\frac{r^2}{r_0^2}\right)\right] \quad \text{for } r > r_c, \tag{5.55}$$

where $a^2 = a_0^2 \exp(-r^2 / r_0^2)$ and $a_0 = eA_0/m\omega c$. In Fig. 5.3, we have plotted $\omega_c/\omega$ as a function of $r/r_0$ for $a_0 = 2$, $r_0\omega/c = 15$, and for $\omega_p^2/\omega^2 = 0.5$ and 0.9. At large values of $r_0$, the magnetic field is maximum on the laser axis and decreases monotonically with $r$ on the scale-length of the laser spot size. At low electron density, when an electron hole is created near the laser axis, the magnetic field is zero up to $r \approx r_e$. It then rises, attains a maximum and falls beyond that (cf. Fig. 5.4). A circularly polarized laser propagating through a plasma at a normalized laser amplitude $a \approx 2$ (laser intensity $I_L \approx 10^{19}$ W/cm$^2$ at 1 μm wavelength) produces an axial magnetic field that has the normalized value $e|B_s| / m\omega \approx 0.4\omega_p^2 / \omega^2$. For $\lambda_L = 1$ μm, $n_0 = n_{cr} \approx 0.9$; this gives $B_s \approx 40$ MG.

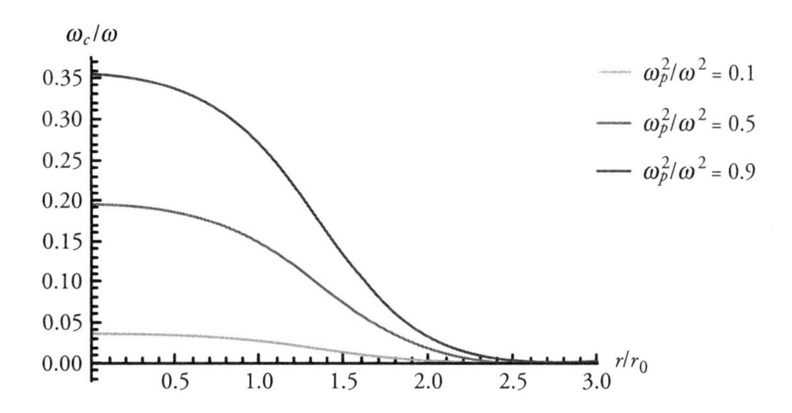


**Fig. 5.3** The normalized magnetic field as a function of the normalized radial distance $r/r_0$ for $a_0 = 2$, and $r_0\omega/c = 15$, with $\omega_p^2 / \omega^2 = 0.1, 0.5$ and $0.9$.

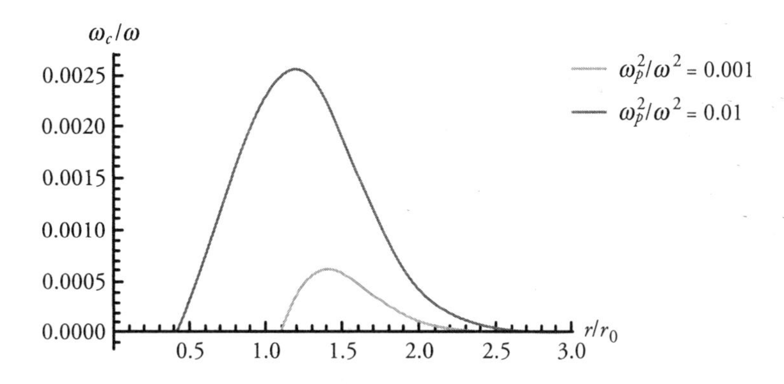


**Fig. 5.4** The normalized magnetic field as a function of the normalized radial distance $r/r_0$ for $a_0 = 2$, and $r_0\omega/c = 15$, $\omega_p^2 / \omega^2 = 0.001$ and $0.01$.

In some experiments on proton acceleration that use $CO_2$ laser on a gas jet target, one encounters a parabolic density profile,

$$\frac{\omega_p^2}{\omega^2} = \begin{cases} \dfrac{\omega_{p0}^2}{\omega^2}\left(1 - \dfrac{z^2}{L_n^2}\right) & \text{if } |z| < L_n, \\ 0 & \text{if } |z| \geq L_n. \end{cases} \tag{5.56}$$

In this case, both the laser amplitude and the density are functions of $z$. Hence, the magnetization $M_z$ is also a function of $z$, being largest at the density peak. If one takes the plasma density to be independent of $r$, the laser intensity is uniform over its spot size $r_0$, and the bound volume current is zero, only the bound surface current is finite at $r = r_0$,

$$K_b = M \times \hat{r} = -\frac{n_0 e c^2 a^2}{2\omega(1+a^2)}\hat{\phi}. \tag{5.57}$$

Using Biot-Savart's law, the magnetic field on the laser axis ($r = 0$, $z$) due to this current can be obtained as

$$B_s = -\hat{z}\frac{m\omega}{e}\frac{\omega_{p0}^2}{2\omega^2}\int_{-L_n}^{L_n}\frac{r_0^2(1 - z'^2 / L_n^2)}{\left[r_0^2 + (z - z')^2\right]^{3/2}}\frac{a^2}{1+a^2}dz'. \tag{5.58}$$

For typical parameters $\omega_{p0}^2/\omega^2 = 0.5$, $L_n/\lambda_L = 20$, $\lambda_L = 2\pi c/\omega$, $r_0/\lambda_L = 3.0$, $a_{00} = 2$ ($I_L \approx 3 \times 10^{16}$ W/cm$^2$ at 10.6 μm wavelength), $\omega_c/\omega$ turns out to be 0.4 at $z/\lambda_L = 0$ and 0.3 at $z/\lambda_L = 10$.

In the case of a laser impinging on an overdense plasma (cf. Section 5.5), the laser field inside the plasma is evanescent and its amplitude falls off with $z$. For a Gaussian beam, it depends on $r$ as well. The bound current due to the laser in the skin layer of the slab is given by Eq. (5.50) with $a$ replaced by $a_T = e|E_{T0}|/m\omega c$ and $n_0$ replaced by $n_e$, the modified electron density due to the ponderomotive force. The laser induced axial static magnetic field in the skin layer at an axial point ($r = 0$, $z$) is

$$B_s(z) = \frac{e}{2\omega\varepsilon_0}\int_0^\infty\int_0^\infty\frac{r^2}{\left[r^2 + (z - z')^2\right]^{3/2}}\frac{\partial}{\partial r}\frac{n_e a_T^2}{1+a_T^2}drdz'$$

$$= -\frac{e}{2\omega\varepsilon_0}\int_0^\infty\int_0^\infty\frac{r[2(z - z')^2 - r^2]}{\left[r^2 + (z - z')^2\right]^{5/2}}\frac{n_e a_T^2}{1+a_T^2}drdz'$$

This integral can be evaluated numerically once we know the space dependence of $a_T^2$ and $n_e$. For a normalized incident laser amplitude $a_{00} = 5$, and for $\omega_p^2 / \omega^2 = 10$ at 1 micron laser wavelength, a magnetic field of the order of 15MG is produced.

## 5.8 Discussion

Relativistic effects strongly modify electron dynamics. In a circularly polarized plane wave, the quiver velocity is transverse and has no higher harmonics. For linear polarization, the quiver velocity has a strong longitudinal component at the second harmonic. In the case of lasers of finite spot size or finite pulse duration, electrons experience a quasi-static ponderomotive force that modifies their density and plasma refractive index. It may also give rise to a host of nonlinear phenomena like self phase modulation, super continuum generation, self-focusing, filamentation and acceleration of charged particles.

A circularly polarized laser of finite spot size gives rise to a self-generated magnetic field via the inverse Faraday effect. The direction of the magnetic field is such that the electron cyclotron motion caused by it is opposite to the circular motion that caused it. At a normalized laser amplitude $a = 1$, the normalized magnetic field, $\omega_c/\omega = n_0/4n_{cr}$. For a 1 μm, $3 \times 10^{18}$ W/cm$^2$ laser in a plasma of $n_0/n_{cr} = 0.8$, this amounts to 20 MG. At higher amplitudes it saturates at 40 MG. For $r_0 < c/\omega_p$ and $a > 1$, the laser drills an ion channel in the plasma. The magnetic field is zero inside the hole, up to the hole boundary, beyond which it rises with $r$, attains a maximum and then falls off. In the case of an overdense plasma, the magnetic field is maximum on the plasma boundary and decreases with distance on the scale length of laser spot size $r_0$.

There are other mechanisms of magnetic field generation,[17–21] for example, crossed density and temperature gradients, ponderomotive force, temperature anisotropy, (the Weibel instability) laser induced current flows, that dominate in different parameter regimes.

## References


1. Sodha, M. S., A. K. Ghatak, and V. K. Tripathi. 1976. "Self Focusing of Laser Beams in Plasmas and Semiconductors." In *Progress in Optics* 13: 169–265. Elsevier.
2. Bauer, D. 2006. *Lecture Notes on the Theory of Intense Laser-Matter Interaction*. Heidelberg, Germany: Max Planck Institut für Kernphysik.
3. Tripathi, V. K., Chuan-Sheng Liu, Xi Shao, Bengt Eliasson, and Roald Z. Sagdeev. 2009. "Laser Acceleration of Monoenergetic Protons in a Self-organized Double Layer from Thin Foil." *Plasma Physics and Controlled Fusion* 51 (2): 024014.
4. Marburger, John H., and Robert F. Tooper. 1975. "Nonlinear Optical Standing Waves in Overdense Plasmas." *Physical Review Letters* 35 (15): 1001.
5. Felber, F. S., and J. H. Marburger. 1976. "Nonlinear Optical Reflection and Transmission in Overdense Plasmas." *Physical Review Letters* 36 (20): 1176.
6. Borghesi, M., A. J. MacKinnon, A. R. Bell, R. Gaillard, and O. Willi. 1998. "Megagauss Magnetic Field Generation and Plasma Jet Formation on Solid Targets Irradiated by an Ultraintense Picosecond Laser Pulse." *Physical Review Letters* 81 (1): 112–5.
7. Gopal, A., S. Minardi, Matthias Burza, Guillaume Genoud, I. Tzianaki, A. Karmakar, Paul Gibbon, M. Tatarakis, Anders Persson, and Claes-Göran Wahlström. 2013. "Megagauss

Magnetic Field Generation by Ultra-short Pulses at Relativistic Intensities." *Plasma Physics and Controlled Fusion* 55 (3): 035002.

8. Wagner, U., M. Tatarakis, A. Gopal, F. N. Beg, E. L. Clark, A. E. Dangor, R. G. Evans et al. 2004. "Laboratory Measurements of 0.7 GG Magnetic Fields Generated during High-Intensity Laser Interactions with Dense Plasmas." *Physical Review E* 70 (2): 026401.
9. Li, C. K., F. H. Séguin, J. A. Frenje, J. R. Rygg, R. D. Petrasso, R. P. J. Town, O. L. Landen, J. P. Knauer, and V. A. Smalyuk. 2007. "Observation of Megagauss-field Topology Changes due to Magnetic Reconnection in Laser-produced Plasmas." *Physical Review Letters* 99 (5): 055001.
10. Abdullaev, A. Sh, and A. A. Frolov. 1981. "Inverse Faraday Effect in a Relativistic Electron Plasma". *Journal of Experimental and Theoretical Physics* 54: 493.
11. Bychenkov, V. Yu, V. I. Demin, and V. T. Tikhonchuk. 1994. "Electromagnetic Field Generation by an Ultrashort Laser Pulse in a Rarefied Plasma". *Journal of Experimental and Theoretical Physics* 78 (1): 62
12. Qiao, Bin, Shao-ping Zhu, C. Y. Zheng, and X. T. He. 2005. "Quasistatic Magnetic and Electric Fields Generated in Intense Laser Plasma Interaction." *Physics of Plasmas* 12 (5): 053104.
13. Deschamps, J., M. Fitaire, and M. Lagoutte. 1970. "Inverse Faraday Effect in a Plasma." *Physical Review Letters* 25 (19): 1330–2.
14. Horovitz, Y., S. Eliezer, A. Ludmirsky, Z. Henis, E. Moshe, R. Shpitalnik, and B. Arad. 1997. "Measurements of Inverse Faraday Effect and Absorption of Circularly Polarized Laser Light in Plasmas." *Physical Review Letters* 78 (9): 1707.
15. Gorbunov, L. M., and R. R. Ramazashvili. 1998. "Magnetic Field Generated in a Plasma by a Short, Circularly Polarized Laser Pulse." *Journal of Experimental and Theoretical Physics* 87 (3): 461–7.
16. Griffiths, David J. 2012. *Introduction to Electrodynamics,* Boston: Addison-Wesley.
17. Sudan, R. N. 1993. "Mechanism for the Generation of $10^9$ G Magnetic Fields in the Interaction of Ultraintense Short Laser Pulse with an Overdense Plasma Target." *Physical Review Letters* 70 (20): 3075–8.
18. Pukhov, A., Z-M. Sheng, and J. Meyer-ter-Vehn. 1999. "Particle Acceleration in Relativistic Laser Channels." *Physics of Plasmas* 6 (7): 2847–54.
19. Tripathi, V. K., and C. S. Liu. 1994. "Self-generated Magnetic Field in an Amplitude Modulated Laser Filament in a Plasma." *Physics of Plasmas* 1 (4): 990–2.
20. Shukla, P. K., and M. Y. Yu. 1984. "Magnetic Field Generation by Powerful Laser Beams." *Plasma Physics and Controlled Fusion* 26 (6): 841.
21. Gradov, O. M., and L. Stenflo. 1983. "Magnetic-field Generation by a Finite-radius Electromagnetic Beam." *Physics Letters A* 95 (5): 233–4.
22. Sun, Guo-Zheng, Edward Ott, Y.C. Lee, and Parvez Guzdar. 1987. "Self-focusing of Short Intense Pulses in Plasmas." *The Physics of Fluids* 30 (2): 526–32.

# 6 Laser Electron Acceleration

## 6.1 Introduction

The acceleration of charged particles by waves is a fascinating field of study. A wave can accelerate co-moving particles when i) it exerts a longitudinal force (parallel to the direction of propagation) on them and ii) its phase velocity is close to the velocity of the particles so that the wave field appears quasi-static to the particles and efficient energy transfer from the wave to the particles can take place. The wave must also be able to overcome diffraction divergence in order to have a long enough interaction length with the particles.

In a linear accelerator (LINAC), a microwave is passed through a cylindrical waveguide loaded with periodic rings to slow down the phase velocity of the wave to a level slightly below the velocity of light in vacuum, $c$. The transverse magnetic (TM) modes of the waveguide inherently possess a longitudinal component of the electric field, and the waveguide prevents the divergence of the wave. Hence, the conditions for the acceleration of a collinearly propagating electron beam are met and the LINAC can yield electron energies in the tens of GeV range. The dimensions of the device, however, are huge and the acceleration process is quite slow. The same is true of storage rings and other accelerators.

By using an intense short pulse laser, it is possible to reduce the acceleration region by orders of magnitude and accelerate electrons at a much higher rate – this can be done by exciting a large amplitude plasma wave[1–11] with a longitudinal electric field, and whose phase velocity is slightly less than $c$. The plasma wave can be guided over a long distance by the laser; hence, it is suitable for electron acceleration. The laser beat wave accelerator (LBWA) and laser wake-field accelerator (LWFA) follow this method.[9–11]

In the LBWA,[4–6] two collinear lasers with frequencies close to each other, $\omega_1$ and $\omega_2$, are injected into an underdense plasma with plasma frequency $\omega_p = \omega_1 - \omega_2$. The lasers exert a longitudinal ponderomotive force (Lorentz force) on the electrons and resonantly drive a plasma wave having a phase velocity equal to the group velocity of the lasers. Due to the resonant interaction, the electrostatic potential of the plasma wave is much larger than the ponderomotive potential.

In the LWFA,[1–3] a single laser pulse of short duration, $\tau \approx 2\pi / \omega_p$, is employed. It drives a plasma wave in the wake of the pulse. The electrostatic potential of the plasma wave is comparable to the ponderomotive potential. Studies using fast rising flat top pulses reveal that a plasma wave is excited at the front of the pulse. It focuses and defocuses the laser periodically, modulating the axial intensity. The modulated laser exerts a ponderomotive force on the electrons enhancing the amplitude of the plasma wave, and giving higher acceleration gradients[12–16].

For laser pulses having a finite spot size and short duration at relativistic intensities ($I_L > 3 \times 10^{18}$ W/cm$^2$ at 1 μm wavelength) yet another fascinating scenario develops. A short laser pulse of spot size $r_0 = c/\omega_p$ and pulse length $c/\omega_p$, propagating through a plasma of plasma frequency $\omega_p$, exerts axial and radial ponderomotive forces on the electrons. As a consequence, a mild electron density increase takes place at the front of the pulse while a fully electron-evacuated ion bubble is created behind the pulse. In the moving frame of the pulse, electrons at the boundary of the bubble move backward and surge toward the stagnation point. The negative charge built up at the stagnation point makes the electrons to come to rest (in the moving frame). They are then pulled into the bubble by the ion space charge field and accelerated to high energies. One may note that the self-injection of electrons is part of this scheme.

In this chapter, we will develop analytical formalisms of laser beat wave and wake-field acceleration of electrons. In Section 6.2, we will study the excitation of a plasma wave by the beating of two collinear laser beams in an underdense plasma, and estimate its saturation amplitude via the frequency detuning introduced due to the relativistic effects of the plasma wave. We will study the laser wake-field excitation of plasma waves by a short pulse laser in Section 6.3. In Section 6.4, we will study single-particle phase space behavior in a large amplitude plasma wave and estimate the energy gain and acceleration length. A semi-analytical description of the bubble regime acceleration of electrons is developed in Section 6.5. In Section 6.6, we discuss some experiments and simulations. Concluding remarks are made in Section 6.7.

## 6.2 Laser Beat Wave Excitation of a Plasma Wave

In a laser beat wave accelerator (LBWA), two laser beams with different frequencies drive a plasma wave resonantly. This plasma wave accelerates a pre-accelerated beam of electrons to high energies. Figure 6.1 shows a schematic of the LBWA, comprising a high density plasma, an electron pre-acceleration mechanism, and a table-top terawatt laser system producing laser beams of two frequencies close to each other, $\omega_1$ and $\omega_2$ and focusing optics. The frequency difference of the lasers equals the electron plasma frequency. The lasers exert an axial ponderomotive force on the electrons, producing a large-amplitude plasma wave. At the same time when the plasma wave is excited, a pre-accelerated electron beam, co-propagating with the lasers, is launched into the plasma wave. It is segments of this electron beam that are accelerated by the plasma wave to high energies.

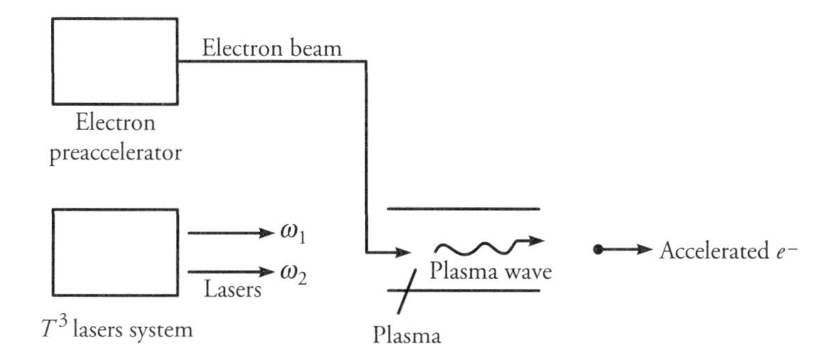


**Fig. 6.1** Schematic of a laser beat wave accelerator (LBWA).

Let the electric and magnetic fields of the linearly polarized lasers propagating through the plasma of equilibrium density $n_0^0$ be

$$\vec{E}_j = \hat{x}A_j e^{-i(\omega_j t - k_j z)},\ \vec{B}_j = \hat{y}k_j E_{jx}/\omega_j,$$

$$k_j \cong \frac{\omega_j}{c}\left(1 - \frac{\omega_p^2}{2\,\omega_1^2}\right),\quad j = 1,\ 2 \tag{6.1}$$

where the frequency difference matches the plasma frequency, $\omega_1 - \omega_2 \approx \omega_p = (n_0^0 e^2/m\varepsilon_0)^{1/2}$, in an underdense plasma so that $\omega_1$, $\omega_1 >> \omega_p$, where $-e$ and $m$ are the electron charge and rest mass, respectively, and $\varepsilon_0$ is the free space permittivity.

The lasers impart oscillatory velocities to the electrons,

$$\vec{v}_1 = \frac{e\vec{E}_1}{mi\,\omega_1},\ \vec{v}_2 = \frac{e\vec{E}_2}{mi\,\omega_2}. \tag{6.2}$$

They also exert a ponderomotive force on the electrons at $\omega = \omega_1 - \omega_2$, $\vec{k} = \vec{k}_1 - \vec{k}_2$,

$$\vec{F}_p = e\nabla\phi_p = -\frac{e}{2}\mathrm{Re}\left(\vec{v}_1 \times \vec{B}_2^* + \vec{v}_2^* \times \vec{B}_1\right), \tag{6.3}$$

where the asterisk denotes the complex conjugate and Re the real part of the quantity. Using $\vec{v}_1$, $\vec{v}_2$, $\vec{B}_1$ and $\vec{B}_2$ in Eq. (6.3), we obtain

$$\phi_p = -\frac{m}{2e}\vec{v}_1 \cdot \vec{v}_2^* = \Phi_p e^{-i(\omega t - kz)}, \tag{6.4}$$

where $\Phi_p = -eA_1A_2^*/(2m\omega_1\omega_2)$. The phase velocity of the ponderomotive force equals the group velocity of the lasers,

$$\omega/k \approx c\left(1-\omega_p^2/2\omega_1^2\right).$$

The ponderomotive force drives a Langmuir wave having the electrostatic potential

$$\phi = \Phi e^{-i(\omega t - kz)} \tag{6.5}$$

where $\Phi$ is complex-valued with a slow space–time dependence. Solving the electron momentum equation under the influence of $\phi$ and $\phi_p$, we obtain the velocity perturbation at $\omega, \vec{k}$, as

$$\vec{v} = -\frac{e\vec{k}\left(\phi+\phi_p\right)}{m\omega}. \tag{6.6}$$

Using $\vec{v}$ in the linearized electron continuity equation, we obtain the density perturbation

$$n = n_0^0\frac{\vec{k}.\vec{v}}{\omega} = -\frac{n_0^0 ek^2}{m\omega^2}\left(\phi+\phi_p\right). \tag{6.7}$$

Using $n$ in the Poisson's equation, $\nabla^2\phi = ne/\varepsilon_0$, we obtain

$$\varepsilon\Phi = \chi_e\Phi_p \tag{6.8}$$

where $\varepsilon = 1 + \chi_e$ is the dielectric constant and $\chi_e = -\omega_p^2/\omega^2$ is the electron susceptibility. In particular, at $\omega \approx \omega_p$, $\varepsilon \to 0$. At this point, $F$ acquires a resonantly large value, much larger than the ponderomotive potential $\Phi_p$. However, Eq. (6.8) is not valid when $\omega = \omega_p$. In this case the equation governing the temporal evolution of the plasma wave amplitude can be deduced from Eq. (6.8) by replacing $\omega$ by $\omega_p + i\partial/\partial t\ \omega_p$ and $\varepsilon$ by $i(\partial\varepsilon/\partial\omega)\partial/\partial t$,

$$\frac{\partial}{\partial t}\Phi = -\,i\frac{\omega_p}{2}\Phi_p. \tag{6.9}$$

Following Rosenbluth and Liu,[4] we express $\Phi = A(t)e^{-i\Psi'(t)}$ and separate the real and imaginary parts of Eq. (6.9) to obtain

$$\dot{A} = \frac{\omega_p}{2}\Phi_p \sin\Psi',$$

$$A\dot{\Psi}' = \frac{\omega_p}{2}\Phi_p \cos\Psi', \tag{6.10}$$

where the overhead dot represents the time derivative. One must remember that $\psi'$ is the phase of the plasma wave with respect to the ponderomotive wave (the beat wave). Since $\psi'$ has positive slope when $-\pi/2 < \psi' < \pi/2$ and negative slope when $\pi/2 < \psi' < 3\pi/2$, the plasma wave quickly phase locks with the beat wave with stationary phase $\psi' = \pi/2$. The stationary phase is reached when $\cos\psi' = 0$. In this situation $\dot{A} > 0$ and the plasma wave amplitude grows linearly with time,

$$A(t) = A(0) + \frac{\omega_p}{2}\Phi_p t. \tag{6.11}$$

As the amplitude increases, the relativistic mass correction becomes significant, introducing a frequency mismatch and leading to phase unlocking and saturation of the amplitude.

Then, the electron momentum equation can be written

$$\frac{\partial}{\partial t}(\gamma v_z) + v_z\frac{\partial}{\partial z}(\gamma v_z) = \frac{e}{m}\frac{\partial}{\partial z}(\phi + \phi_p), \tag{6.12}$$

where $\gamma = \left(1 - v_z^2/c^2\right)^{-1/2} \approx 1 - v_z^2/2c^2$ is the relativistic gamma-factor in the limit $v_z^2 << c^2$. To the lowest order $v_z$ is obtained from the real part of Eq. (6.6). Using this value of $v_z$ in the nonlinear term of Eq. (6.12), we obtain $v_z$ to the next order,

$$v_z = -\frac{ek}{m\omega}(\phi + \phi_p) + \frac{3}{8c^2}\frac{e^3k^3}{m^3\omega^3}\phi^2\phi^*. \tag{6.13}$$

where we have neglected $\phi_p$ in the nonlinear term as $\phi_p << \phi$ when $\omega \sim \omega_p$. Using $v_z$ in the continuity equation, we obtain the density perturbation as

$$n = n_0^0\frac{kv_z}{\omega} = -\frac{n_0^0k^2e}{m\omega^2}(\phi + \phi_p) + \frac{3n_0^0e^3k^4}{8c^2m^3\omega^4}\phi^2\phi^*, \tag{6.14}$$

where we have neglected the nonlinearity arising in the continuity equation. Using in Poisson's equation, we obtain

$$\varepsilon\Phi = -\chi_e\Phi_p - \frac{3}{8}\frac{e^3k^2\Phi^2\Phi^*}{m^2\omega^2c^2} \tag{6.15}$$

which by expressing $\omega = \omega_p + i\partial/\partial t$ leads to

$$\frac{\partial \Phi}{\partial t} = -i\frac{\omega_p}{2}\Phi_p + i\alpha\Phi^2\Phi^*, \tag{6.16}$$

where $\alpha = 3e^2k^2/16m^2\omega c^2$. Writing F = $A(t)\exp(-i\psi')$ and separating real and imaginary parts gives

$$\dot{A} = \frac{\omega_p}{2}\Phi_p \sin\psi', \tag{6.17}$$

$$A\dot{\psi}' = \frac{\omega_p}{2}\Phi_p \cos\psi' - \alpha A^3. \tag{6.18}$$

We seek a solution of these equations around $\psi' = \pi/2$, by writing $\psi' = \pi/2 - f$ with $f << \pi/2$. Dividing the equation for $f$ by $\dot{A}$ and using Eq. (6.17), we obtain

$$A\frac{df}{dA} \equiv -f + \frac{2\alpha}{\omega_p\Phi_p}A^3,$$

$$f = \frac{\alpha}{2\omega_p\Phi_p}A^3$$

when $A(0) << \Phi_p$. Hence, $\dot{A} = \omega_p\Phi_p(2 - f^2)/4$. The growth stops when $f = \sqrt{2}$,

$$A = \left(2\sqrt{2}\omega_p\Phi_p/\alpha\right)$$

or

$$|v_z|/c = \left(32\sqrt{2}v_1v_2^*/c^2\right)^{1/3}. \tag{6.19}$$

The oscillatory electron velocity due to the plasma wave is larger than $v_1$, $v_2$, which justifies our neglect of relativistic mass corrections due to $v_1$, $v_2$.

This level of saturation by relativistic detuning of resonance is much lower than that due to plasma wave breaking.

## 6.3 Laser Wake-field Excitation of a Plasma Wave

A laser wake-field accelerator (LWFA) comprises a gas jet target (cf. Fig. 6.2), which is converted into a plasma by a pre-pulse. After the passage of the pulse, the plasma expands radially outward creating an electron density profile, which is minimum on the axis. Such a channel is suitable for guiding the laser and overcoming the diffraction divergence. An intense short pulse laser is launched into the channel with the pulse duration of the order of a plasma period, $\tau \sim 2\pi\omega_p^{-1}$. Here we ignore 2D off-axis effects and consider the plasma wave excitation in a uniform plasma by a laser pulse uniform in the *x* and *y* axes. The electric field of the laser pulse is

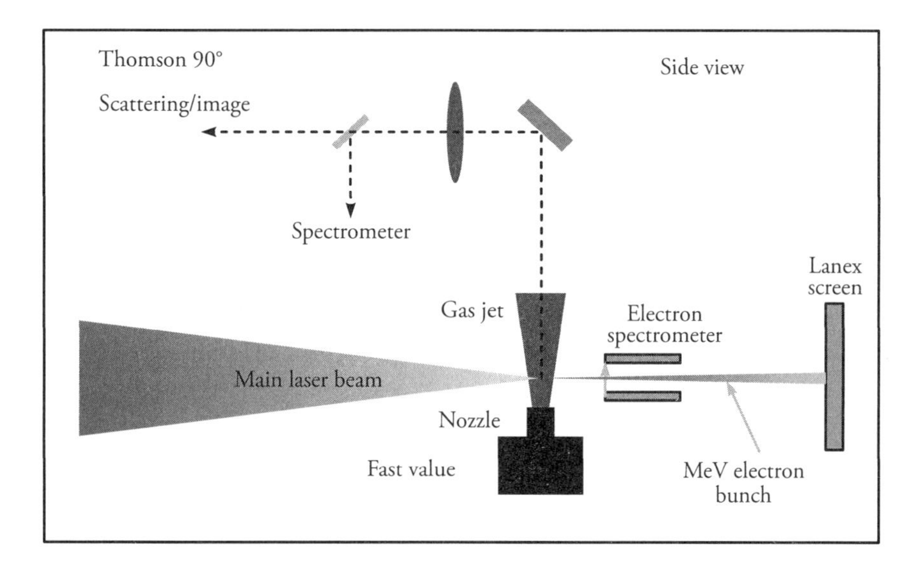


**Fig. 6.2** Schematic of a laser wake-field accelerator (LWFA). The energetic electrons are produced in the plasma itself or may be launched from outside. (From Gizzi et al.[29])

$$\vec{E}_1 = \hat{x}A_1(t,\, z)e^{-i(\omega_1 t\, -\, k_1 z)} \tag{6.20}$$

$$A_1^2 = A_{10}^2 e^{-(t\, -\, z/v_g)^2/\tau^2}$$

where $v_g = c\left(1-\omega_p^2/2\omega_1^2\right)$ is the group velocity. The pulse imparts an oscillatory velocity on the electrons, with $\vec{v}_1$ given by Eq. (6.2). It also exerts a ponderomotive force with the ponderomotive potential

$$\phi_p = -\frac{m}{2e}\left|v_1\right|^2 = -\frac{eA_{10}^2}{2m\omega_1^2}e^{-(t-z/v_z)^2/\tau^2}. \tag{6.21}$$

The ponderomotive force drives a plasma wave of potential $\phi$. The electron response to $\phi$ and $\phi_p$ is governed by the momentum and continuity equations

$$\frac{\partial v_z}{\partial t} = \frac{e}{m}\frac{\partial}{\partial z}\left(\phi + \phi_p\right), \tag{6.22}$$

$$\frac{\partial n}{\partial t} + \frac{\partial}{\partial z}\left(n_0^0 v_z\right) = 0. \tag{6.23}$$

Differentiating Eq. (6.23) with respect to time and using Eq. (6.22) we obtain

$$\frac{\partial^2 n}{\partial t^2} + \frac{n_0 e}{m}\frac{\partial^2}{\partial z^2}\left(\phi + \phi_p\right) = 0. \tag{6.24}$$

Since Poisson's equation gives $ne = \varepsilon_0 \partial^2\phi / \partial z^2$, we obtain

$$\frac{\partial^2 \phi}{\partial t^2} + \omega_p^2 \phi = -\omega_p^2 \phi_p. \tag{6.25}$$

Introducing a new set of variables $\xi = t - z / v_g$ and $\eta = z$ this equation takes the form

$$\frac{\partial^2 \phi}{\partial \xi^2} + \omega_p^2 \phi = \frac{\omega_p^2 e A_{10}^2}{2m\omega_1^2} e^{-\xi^2/\tau^2}, \tag{6.26}$$

giving the wake potential

$$\phi = \frac{\omega_p e A_{10}^2}{2m\omega_1^2} \int_{-\infty}^{\xi} \sin\left((\xi - \xi')\omega_p\right) e^{i\omega_1\xi} e^{-(\xi')^2/\tau^2} d\xi'. \tag{6.27}$$

In the wake of the laser pulse, as $\xi \to \infty$, the plasma wave potential attains the value

$$\phi = \frac{\omega_p e A_{10}^2 \tau\sqrt{\pi}}{2\, m\, \omega_1^2} e^{-\omega_p^2\tau^2/4} \sin\left(\omega_p t - k_p z\right), \tag{6.28}$$

where $k_p = \omega_p / v_g$. The amplitude of the plasma wave maximizes for $\tau = \sqrt{2}/\omega_p$,

$$|\phi|_{\max} = \frac{e A_{10}^2 \sqrt{2\pi}}{2\, m\, \omega_1^2} e^{-1/2} \approx 0.76 \frac{e A_{10}^2}{m\omega_1^2}. \tag{6.29}$$

which is of the same order as the peak value of the ponderomotive potential.

In the strongly nonlinear regime, the relativistic effects lead to a saw tooth-shape of the wake electric field.[3]

The plasma wave thus excited in the wake of the laser pulse lasts for quite some time. During this period, the wave can be employed to accelerate a co-moving electron beam to high energies.

## 6.4 Electron Acceleration

We next study the acceleration of energetic electrons by the large amplitude plasma waves. These electrons can be externally injected into the plasma wave as a beam, or could be self-injected into the plasma wave by some internal mechanism. We can write the electric field of the plasma wave as

$$\vec{E} = -\hat{z}E_0 \sin\left(\omega t - kz + \psi_0\right). \tag{6.30}$$

The electron energy equation under this field is

$$mc^2 \frac{d\gamma}{dt} = -e\vec{E} \cdot \vec{v}, \tag{6.31}$$

where $\gamma = \left(1 + p^2/m^2c^2\right)^{1/2}$ is the relativistic gamma factor and $\vec{p} = m\gamma\vec{v}$ is the electron momentum. We ignore the $x$–$y$ motions and introduce

$$\psi = kz - \omega t - \psi_0 \tag{6.32}$$

to obtain the system of differential equations for the electron,

$$\frac{d\gamma}{dt} = -\frac{eE_0}{mc}\left(1 - \frac{1}{\gamma^2}\right)^{1/2} \sin\psi \tag{6.33}$$

$$\frac{d\psi}{dt} = ck\left(1 - \frac{1}{\gamma^2}\right)^{1/2} - \omega, \tag{6.34}$$

where $\psi$ is the phase of the wave as seen by the electron. Dividing Eq. (6.33) by Eq. (6.34) we obtain

$$\left[1 - \beta \frac{\gamma}{\left(\gamma^2 - 1\right)^{1/2}}\right] \frac{d\gamma}{d\psi} = -A\sin\psi,$$

which can be integrated as

$$F(\gamma) \equiv \gamma - \beta\left(\gamma^2 - 1\right)^{1/2} = A\cos\psi + c_1, \tag{6.35}$$

where we denoted $\beta = \omega / kc$ and $A = eE_0 / mc^2 k$. The constant of integration is $c_1 = \gamma_0 - \beta \left(\gamma_0^2 - 1\right)^{1/2} - A \cos \psi_0$, and $(\gamma_0 - 1)mc^2$ is the electron kinetic energy at the initial phase $\psi_0$ (at $t = 0$, $z = 0$). We recall that the phase speed $w/k$ of the plasma wave equals the group speed of the laser pulse, which is less than $c$. Hence the second term on the left-hand side is smaller than the first one, and $F(\gamma)$ is positive definite. In Fig. 6.3, $F(\gamma)$ is plotted for different values of $\beta$. One may notice that for one value of $F$, and hence for one value of $\psi$, there are two values of $\gamma$. These values coalesce when $dF / d\gamma = 0$, i.e.,

$$\gamma = \gamma_m = \left(1 - \beta^2\right)^{-1/2}, \; F_{\min} = \left(1 - \beta^2\right)^{1/2}. \tag{6.36}$$

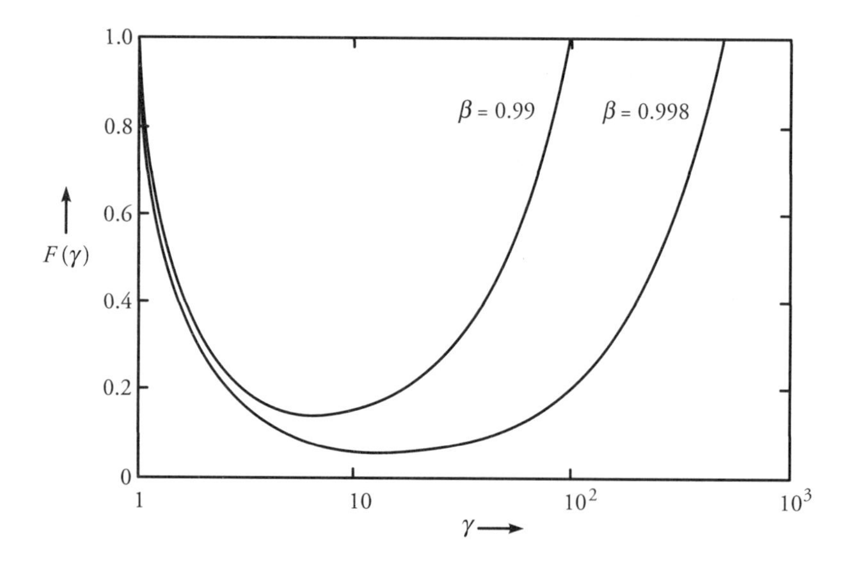


**Fig. 6.3** Variation of $F(\gamma)$ with $\gamma$ for different values of $\beta = \omega/kc$.

Thus the left-hand side of Eq. (6.35) is positive definite, greater than $F_{\min}$. Therefore, the right-hand side is also positive definite. For $c_1 < A + F_{\min}$, all values of $\psi$ are not accessible, and hence, this limit represents trapped electrons. The opposite limit $c_1 > A + F_{\min}$ represents untrapped electrons.

Figure 6.4 shows the phase space trajectories ($\gamma$ vs $\psi$) of the electrons plotted from Eq. (6.35). The separatrix corresponds to $c_1 = A + (1 - \beta^2).^{1/2}$ The trapped particles have closed curves inside the separatrix, while the untrapped particles have trajectories outside the separatrix. The trapped electrons near the bottom of the trajectory, in course of their interaction with the plasma wave, gain large amounts of energy and move to the top of the curve. If a horizontal line $F = F(\gamma_0)$ is drawn in the graph in Fig. 6.3, it will cut the $F-\gamma$ curve at two points. The span between the two cuts gives the maximum energy the electron can gain from the plasma wave. As $\beta$ approaches unity, the energy gain becomes large.

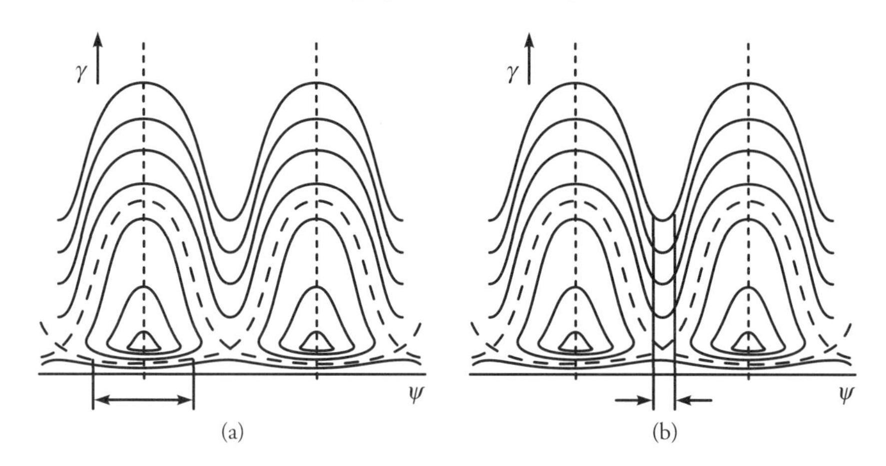


Fig. 6.4 Phase space behavior of trapped and untrapped electrons: (a) acceleration of trapped electrons, (b) acceleration of passing electrons. (After Bobin[17])

The electrons on passing trajectories, can also be accelerated to high energies (c.f. 6.4b).

## 6.4.1 Acceleration energy and length

Expressing $c_1$ in terms of $\gamma_0$ and $\psi_0$, Eq. (6.35) can be written as

$$\gamma-\gamma_0-\beta\left[\left(\gamma^2-1\right)^{1/2}-\left(\gamma_0^2-1\right)\right]=A\left(\cos\psi-\cos\psi_0\right). \tag{6.37}$$

The largest value of the right-hand side is 2A. If we assume $\gamma, \gamma_0 >> 1$, this equation gives

$$\gamma-\gamma_0\approx\frac{2eE_0}{mc\omega_p}\frac{\beta}{1-\beta} \tag{6.38}$$

giving the maximum energy gain

$$W_A=mc^2\left(\gamma-\gamma_0\right)=\frac{4eE_0c}{\omega_p}\gamma_r^2, \tag{6.39}$$

where $\gamma_r=\left(1-\beta_r^2\right)^{-1/2}$ is the Lorentz factor associated with the phase velocity of the plasma wave. If one assumes that the maximum value of plasma wave amplitude is limited (in the wake-field accelerator) by wave breaking, corresponding to $n/n_0^0 \sim 0.5$, i.e., $E_0 \approx mc\omega_p/2e$, one obtains

$$W_{A\max} \cong 2m_0c^2\gamma_r^2, \tag{6.40}$$

The acceleration length $L_A$ is given by

$$L_A \cong W_A / eE_0 = 2\gamma_r^2 c / \omega_p. \tag{6.41}$$

Akhiezer and Polovin[18] and Esarey and Pilloff[19] have found a higher wave breaking limit of relativistically nonlinear plasma waves,

$$E_{WB} = \sqrt{2}\left(\gamma_r - 1\right)^{1/2} E_0. \tag{6.42}$$

Such a plasma wave can accelerate electrons to much higher energy.

## 6.5 Bubble Regime Acceleration

At high intensity, the ponderomotive force on the electrons due to a short pulse laser of small spot size acquires a large value and causes complete electron evacuation from the region just behind the laser pulse. This creates an ion bubble moving with the group velocity of the laser. Outside the bubble, the electrostatic potential nearly balances the ponderomotive potential

$$\phi \approx -\phi_p = \frac{mc^2}{e}\left[(1 + a^2/2)^{1/2} - 1\right], \tag{6.43}$$

where $a = eA_L/mw_Lc$, with $A_L$ being the amplitude of the laser of frequency $\omega_L$. Using $\phi$ in Poisson's equation, $\nabla^2\phi = (e/\varepsilon_0)\left(n_e - n_0^0\right)$, we obtain the modified electron density

$$n_e = n_0^0\left(1 + \frac{c^2}{\omega_p^2}\nabla^2\left(1 + \frac{a^2}{2}\right)\right)^{1/2}. \tag{6.44}$$

The boundary of the bubble is given by $n_e = 0$, i.e.,

$$\nabla^2\left(1 + \frac{a^2}{2}\right)^{1/2} = -\frac{\omega_p^2}{c^2} \tag{6.45}$$

For a Gaussian pulse, $a^2 = a_{00}^2 e^{-r^2/r_0^2} e^{-\left(t - z/v_g\right)^2/\tau^2}$, Eq. (6.45) gives

$$\frac{r^2}{r_0^2} + \frac{\left(z - v_g t\right)^2 r_0^2}{v_g^4 \tau^4} = \frac{1 + a^2/2}{1 + a^2/4} + \left(1 + \frac{r_0^2}{2v_g^2\tau^2}\right) - \frac{\omega_p^2 r_0^2}{c^2}\frac{\left(1 + a^2/2\right)^{3/2}}{\left(1 + a^2/4\right)a^2}, \tag{6.46}$$

describing an ellipsoid-shaped bubble moving with the group velocity $v_g$. When $r_0 = v_g\tau$, it gives a spherical bubble. Figure 6.5 shows a typical bubble of almost spherical shape seen in PIC (particle in cell) simulations.

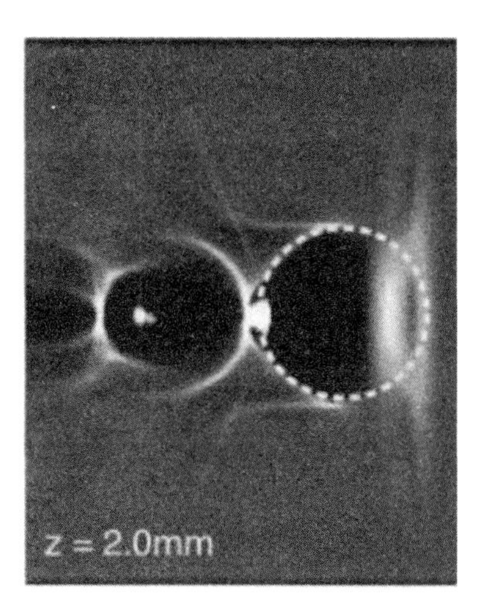


Fig. 6.5 Bubble regime acceleration. (From Lu et al.[21])

The electrons at the periphery of the bubble move backward (in the moving frame of the bubble). They surge towards the stagnation point at the back of the bubble and merge to create a small electron sphere. This sphere slows down the boundary electrons with respect to the moving frame and brings them to rest, after which they are pulled by the ions inside the bubble and acquire very high energies with respect to the lab frame.

### 6.5.1 Energy gain

Consider a moving bubble (cf. Fig. 6.5) of ion density $n_0$ and radius $R$ moving with velocity $v_g\hat{z}$. At $t = 0$, the center of the bubble is at the origin $z = 0$. At that instant, the electric field at $z$ (on the bubble axis) is $E_z = n_0 ez/3\varepsilon_0$. At time $t$, the field at $z$ is

$$E_z = \frac{n_0 e}{3\varepsilon_0}(z - v_g t). \tag{6.47}$$

An electron enters the ion bubble from the stagnation point at time $t = 0$. Let its position at time $t$ be $z_p$ and velocity $v_z\hat{z}$. At $t = 0, z_p = -R, v_z = v_g$. When $t > 0$, the energy equation for the electron is

$$mc^2\frac{d\gamma}{dt} = -eE_z v_z = -\frac{n_0 e^2}{3\varepsilon_0}(z_p - v_g t)v_z.$$

Introducing a new variable $\xi = z_p - v_g t$ and writing $v_z = c(\sqrt{\gamma - 1})/\gamma$, this equation reduces to

$$\left(1 - \frac{v_g}{c}\frac{\gamma}{\sqrt{\gamma^2 - 1}}\right)\frac{d\gamma}{d\xi} = -\frac{\omega_p^2}{3c^2}\xi, \tag{6.48}$$

giving

$$\gamma - \frac{v_g}{c}\sqrt{\gamma^2 - 1} = -\frac{\omega_p^2 \xi^2}{6c^2} + C_1, \tag{6.49}$$

where $\omega_p = (n_0 e^2/m\varepsilon_0)^{1/2}$ and the constant of integration on using the initial conditions turns out to be $C_1 = \gamma_g - (v_z/c)\sqrt{\gamma_g^2 - 1} + \omega_p^2 R^2/6c^2$ with $\gamma_g = (1 - v_g^2/c^2)^{-1/2}$. When the electron reaches the center $\xi = 0$ (where the field vanishes), we denote $\gamma = \gamma_M$, governed by

$$\gamma_M - \frac{v_g}{c}\sqrt{\gamma_M^2 - 1} = \frac{\omega_p^2 R^2}{6c^2} + \gamma_g - \frac{v_g}{c}\sqrt{\gamma_g^2 - 1}. \tag{6.50}$$

For $\gamma_M^2 >> 1, \gamma_g^2 >> 1$, this gives the electron energy on reaching the center of the ion bubble,

$$E_e = mc^2(\gamma_M - 1) = mc^2\left((\gamma_g - 1 + \frac{\omega_p^2 R^2}{3c^2}\gamma_g^2)\right). \tag{6.51}$$

This energy estimate compares well with that of Lu et al.[20,21], $E_e = 2mc^2\omega_L^2 a_{00}/3\omega_p^2$, when one takes $\omega_p R/c = 2\sqrt{a_{00}}$ and $\gamma_g = \omega_L/\omega_p$. For $\omega_p R/c \simeq 1.7, \gamma_g \approx 40$ this gives an electron energy of 800 MeV, comparable to the energy gain observed in PIC simulations.

## 6.6 Experiments and Simulations

A major milestone in laser-driven electron acceleration was achieved in 2004 when groups from the Rutherford Appleton Laboratory,[22] the Lawrence Berkeley National Laboratory,[23] and Laboratoire d'Optique Appliquee[24] independently reported experimental results producing electron bunches of $\sim 100$ MeV with only a small percent energy spread and very low divergence. In this section, we summarize the parameters used in these experiments.

Mangles et al.[22] employed a laser of energy 0.5 J, intensity $2.5 \times 10^{18}$ W/cm$^2$, pulse length 40 fs, and spot size 12.5 μm, using a 2 mm long gas jet plume with a plasma density of $2 \times 10^{19}$ cm$^{-3}$. A 78 MeV electron bunch with 3% energy spread, and 22 pC charge was observed.

Faure et al.[24] used a pulse of 1 J energy, 33 fs pulse length, $3.2 \times 10^{18}$ W/cm$^2$ intensity, and 10.5 μm spot size on a 3 mm gas jet with a plasma density of $6 \times 10^{18}$ cm$^{-3}$. They observed an electron bunch of 170 MeV energy, 24% energy spread, and 500 pC charge. Geddes et al.[23]

employed a laser pulse of 9 TW power, 55 fs pulse length, $10^{19}$ W/cm$^2$ intensity, and 4.2 μm spot size on a preformed plasma channel created on a 2 mm gas jet with an on-axis electron density of $2 \times 10^{19}$ cm$^{-3}$. They observed 86 MeV energy bunches with 4% energy spread, 3 mrad divergence and $2\times10^9$ electrons. In some experiments with varied parameters, 170 MeV bunches with $10^9$ electrons were also observed.

These experiments demonstrate the efficacy of self-injected bubble regime acceleration of mono energy electrons. There are, however, three crucial issues in laser wake-field and bubble regime acceleration schemes: the problem of increasing the propagation length of the laser by overcoming diffraction divergence, the problem of increasing the dephasing length of trapped electrons (beyond which they slip into the retarding phase of the plasma wave), and the injection of electrons into the accelerating phase of the wave or the bubble. Subsequent experiments have addressed these issues. PIC simulations have demonstrated electron acceleration beyond GeV energy. Experiments have also made steady progress. Leemans et al.[25] have reported 4 GeV electron bunches with 6% energy spread, 0.3 m rad divergence and 6 pC charge from a 9 cm long capillary discharge plasma waveguide with electron density ~ $7\times10^{17}\,\text{cm}^{-3}$ using a laser of energy 16 J, power 0.3 PW, pulse length 40 fs, and wavelength 0.8 μm. Very high energy electrons in the range 100–250 MeV may have the potential use in radiotherapy of cancers due to their improved dosimetry properties compared to conventional methods.[26]

## 6.7 Discussion

The acceleration schemes mentioned in this chapter employ low density plasmas with densities of the order 1% of the critical density. Experiments and PIC simulations at higher densities reveal a different scenario. At densities of about 10% of the critical density and using an intensity ≥ $10^{18}$ W/cm$^2$ at 1 μm wavelength, Gahn et al.[27] observed strong azimuthal magnetic fields (~100 MG) associated with the flow of energetic electrons. The electrons execute transverse betatron oscillations in these fields. The laser magnetic field beats with this motion to exert a longitudinal Lorentz force on electrons. When the Doppler shifted laser frequency equals the betatron frequency, the electrons get a sharp boost in their energy.[28] This scheme of direct laser acceleration is especially relevant to fast ignition laser driven fusion that relies on generation and penetration of multi-MeV electrons into the shock-compressed core.

Katsouleas and Dawson[30] put forth the concept of surfatron acceleration, to avoid the accelerating electrons from going into the decelerating phase of the plasma wave. They suggested the superimposition of a uniform magnetic field transverse to the wave vector on the plasma wave. The magnetic field deflects the particles across the wave front, thereby preventing them from outrunning the wave. One may visualize this phenomenon by considering a trapped electron moving with velocity $v_z\hat{z} \approx (\omega_p / k_z)\hat{z}$, i.e., the phase velocity of the wave. As there exists a static magnetic field $B_s\hat{y}$, the electron experiences a force $ev_zB_s\hat{x}$ and acquires the velocity $v_x$ in the $x$ direction, which in turn gives rise to a retarding force $F_z = -ev_xB_s$ in the negative $z$ direction. Due to relativistic effects, the velocity $v_x$ is limited by the speed of light $c$. If the electric force $-eE_z$ on the electron due to the wave is sufficient to overcome the retarding force $E_z = -ecB_{\varsigma}$, the electron would remain trapped and get accelerated in the $x$ direction without limit. However, the finite transverse extent of the plasma wave limits this acceleration mechanism.

## References


1. Tajima, T., and J. M. Dawson. 1979. "Laser Electron Accelerator." *Physical Review Letters* 43 (4): 267.
2. P. Sprangle, E. Esarey, A. Ting, and G. Joyce. 1988. "Laser Wakefield Acceleration and Relativistic Optical Guiding." *Applied Physics Letters* 53: 2146.
3. Berezhiani, V. I., and I. G. Murusidze. 1990 "Relativistic Wake-field Generation by an Intense Laser Pulse in a Plasma." *Physics Letters A* 148(6–7): 338–40.
4. Rosenbluth, M. N., and C. S. Liu. 1972. "Excitation of Plasma Waves by Two Laser Beams." *Physical Review Letters* 29 (11): 701.
5. Joshi, C., W. B. Mori, T. Katsouleas, J. M. Dawson, J. M. Kindel, and D. W. Forslund. 1984. "Ultrahigh Gradient Particle Acceleration by Intense Laser-driven Plasma Density Waves." *Nature* 311 (5986): 525.
6. Clayton, C. E., K. A. Marsh, A. Dyson, M. Everett, A. Lal, W. P. Leemans, R. Williams, and C. Joshi. 1993. "Ultrahigh-gradient Acceleration of Injected Electrons by Laser-excited Relativistic Electron Plasma Waves." *Physical Review Letters* 70 (1): 37.
7. Clayton, C. E., M. J. Everett, A. Lal, D. Gordon, K. A. Marsh, and C. Joshi. 1994. "Acceleration and Scattering of Injected Electrons in Plasma Beat Wave Accelerator Experiments." *Physics of Plasmas* 1(5): 1753–60.
8. Malka, G., J. Fuchs, F. Amiranoff, S. D. Baton, R. Gaillard, J. L. Miquel, H. Pépin, et al. 1997. "Suprathermal Electron Generation and Channel Formation by an Ultrarelativistic Laser Pulse in an Underdense Preformed Plasma." *Physical Review Letters* 79 (11): 2053.
9. Esarey, Eric, Phillip Sprangle, Jonathan Krall, and Antonio Ting. 1996. "Overview of Plasma-based Accelerator Concepts." *IEEE Transactions on Plasma Science* 24 (2): 252–88.
10. Bingham, R., J. T. Mendonca, and P. K. Shukla. 2004. "Plasma Based Charged-particle Accelerators" *Plasma Physics and Controlled Fusion* 46 (1): R1–R23.
11. Esarey, E., C. B. Schroeder, and W. P. Leemans. 2009. "Physics of Laser-Driven Plasma-based Electron Accelerators." *Reviews of Modern Physics* 81 (3): 1229.
12. Antonsen Jr, T. M., and P. Mora. 1992. "Self-focusing and Raman Scattering of Laser Pulses in Tenuous Plasmas." *Physical Review Letters* 69 (15): 2204.

    Antonsen Jr, T. M., and P. Mora. 1993. "Self-focusing and Raman Scattering of Laser Pulses in Tenuous Plasmas." *Physics of Fluids B: Plasma Physics* 5 (5): 1440.
13. Esarey, Eric, Phillip Sprangle, Jonathan Krall, Antonio Ting, and Glenn Joyce. 1993. "Optically Guided Laser Wake-field Acceleration." *Physics of Fluids B: Plasma Physics* 5 (7): 2690–7.
14. Sprangle, Phillip, Eric Esarey, Jonathan Krall, and G. Joyce. 1992. "Propagation and Guiding of Intense Laser Pulses in Plasmas." *Physical Review Letters* 69 (15): 2200.
15. Andreev, N. E., L. M. Gorbunov, V. I. Kirsanov, A. A. Pogosova, and R. R. Ramazashvili. 1994. "The Theory of Laser Self-resonant Wake Field Excitation." *Physica Scripta* 49 (1): 101.

16. Liu, C. S., and V. K. Tripathi. 1995 "Thermal Effects on Coupled Self-focusing and Raman Scattering of a Laser in a Self-consistent Plasma Channel." *Physics of Plasmas* 2 (8): 3111–4.
17. J. L. Bobin. 1987. in *Proceedings of the ECFA-CAS/CEFN-In-2P3-IRF/CEA-EPS Workshop*, Brsay, Italy. Edited by S. Turnet. Geneva: CERN, vol. 1, p. 58.
18. Akhiezer, Aleksander Ilyich, and R. V. Polovin. 1956 "Theory of Wave Motion of an Electron Plasma." *Soviet Phys. JETP* 3: 696.
19. Esarey, Eric, and Mark Pilloff. 1995 "Trapping and Acceleration in Nonlinear Plasma Waves." *Physics of Plasmas* 2 (5): 1432–6.
20. Lu, Wei, Chengkun Huang, Miaomiao Zhou, W. B. Mori, and T. Katsouleas. 2006. "Nonlinear Theory for Relativistic Plasma Wakefields in the Blowout Regime." *Physical Review Letters* 96 (16): 165002.
21. Lu, Wei, M. Tzoufras, C. Joshi, F. S. Tsung, W. B. Mori, J. Vieira, R. A. Fonseca, and L. O. Silva. 2007 "Generating Multi-Gev Electron Bunches Using Single Stage Laser Wakefield Acceleration in a 3D Nonlinear Regime." *Physical Review Special Topics-Accelerators and Beams* 10 (6): 061301.
22. Mangles, Stuart PD, C. D. Murphy, Zulfikar Najmudin, Alexander George Roy Thomas, J. L. Collier, Aboobaker E. Dangor, E. J. Divall et al. 2004 "Monoenergetic Beams of Relativistic Electrons from Intense Laser–Plasma Interactions." *Nature* 431 (7008): 535.
23. Geddes, C. G. R., Cs Toth, J. Van Tilborg, E. Esarey, C. B. Schroeder, D. Bruhwiler, C. Nieter, J. Cary, and W. P. Leemans. 2004 "High-quality Electron Beams from a Laser Wakefield Accelerator Using Plasma-channel Guiding." *Nature* 431 (7008): 538.
24. Faure, Jérôme, Yannick Glinec, A. Pukhov, S. Kiselev, S. Gordienko, E. Lefebvre, J-P. Rousseau, F. Burgy, and Victor Malka. 2004 "A Laser–Plasma Accelerator Producing Monoenergetic Electron Beams." *Nature* 431 (7008): 541.
25. Leemans, W. P., A. J. Gonsalves, H-S. Mao, K. Nakamura, C. Benedetti, C. B. Schroeder, Cs Tóth et al. 2014 "Multi-GeV Electron Beams from Capillary-Discharge-Guided Subpetawatt Laser Pulses in the Self-trapping Regime." *Physical Review Letters* 113 (24): 245002.
26. Subiel, Anna, V. Moskvin, Gregor H. Welsh, Silvia Cipiccia, D. Reboredo, P. Evans, M. Partridge et al. 2014. "Dosimetry of Very High Energy Electrons (VHEE) for Radiotherapy Applications: Using Radiochromic Film Measurements and Monte Carlo Simulations." *Physics in Medicine & Biology* 59(19) : 5811.
27. Gahn, C., G. D. Tsakiris, A. Pukhov, J. Meyer-ter-Vehn, G. Pretzler, P. Thirolf, D. Habs, and K. J. Witte. 1999 "Multi-MeV Electron Beam Generation by Direct Laser Acceleration in High-density Plasma Channels." *Physical Review Letters* 83 (23): 4772.
28. Tsakiris, G. D., C. Gahn, and V. K. Tripathi. 2000 "Laser Induced Electron Acceleration in the Presence of Static Electric and Magnetic Fields in a Plasma." *Physics of Plasmas* 7 (7): 3017–30.
29. Gizzi, Leonida Antonio, Carlo Benedetti, Carlo Alberto Cecchetti, Giampiero Di Pirro, Andrea Gamucci, Giancarlo Gatti, Antonio Giulietti et al. 2013 "Laser-Plasma Acceleration with FLAME and ILIL Ultraintense Lasers." *Applied Sciences* 3 (3): 559–80.
30. Katsouleas, T., and J. M. Dawson. 1983 "Unlimited Electron Acceleration in Laser-Driven Plasma Waves." *Physical Review Letters* 51 (5): 392.

# 7 Laser Acceleration of Ions

## 7.1 Introduction

There has been increasing interest in high energy ion beams since the invention of the cyclotron in 1938. This is mainly due to the wide ranging applications of these beams, including their use in probing the inner structure of matter,[1] hadron therapy of cancer,[2] isotope production for positron emission tomography,[3] and in recent years, fast ignition fusion.[4] While conventional sources of beams like the synchrotron, cyclotron, pinch reflex diodes, magnetically insulated diodes, etc., are well established, they are expensive and bulky. Therefore, there is a quest for accelerators that are compact. A possible way to achieve this is by using intense short pulse laser--plasma interaction to produce quasi monoenergetic high energy ion beams.

In target normal sheath acceleration[5–9] (TNSA), one employs a linearly polarized laser on a solid thin foil of a few micron thickness with hydrogen contamination or adsorption on the rear side. The laser quickly converts the foil into an overdense plasma and penetrates into the skin layer where the laser field falls off rapidly with depth. This laser exerts a strong ponderomotive force on the electrons with second harmonic and quasi-static components. The former heats the electrons to a temperature of the order of the ponderomotive energy $T_h \sim e|\phi_p|$; while the latter pushes the electrons forward, forming a hot electron sheath on the rear side of the target. The sheath accelerates the protons adsorbed.

TNSA has been studied quite extensively in the last two decades. It is effective in producing ions below 100 MeV. Ceccotti et al.[10] have seen significant improvement in TNSA proton energy gain with targets having surface grating due to the mode conversion of lasers into surface plasma waves. Wagner et al.[11] achieved a proton cutoff energy of 85 MeV with particle numbers about $10^9$. Poole et al.[12] experimentally showed a scaling relation between the target thickness and the maximum proton energy. Nakatsutsumi et al.[13] determined from their experiments that the self-generated magnetic field degrades the proton cutoff energy in the TNSA process.

In radiation pressure acceleration[14–19] (RPA), the basic idea is that laser light carries momentum. A sub-micron thick foil is normally impinged by a circularly polarized laser. The laser front ionizes the foil into an overdense plasmaand exerts a forward ponderomotive force on the electrons. As the electrons move forward, they leave behind an ion space charge. A quasi-steady state is realized, after the electrons have moved a distance $\Delta$, where the space charge force on the electrons balances the ponderomotive force. If the foil thickness $l \sim \Delta$, a self-organized double layer, comprising a compressed electron layer and an extended ion layer, is formed. The continued laser irradiance accelerates the double layer as a single unit, effectively converting the momentum of the laser light to momentum of the foil. RPA has shown the potential to produce hundreds of MeV protons using laser intensities $I_L \geq 10^{22}$ W/cm$^2$ at 0.8 μm wavelength.

Henig et al.[19] reported the first experimental results on the RPA of ions from 2.9 – 40nm thick diamond like carbon (DLC) foils with hydrogen adsorbed on the rear. A circularly polarized laser of intensity ~ $5 \times 10^{19}$ W/cm$^2$ at 0.8 μm wavelength ($a_{00} = 3.5$), pulse duration 45 fs, pulse contrast ratio $10^{-11}$ and focal spot radius 1.8 μm was normally impinged on the foil. For an optimum thickness of 5.3 ± 1.3 nm, protons of 10 MeV and $C^{6+}$ ions of 30 MeV energy were observed. This thickness is 1.6 times more than the optimal thickness given in formula (7.17) below with $n_0^0 / n_{cr} = 500$, however, this is reasonable given the two ion species and 3-D character of the experiment. Kar et al.[18] observed similar energies but higher fluxes at $3 \times 10^{20}$ W/cm$^2$. Bin et al.[20] also achieved a beam of $C^{6+}$ ions with an energy peak of 15 MeV/u by attaching near-critical-density carbon-nanotube foams to the DLC target. Palaniyappan et al.[21] used different mixtures of aluminum and carbon to make thin target foils and obtained a maximum peak energy of 18 MeV/u. Scullion et al.[22] performed a series of experiments with target foils with thickness ranging from 10 nm to 100 nm irradiated by linearly and circularly polarized laser beams and achieved a $C^{6+}$ ion beam with peak energy 20 MeV/u while using a circularly polarized laser on a foil of thickness 10 nm.

One of the first PIC simulation results of RPA using circularly polarized laser in the phase stable regime, in 1D, was reported by Yan et al.[14] For 1D PIC simulations, they reported proton energies ~ 400 MeV with a narrow energy spread at $n_0/n_{cr} = 10$, $a_{00} = 5$, pulse duration $\tau_L = 500\ T_L$, and $l = 0.2\lambda_L$, where $T_L$ is the laser period. Tripathi et al.[16] explained these results using an analytical framework. Eliasson et al.[17] carried out Vlasov simulations and observed the phase space bunching of ions. Qiao et al.[15], and many others[23–25] have carried extensive 2D and 3D PIC simulations of one and two ion species foils. It is noted that the Rayleigh–Taylor instability in a two ion species foil is primarily limited to the heavier ion species; hence, protons can be accelerated during longer times than the growth time of the Rayleigh–Taylor instability.

Experiments have also been conducted on gas jet targets using laser micropulses with circular and linear polarizations.[26,27] Steepened plasma density profile and radiation pressure-driven double layers or shock fronts moving through a lower density plasma driven by radiation pressure have been reported. The upstream plasma protons suffer reflection from the moving front acquiring a velocity twice that of the front. This acceleration is referred to as hole boring acceleration. With linear polarization, the plasma front has a high temperature and the method is known as shock acceleration.

In this chapter, we will develop analytical models for TNSA and RPA. In Section 7.2, we will study TNSA by a normally impinged laser following the same procedure as Mora.[9] In

Section 7.3, we will study TNSA by a laser driven surface plasma wave over a rippled target. In Section 7.4 we will study RPA, giving details of the formation of the self-organized double layer, ion trapping, and acceleration of the double layer. In Section 7.5, we address the 2D effects on the laser irradiated foil and discuss its Rayleigh–Taylor instability. In Section 7.6, we present a model for hole boring proton acceleration by a circularly polarized laser in a gas jet. Finally, some concluding remarks are given in Section 7.7.

## 7.2 Target Normal Sheath Acceleration (TNSA)

In TNSA (cf. Fig. 7.1) a laser impinging on a foil creates a hot sheath of electrons on the rear side that accelerates protons.

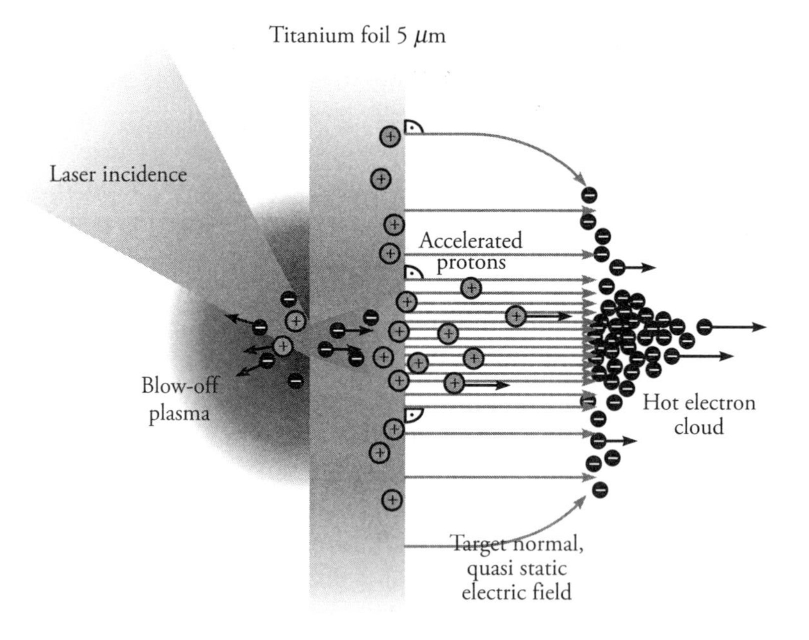


Fig. 7.1 Schematic of a laser-irradiated thick foil target. The electrons ($e^-$) at the front are heated and pushed by the laser to the rear side, forming an electrostatic sheath that accelerates the protons. (From Ledingham et al.[2])

Consider a foil that is turned into an overdense plasma of electron density $n_0$ normally irradiated by a linearly polarized laser,

$$\vec{E} = \hat{x}A_0 \exp\left[-i\left(\omega t - \omega z / c\right)\right]. \tag{7.1}$$

The laser penetrates into the skin layer with field $\vec{E}_T$ as discussed in Chapter 5. In the linear theory (valid for $a_0 = eA_0/m\omega c << 1$), the ratio of the amplitude of the transmitted wave at the boundary to the amplitude of the incident wave is $2\omega/\omega_p$, where $\omega_p = (n_0^0 e^2/m\varepsilon_0)^{1/2}$ is the plasma frequency, $-e$ and $m$ are the electron charge and rest mass, and $\varepsilon_0$ is the free space permittivity. At higher intensities this ratio decreases gradually. The transmitted field imparts an oscillatory velocity to the electrons and exerts a ponderomotive force having a quasi-static component, $\vec{F}_p = \hat{z}e\partial\phi_p/\partial z$, with $\phi_p = -(mc^2/e)(\gamma - 1)$, $\gamma = (1 + a_T^2/2)^{1/2}$, $a_T = e|E_T|/m\omega c$ and a second harmonic component $\vec{F}_{p2\omega}$ (of similar magnitude) that heats the electrons to a temperature $T_h \approx e|\phi_p|$. The quasi-static component pushes the hot electrons to the rear side of the foil.

At time $t = 0$, the proton density of the plasma can be taken to be

$$n_i = n_0 \text{ for } 0 < z < l$$

$$= 0 \text{ for } z > l. \tag{7.2}$$

The hot electron density may be taken to follow the Boltzmann distribution

$$n_h = n_{0h} e^{e\phi/T_h} \tag{7.3}$$

where $\phi$ is the electrostatic potential in the sheath region, satisfying Poisson's equation

$$\frac{\partial^2 \phi}{\partial z^2} = (n_h - n_i)e/\varepsilon_0. \tag{7.4}$$

Numerical integration of Eq. (7.4) at $t = 0$ from the ion surface at $z = l$ to $z = \infty$ yields the electric field at $z = l$ as[9]

$$E_{\text{front},0} \approx E_0 \approx n_{0h} e\lambda_D / \varepsilon_0 \approx (n_{0h} T_h/\varepsilon_0)^{1/2} \tag{7.5}$$

where $\lambda_D = (T_h\varepsilon_0/n_{0h}e^2)^{1/2}$ is the initial Debye length.

For $t > 0$, the motion of the protons (on the rear side) is governed by the momentum and continuity equations

$$\frac{\partial v_i}{\partial t} + v_i \frac{\partial}{\partial z} v_i = -\frac{e}{m_i}\frac{\partial \phi}{\partial z}, \tag{7.6a}$$

$$\frac{\partial n_i}{\partial t} + \frac{\partial}{\partial z}(n_i v_i) = 0. \tag{7.6b}$$

Physically we expect the hot electrons to move out to $z > 1$ upto a Debye length and create a space charge field $E_s = -\partial\phi/\partial z$ that pulls the ions. As the ions move out, the electrons move farther. Thus at larger $z$, electrons and ions arrive at a later time. We introduce a self-similarity variable,[9] $\xi = (z - l)/c_s t$, where $c_s = (T_h/m_i)^{1/2}$ is the velocity of sound. Using $\partial/\partial t = -(\xi/t)\, d/d\xi$ and $\partial/\partial z = (1/c_s t)\, d/d\xi$ in Eqs. (7.6a) and (7.6b), we obtain

$$\left(v_i - c_s\xi\right)\frac{dv_i}{d\xi} = \frac{e}{m_i}\frac{d\phi}{d\xi}, \tag{7.7a}$$

$$\left(v_i - c_s\xi\right)\frac{dn_i}{d\xi} = -n_i\frac{dv_i}{d\xi}. \tag{7.7b}$$

These equations combine to give

$$\left(v_i - c_s\xi\right)^2\frac{1}{n_i}\frac{dn_i}{d\xi} = \frac{e}{m_i}\frac{d\phi}{d\xi}. \tag{7.8}$$

Assuming quasi-neutrality, $n_i \approx n_h = n_{h0}\exp(e\phi/T_h)$, this equation yields $(v_i - c_s\xi)^2 = c_s^2$, or

$$v_i = c_s\left(1+\xi\right) = c_s + \left(z - l\right)/t. \tag{7.9}$$

Using this in Eq. (7.7a), we obtain

$$\phi = -\left(T_h/e\right)\xi,\; E_s = E_0/\left(\omega_{\rm pi}t\right), \tag{7.10}$$

where $\omega_{\rm pi} = (n_{0h}e^2/m_i\varepsilon_0)^{1/2}$ is the ion plasma frequency.

The quasi-neutrality approximation is not valid in the begining when density scale length $c_s t$ is smaller than the initial Debye length $\lambda_{D0}$ that is, for $\omega_{\rm pi}\, t < 1$. It is also not valid, even for $\omega_{\rm pi} t >> 1$, in the low density part of the plasma where the local Debye length $\lambda_D = \lambda_{D0}(n_{0h}/n_h)^{1/2}$ equals the density scale length, $c_s t$. This position corresponds to

$$\left(z - l\right)/c_s t = 2\log_e\left(\omega_{\rm pi}t\right) - 1. \tag{7.11}$$

At this position, there is a well-defined ion front with front velocity

$$v_{\rm front} \approx 2c_s\log_e\left(\omega_{\rm pi}t\right). \tag{7.12}$$

A numerical evaluation gives the location and velocity of the ion front as

$$z_{\rm front} = 1 + 2\sqrt{5.4}\lambda_{D0}\left[\tau\log\left(\tau + \sqrt{\tau^2+1}\right) - \sqrt{\tau^2+1} + 1\right]$$

$$v_{\text{front}} \approx 2c_s \log_e\left(\tau + \sqrt{\tau^2 + 1}\right) \tag{7.13}$$

where $\tau = \omega_{\text{pi}} t / \sqrt{5.4}$. The ions at the front acquire a velocity a few times the acoustic speed and an energy an order ofmagnitude larger than the ponderomotive energy of the electrons.

## 7.3 TNSA by Surface Plasma Wave

Ceccotti et al.[10] have recently reported significant improvement in TNSA proton energy gain when the target had a surface grating. The energy gain is maximum at an angle of incidence at which a p-polarized laser is mode converted into a surface plasma wave (SPW). It is inferred that the SPW is responsible for the enhanced TNSA. Liu et al.[28] have developed an analytical model for this mechanism. Following their approach we have given the mode structure of a large amplitude surface plasma wave, including relativistic and ponderomotive nonlinearities, in Appendix 7.1. One may note that the nonlinearities cause stronger localization of the SPW fields near the interface; however, the propagation constant is not affected significantly.

A surface ripple $z = h\cos(qx)$, of wave number $q$ and depth $h$ resonantly couples a p-polarized laser, obliquely incident on the plasma at an angle of incidence $\theta_i$ (cf. Fig. 7.2), to the SPW when the sum of the ripple wave number and the laser wave number along $\hat{x}$ equals the SPW wave number

$$k_x = q + \frac{\omega}{c}\sin\theta_i = \frac{\omega}{c}\left(\frac{\varepsilon}{1+\varepsilon}\right)^{1/2}; \; \varepsilon = 1 - \frac{\omega_p^2}{\omega^2},$$

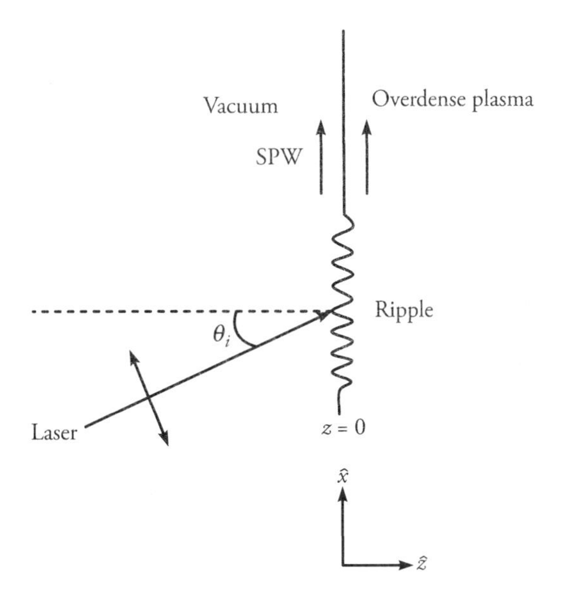


Fig. 7.2 Schematic of surface plasma wave propagation over an overdense plasma–vacuum interface. The surface ripple couples the laser to SPW.

where $\omega_p$ is the plasma frequency. In the limit when nonlinear effects are weak, the ratio of the SPW amplitude to the laser amplitude (following Chapter 4) can be written as

$$\left|\frac{A_1}{A_L}\right| = \frac{\omega^2 h r_0}{c^2} \frac{\left(\Omega_p^2 - 1\right)^{3/2}}{\left(\Omega_p^2 - 2\right)^2} \frac{1 + \sin\theta_i \left(\dfrac{\Omega_p^2 - 2}{\Omega_p^2 - 1}\right)^{1/2}}{\left[1 + \dfrac{\left(\Omega_p^2 - 1\right)^2}{\Omega_p^2 \sec^2\theta_i - 1}\right]^{1/2}}, \tag{7.14}$$

where $\Omega_p = \omega_p/\omega$. This expression is valid when the normalized SPW amplitude $a_0 < 1$. For $a_0 \geq 1$ a rough estimate of the power conversion efficiency can be obtained by replacing $\omega_p$ with $\omega_p / \left(1 + a_0^2/2\right)^{1/4}$.

The evanescent SPW field in the skin layer exerts a strong ponderomotive force on electrons, as in the aforementioned case and sends them to the rear. The hot electron temperature can be taken as $T_h \approx mc^2\left[\left(1 + a_0^2/2\right)^{1/2} - 1\right]$ and the sheath accelerated proton energies are of the order of $\varepsilon_i \approx 4T_h$. The energy gain is enhanced due to the enhancement in field amplitude via resonant mode conversion of laser into the SPW. For the parameters of Ceccotti et al.'s[10] experiment, $\omega_p/\omega = 10$, $\lambda_L = 0.8$ μm, $I_L = 2 \times 10^{19}$ W/cm$^2$ ($a_L = eA_L/m\omega c \approx 3$), $r_0 = 5$ μm, grating period $2\pi/q = 1.6$ μm, resonant angle of incidence $\theta_i = 30°$, laser amplitude transmission coefficient $|T| \approx 0.45$. Hence, $|T|a_L < 1$ and one may employ the linear coupling estimate. For $h = 150$ nm, $|A_1/A_L| \approx 0.9$. However, as far as the ponderomotive push on the electrons in the skin layer is concerned, a more relevant quantity is the ratio of the SPW field to the laser field in the skin layer $A_1/|T|A_L$; its value for the mentioned parameters is 2. Experimentally Ceccotti et al.[10] have observed field enhancement manifest in ion cutoff energy enhancement by a factor ~ 2.5. Further, a major advantage of the SPW is that it covers a much larger surface area than the cross-sectional area of the laser. Hence, a larger area of the proton layer would undergo target normal sheath acceleration.

## 7.4 Radiation Pressure Acceleration

Experiments on RPA employ an ultrathin foil of diamond-like carbon or other material, with a significant percentage of hydrogen adsorbed on it. A circularly polarized laser pulse normally irradiates the foil and instantly turns it into an overdense plasma. For the sake of simplicity we consider a thin slab of hydrogen plasma of thickness $l$ and density $n_0$, localized between $0 < z < l$ (cf. Fig. 7.3). The incident laser field (for $z < 0$) is $\vec{E}_1 = A_0\left(\hat{x} + i\hat{y}\right)\exp[-i\left(\omega t - \omega z/c\right)]$. Inside the plasma, the laser field is evanescent,

$$\vec{E}_T = \left(\hat{x} + i\hat{y}\right) A_T(z) \exp(-i\omega t), \tag{7.15}$$

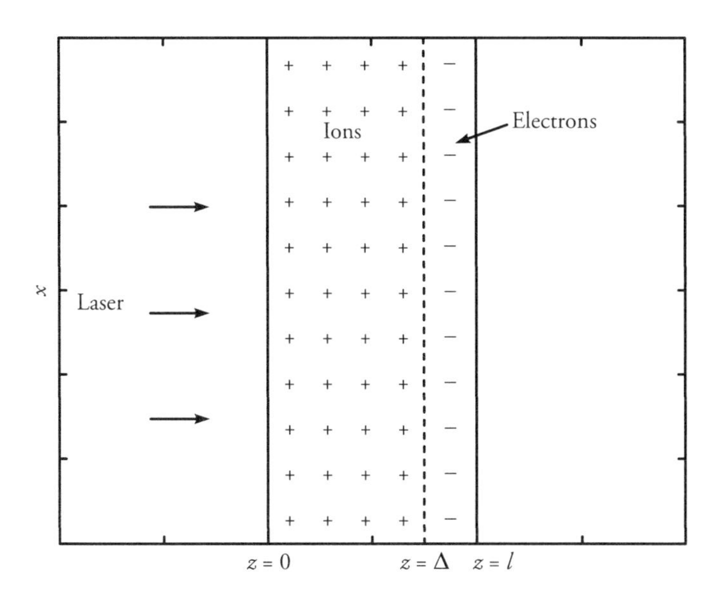


**Fig. 7.3** Schematic of a laser-irradiated overdense plasma thin foil. The electrons are pushed to the rear side of the foil.

with a normalized amplitude $a_T = eA_T/m\omega c$ governed by Eq. (5.37).The field falls off rapidly with distance. It exerts a ponderomotive force $\vec{F}_p = -\hat{z}mc^2\partial(1+a_T^2)^{1/2}/\partial z$ on the electrons that pushes them forward, creating an electron depleted region up to $z = \Delta$ and a compressed electron layer with a density profile given by Eq. (5.35),

$$\frac{n_e}{n_0} = 1 + \frac{c^2}{\omega_p^2}\frac{\partial^2}{\partial z^2}\left(1+a_T^2\right)^{1/2}, \tag{7.16}$$

The electron depleted region ($0 < z < \Delta$) has ion space charge electric field $\vec{E}_s = \hat{z}n_0 ez/\varepsilon_0$. At $z = \Delta$, the space charge field balances the ponderomotive force. In the electron layer, the ponderomotive force falls off rapidly, and so does the space charge field. In the quasi-steady state, the total radiation pressure force on electrons per unit area equals the electrostatic force on $n_0 l$ electrons, $n_0 leE_s \approx 2I_0/c$. Choosing the width of the foil $l = \Delta$, this condition gives the optimal foil thickness[16]

$$l \cong a_{00}\frac{n_{\rm cr}}{n_0}\frac{\lambda_L}{\pi\sqrt{2}}, \tag{7.17}$$

where $n_{cr} = (\omega^2 m\varepsilon_0/e^2)^{1/2}$ is the critical density at frequency $\omega$, and $a_{00}$ is the normalized laser amplitude. In foils like DLC (diamond-like carbon), where the ion charge is $Z_i e$ and the ion density is $n_{0i}$, the optimal thickness is still given by Eq. (7.17) with $n_0$ replaced by $Z_i n_{0i}$.

### 7.4.1 Ion trapping in the self-organized double layer

Once the electrons are pushed to the rear of the foil, their density is compressed. The electron density profile $n_e(z)$ can be deduced using the numerical solution of Eq. (5.37) in Eq. (7.16). The electrostatic potential $\phi_s$ in the electron depleted front region (deduced from the space charge field)is $\phi_s = -n_0^0 ez^2 / 2\varepsilon_0 + c_1$, whereas inside the electron layer $\phi_s = -\phi_p = (mc^2/e)\left[(1+a_T^2)^{1/2} - 1\right]$. The constant $c_1$ is evaluated by demanding the continuity of $\phi_s$ at the front of the electron layer. In Figs. 7.4a and 7.4c we have plotted $n_e$ and $\phi_s$ as a function of $z$. The electron density is zero in large parts of the foil. At the point where $\phi_s = \phi_p$, the density rises sharply and then falls off. The electrostatic potential falls off monotonically. The ions, in the accelerated frame of the foil (moving with velocity $v_f$), also experience an inertial force $-m_i dv_f/dt$, which tends to confine them by reflecting them back to the rear surface. Figure 7.4d shows the effective potential $\phi_{tot} = \phi_s + \phi_{acc}$, where $\phi_{acc} = (m_i/e)(dv_f/dt)z$, $m_i$ is the ion mass, and $dv_f/dt$ is the foil acceleration due to radiationpressure. Thus the ions can be stably trapped in

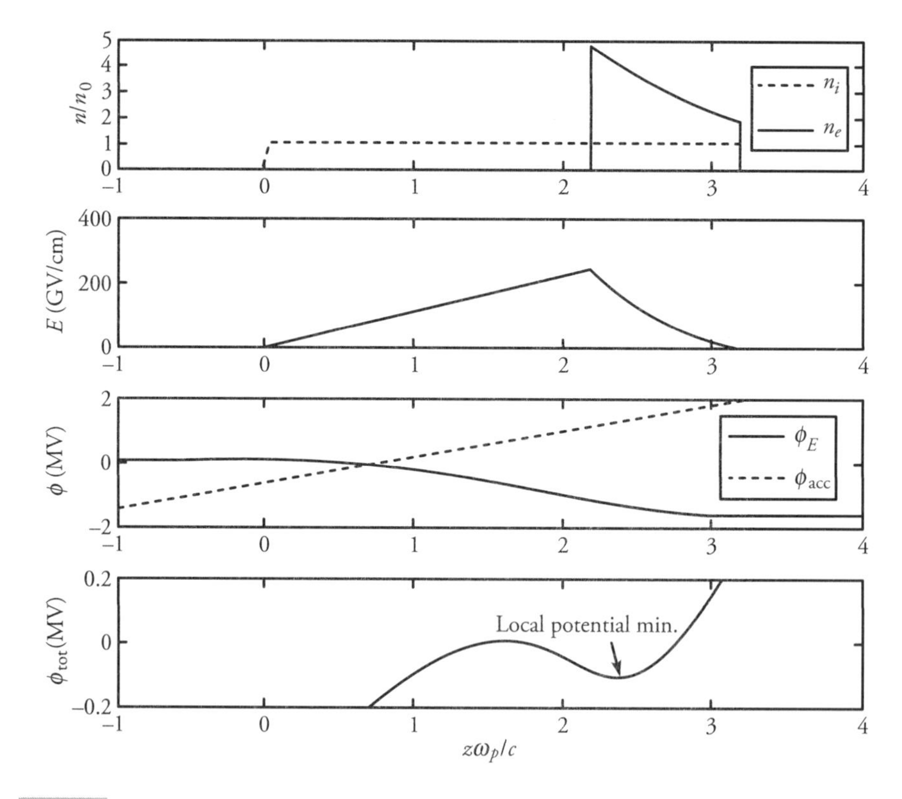


Fig. 7.4 The electron and ion number densities (top panel), the electric field (second panel), the electrostatic and acceleration potentials (third panel) and the effective/total potential in the accelerated frame (bottom panel), as functions of $z$, for $\omega_p^2/\omega^2 = 10$ and $a_{00}$ =5. The total potential has a local minimum which traps ions in the accelerated frame. (From Tripathi et al.[16])

the minimum of $\phi_{tot}$ formed by the combined forces of electron sheath and the inertial force of the accelerated frame. Together, they form a double layer, which is accelerated by the radiation pressure (shown to be equivalent to the ponderomotive force in Chapter 5). The ions that are accelerated to the compressed electron layer and overshoot decrease their momentum inside the trap and are pulled back to the potential well.

### 7.4.2 Acceleration of the double layer

The radiation pressure decreases with increasing foil velocity $v_f\hat{z}$. An expression for the radiation pressure on a moving mirror was first given in its general form by Albert Einstein in 1905 as[29]

$$F = \frac{2I_0}{c}\frac{(\cos\phi - v_f / c)^2}{1 - v_f^2 / c^2}, \tag{7.18}$$

where $\phi$ is the angle of incidence. For normal incidence ($\phi = 0$) the radiation pressure reduces to

$$F = \frac{2I_0}{c}\frac{1 - v_f / c}{1 + v_f / c} \tag{7.19}$$

Thus, for normal incidence of the laser, the relativistic equation of motion, on employing the areal mass density of the foil as $n_0 m_i l$ and laser intensity $I_0 = \varepsilon_0 A_0^2 c$, takes the form

$$\frac{d}{dt}\left(\gamma_f v_f\right) = \gamma_f^3 \frac{dv_f}{dt} = g\frac{1 - v_f/c}{1 + v_f/c}, \tag{7.20}$$

where $\gamma_f = \left(1 - v_f^2/c^2\right)^{-1/2}$, $g = 2I_o / (m_i n_0 lc) = 2\sqrt{2}\pi a_{00} c^2 / m_i \lambda_L$.

At $a_{00} = 20$, $\lambda_L = 1$ μm ($I_0 \sim 10^{21}$ W/cm$^2$), and $m_i/m = 1836$. The acceleration is huge, $g \approx 10^{22}$ m/s$^2$. In the non-relativistic limit, Eq. (7.20) gives

$$v_f = gt, \tag{7.21}$$

giving the ion energy $\varepsilon_i = m_i g^2 t^2/2$. The energy scales with the square of the pulse duration and laser intensity. As $v_f$ approaches the relativistic range, one obtains from Eq. (7.20)

$$v_f/c = \frac{G^2 - 1}{G^2 + 1}, \tag{7.22}$$

where

$$G = \frac{-2^{2/3} + \left(h + \sqrt{4+h^2}\right)^{2/3}}{2^{1/3}\left(h + \sqrt{4+h^2}\right)^{1/3}}, \quad h = 6gt/c + 4. \tag{7.23}$$

An equivalent solution in a different form is given by Robinson et al.[30] The position of the foil can be expressed in terms of $v_f/c$ as

$$z(t) = \frac{c^2}{6g}\left[\left(\frac{1+v_f/c}{1-v_f/c}\right)^{3/2} - 3\left(\frac{1+v_f/c}{1-v_f/c}\right)^{3/2} + 2\right] \tag{7.24}$$

and the ion energy as

$$\varepsilon_i = m_i c^2[(1 - v_f^2/c^2)^{-1/2} - 1]. \tag{7.25}$$

In Fig. 7.5, $\varepsilon_i$ has been plotted as a function of $t/T_L$. For $gt/c < 1$, $\varepsilon_i$ scales as $t^2$ with time $t$. However, later in time, for $gt/c >> 1$, $\varepsilon_i$ scales as $t^{1/3}$, which is consistent with the scaling observed in PIC (particle in cell) simulations.[23,27]

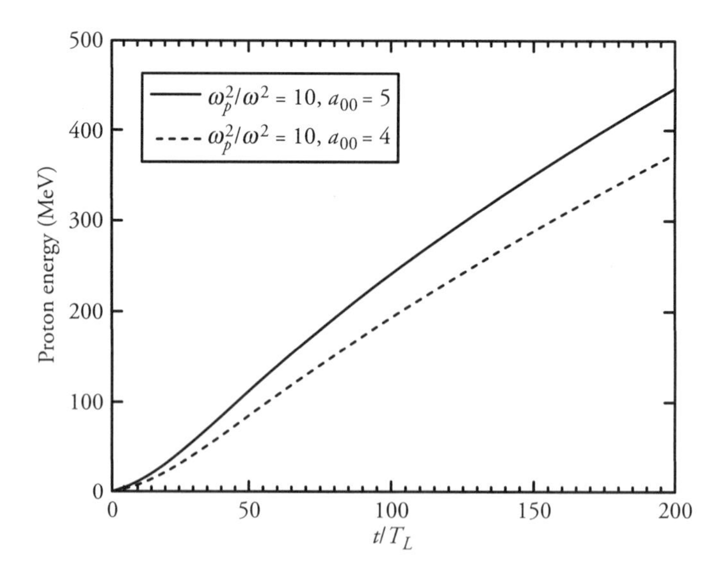


**Fig. 7.5** The ion energy as a function of time for $a_{00} = 4$ and $a_{00} = 5$ with $\omega_p^2/\omega^2 = 10$. (From Tripathi et al.[16])

Eliasson et al.[17] observed phase space bunching of ions in their Vlasov simulations of 1D RPA. In order to provide a balance between the inertial force acting backward and the

Coulomb force acting forward, it is necessary that a portion of the ions are left behind. The fraction of trapped ions is estimated to be[17]

$$\frac{N_{tr}}{N_0} = 1 - \frac{\varepsilon_0 F}{e^2 N_0^2},$$

where $N_{tr}$ is the integrated number density of trapped ions (along $z$) and $N_0$ is the total (trapped and un-trapped) integrated number density of ions $F$ is the radiation pressure. The un-trapped ions experience a Coulomb expansion and form a tail behind the double layer, clearly seen extending to the left of the foil in Fig. 7.6b. As the double layer is accelerated to higher speeds, the RPA decreases due to the Doppler-shifted radiation pressure force, and a portion of the ions left behind catch up with the compressed electron layer (cf. Fig. 7.6b).

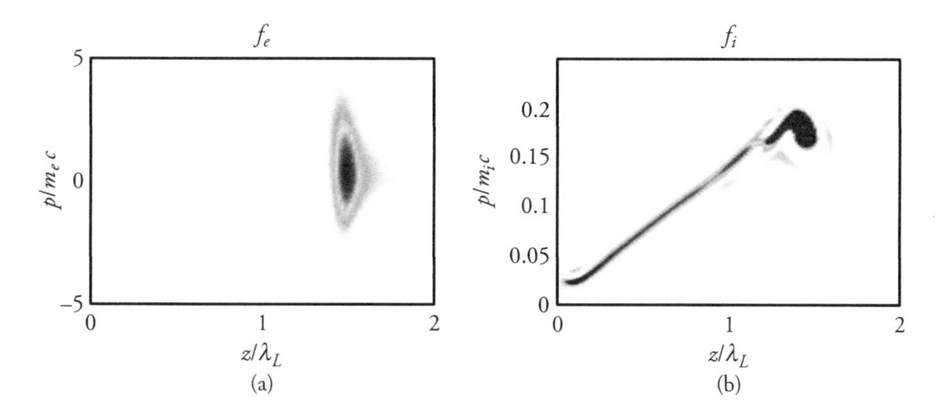


**Fig. 7.6** Simulation results showing (a) the electron and (b) ion distribution function at $t = 16\ T_L$. (From Eliasson et al.[17])

## 7.5 2D Effects: Rayleigh–Taylor Instability

The laser employed in the RPA of ions has a finite spot size and non-uniform (often Gaussian) distribution of intensity across the spot. As a result the foil acquires, curvature. The portion of the foil whose surface normal makes an angle $\theta$ with the laser axis $(\hat{z})$, experiences a diminishing of the radiation pressure force by a factor $\cos^2\theta$, the force is directed along the surface normal. Thus, the force tends to diverge the foil, and brings angular divergence and energy spread to the ions.

A more severe problem arises due to the ripple or kinks in the surface of the foil. On a rippled surface, the radiation pressure force is non-uniform and directed along the local surface normal. As a result the surface ripple grows as the foil moves forward. This is known as the Rayleigh–Taylor instability (RTI). Ott[31] theoretically studied the nonlinear development of the Rayleigh–Taylor instability of a thin foil subject to a pressure difference between the two sides, while Pegoraro and Bulanov[32] extended the study to the relativistic regime, theoretically and with PIC simulations. The effect of the diffraction pattern of the reflected wave was taken into account in recent studies.[33,34] The studies showed that due to a plasmonic resonance, the growth rate of the Rayleigh–Taylor instability has a sharp maximum at perturbation wave numbers near the laser wave number. In this chapter, we deduce only the linear growth rate of Rayleigh–Taylor instability. We exclude diffraction effects, treating the foil of thickness $l_0$, ion density $n_0$ and ion mass $m_i$ as a single fluid. Initially the foil is located at $z = 0$ with a small surface ripple (cf. Fig. 7.7)

$$z_0 = be^{iqx_0}, \tag{7.26}$$

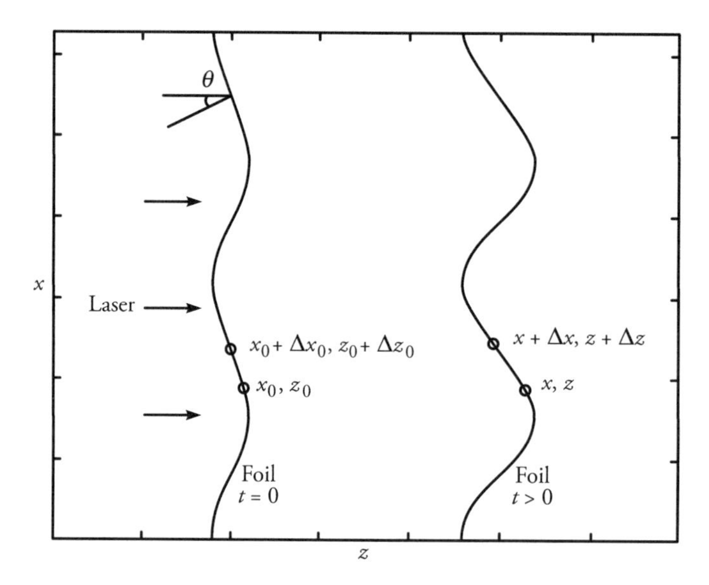


**Fig. 7.7** A rippled foil irradiated by a laser. The perturbation grows as the foil propagates.

where $x_0$, $z_0$ is any initial point on the foil. It is irradiated by a laser of intensity $I_0$ from the left. The point $x_0$, $z_0$ evolves into $x$, $z$, and so does a neighboring point $x_0 + \Delta x_0$, $z_0 + \Delta z_0$ into $x + \Delta x$, $z + \Delta z$, where

$$\Delta x = \frac{\partial x}{\partial x_0}\Delta x_0, \quad \Delta z = \frac{\partial z}{\partial x_0}\Delta x_0. \tag{7.27}$$

The initial length of the segment between $x_0$ and $x_0 + \Delta x_0$ is $ds_0 = (1 + (\partial z_0/\partial x_0)^2)^{1/2}\Delta x_0$; later it becomes $ds = (1 + (\partial z/\partial x)^2)^{1/2}\Delta x$. The width of the segment $l$ changes to maintain mass conservation.

$$l_0 ds_0 = l ds. \tag{7.28}$$

The radiation pressure force on this segment of length $ds$ and $y$-width unity is $ds(2I_0/c)\cos^2\theta(\cos\theta\,\hat{z}+\sin\theta\,\hat{x})$, where $\theta$ is the angle the surface normal makes with the $z$-axis. The equation of motion of the segment in component form is

$$n_0 l_0 m_i ds_0 \frac{d^2x}{dt^2} = \frac{2\,I_0}{c} ds\cos^2\theta\sin\theta \tag{7.29a}$$

$$n_0 l_0 m_i ds_0 \frac{d^2z}{dt^2} = \frac{2I_0}{c} ds\cos^2\theta\cos\theta. \tag{7.29b}$$

The tangent to the surface makes an angle $\alpha = \tan^{-1}(\Delta x/\Delta z)$ with respect to the $z$-axis. Hence the surface normal would make an angle $\theta = -\cot^{-1}(dx/dz)$. Thus, $\sin\theta = -\Delta z/ds$, $\cos\theta = \Delta x/ds$. For $bq << 1$, the aforementioned equations take the form

$$\frac{d^2x}{dt^2} = -g\frac{\partial z}{\partial x_0}, \tag{7.30}$$

$$\frac{d^2z}{dt^2} = g\frac{\partial x}{\partial x_0}, \tag{7.31}$$

where $g = 2I_0/(m_i n_0 l_0 c)$. If the surface had no ripple, the $x$ coordinate of any segment will remain constant ($x = x_0$) while the $z$ coordinate of all segments would remain the same ($\partial z/\partial x_0 = 0$). Thus $d^2z/dt^2 = g$ and

$$z = \frac{1}{2}gt^2. \tag{7.32}$$

We perturb this equilibrium by the ripple

$$z = \frac{1}{2}gt^2 + z_1 \tag{7.33}$$

$$x = x_0 + x_1 \tag{7.34}$$

and take $z_1,\ x_1 \sim e^{-i(\Omega t - qx_0)}$. Using these in Eqs. (7.30) and (7.31), we get

$$-\Omega^2 x_1 = -giqz_1$$

$$-\Omega^2 z_1 = giqx_1$$

leading to the dispersion relation

$$\Omega^4 = q^2 g^2, \tag{7.35}$$

This gives a purely growing mode with growth rate

$$\Gamma = (qg)^{1/2}. \tag{7.36}$$

Shorter wavelength perturbations grow faster. By taking into account the laser diffraction effects of the scattered light,[33] it has been shown that the growth rate has a sharp maximum when the perturbation wave number equals the laser wave number, $q = k_L$. The general dispersion relation for a normal incident circularly polarized light is[33]

$$\Omega^4 - q^2 g^2 = i\frac{\Omega^2 g}{2}\sum_{\pm}\frac{q^2 + 2k_{z\pm}^2}{k_{z\pm}}, \tag{7.37}$$

where the right-hand side includes the effects of diffraction. Here $k_{z\pm} = \sqrt{(k_L \pm \Omega / c)^2 - q^2}$ is the complex-valued normal wave vector component of the reflected sideband of the laser light, where the branch of the square root should be chosen such that its imaginary part is negative when the imaginary part of $\Omega$ is positive. A solution of the dispersion relation is given in Fig. 7.8 for $g/(c^2 k_L) = 0.003$, showing the sharp maximum in the growth rate for $q \approx k_L$. In general, the Rayleigh–Taylor instability grows over a time $\approx \sqrt{\lambda_L / g}$. This is a serious limitation on RPA, limiting the energy gain. In one growth period ions gain energy $\approx \sqrt{2\pi a_{00}} mc^2$ which is of the order of 100 MeV at $a_{00} \approx 50 (I_L \sim 10^{22}\ \mathrm{W/cm^2})$.

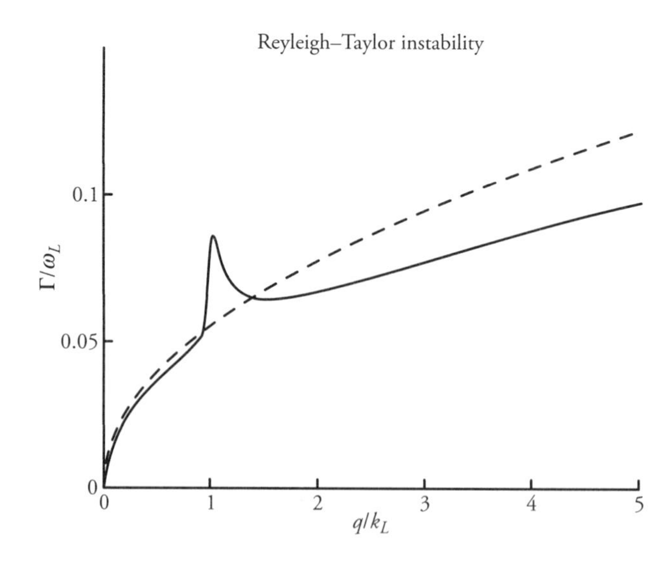


**Fig. 7.8** The growth rate of the standard Rayleigh–Taylor instability (dashed line) and including diffraction effects (solid line) for circularly polarized laser light. The diffraction effects lead to a plasmonic resonance and a maximum of the instability at $q \approx k_L$. (After Eliasson.[33])

When there are two ion species targets, for example, $C^{6+}$ plasma with 10% protons, PIC simulations have revealed the formation of quasi-monoenergetic protons through a combination of RPA and shielded Coulomb repulsion.[35,36] Since the proton layer has a larger charge-to-mass ratio, it gets detached from the heavier $C^{6+}$ layer and the growth of the Rayleigh–Taylor instability has a weaker effect on the proton acceleration (cf. Fig. 7.9).

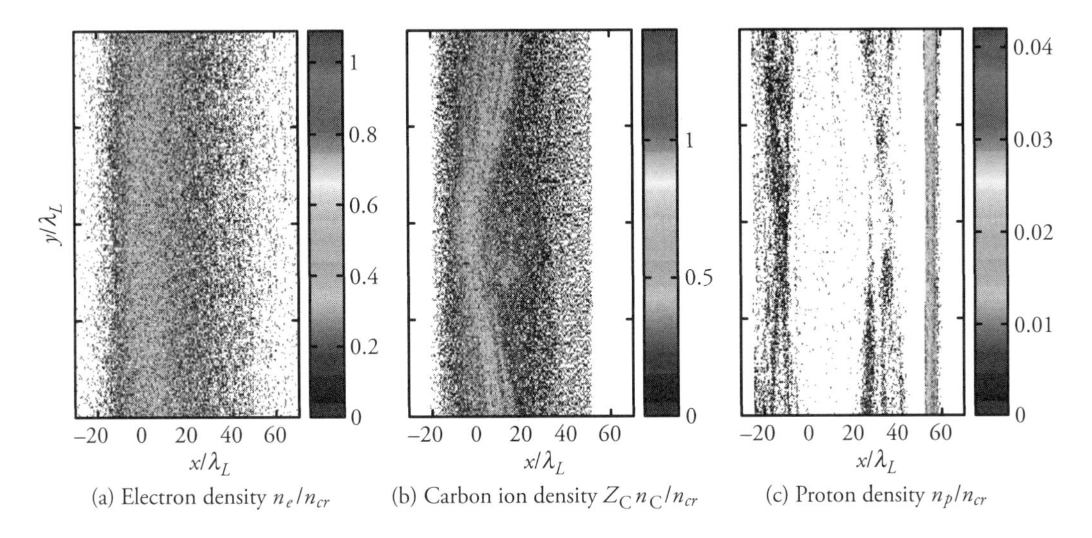


**Fig. 7.9** A planar foil of diamond-like carbon with absorbed hydrogen on the rear side, when irradiated by a laser, evolves into a proton layer (c) detached from the $C^{6+}$ layer (b). The proton layer remains quasi-monoenergetic after the Rayleigh–Taylor instability has been taken place. (After Liu et al.[35])

## 7.6 Hole Boring and Shock Acceleration in Gaseous Targets

A double layer accelerated by radiation pressure and moving through an underdense hydrogen plasma offers an interesting way to increase proton energy. For the double layer moving with velocity $v_f\hat{z}$, the upstream protons appear to be moving toward it with velocity $-v_f\hat{z}$. On reflection from the layer, they acquire velocity $v_f\hat{z}$ with respect to the layer. Hence in the lab frame their velocity is $2v_f\hat{z}$ (in the non-relativistic limit) and their energy is $2m_pv_f^2$, where $m_p$ is the mass of the proton. Macchi et al.[37] had noted such a feature in their 1D PIC simulations of $2\lambda_L$ thick proton plasma of $n_0/n_{cr} = 5$ irradiated by a circularly polarized laser of wavelength $\lambda_L$ and normalized amplitude $a_{00} = 2$. The ion density showed a compressed peak, at the front of the compressed electron layer, moving with velocity $0.02c$ and another peak with velocity $0.04c$. The secondary peak was attributed to the reflection of upstream ions from the moving ion front or shock front. Haberberger et al.[26] conducted experiments with $CO_2$ laser micropulses of 3 ps duration separated by 18 ps and intensity $I_L \approx 6.6 \times 10^{16}$ W/ cm$^2$ ($a_{00} \approx 1.5 - 2.5$) impinging on a hydrogen jet of density peak 3 to 5 times $n_{cr}$. They observed that *i*) the plasma density profile, at the peak of the laser pulse was strongly peaked on the front side and fell off gradually on the rear side, presumably due to laser ponderomotive force, and ii) mono-energy ions of 22 MeV energy with 1% energy spread and an angular divergence one-tenth that of the laser were generated. PIC simulations revealed the formation of an electrostatic shock. He et al.[38] observed shock acceleration of ions in plasmas with densities closer to the critical density.

Tripathi et al.[39] have developed an analytical model for radiation pressure hole boring acceleration of protons. We will present essential features of this model. We consider a circularly polarized laser irradiating a gas jet plasma of peak density close to relativistic critical density. A double layer is presumably created by the ponderomotive force steepening of the density profile.

In course of time, we will have an extended ion-electron double layer, of areal mass density $n_0 m_p l$ ($n_0$ and $l$ being the initial ion density and thickness of the layer, and $m_p$ the ion (proton) mass), irradiated by a laser of intensity $I$ (cf., Fig. 7.10). It experiences $2I/c$ radiation pressure force by the laser and loses $2n_u m_p v_f^2$ momentum to the upstream protons (of density $n_u$) per unit area per second. Hence, the equation of motion is

$$n_0 m_p l \frac{dv_f}{dt} = \frac{2I}{c} - 2n_u m_p v_f^2. \tag{7.38}$$

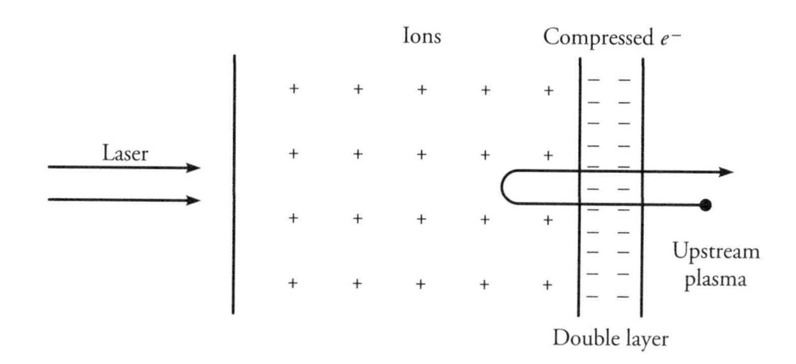


Fig. 7.10 Reflection of the down-stream ions from an extended double layer, accelerated by laser radiation pressure. (From Tripathi, et al.[39])

Using $dv_f/dt = v_f dv_f/dz$ one may write Eq. (7.38) as

$$\frac{dv_f^2}{dz} + \frac{4n_u(z)}{n_0 l} v_f^2 = \frac{2a_0^2 c^2}{l} \frac{n_{cr} m}{n_0 m_p} \tag{7.39}$$

where $a_0$ is the normalized laser amplitude. When laser amplitude is constant in time, this equation can be integrated to give

$$v_f^2 = \left[ v_{f0}^2 + \frac{2a_0^2 c^2}{l} \frac{n_{cr} m}{n_0 m_p} \int_0^z e^{h(z)} dz \right] e^{-h(z)}, \tag{7.40}$$

$$h(z) = \frac{1}{n_0 l}\int_0^z n_u(z)dz,$$

where $v_{f0} = v_f$ at $z = 0$.

For a constant density upstream Eq. (7.40) takes the form

$$v_f^2 = v_{f0}^2 + \frac{a_0^2 c^2}{2}\frac{n_{cr} m}{n_u m_p}\left(1 - e^{-4n_u z/n_0 l}\right). \tag{7.41}$$

One may notice the enhancement in velocity due to the reduction in upstream plasma density $n_u$. For $n_u/n_{cr} = 0.1$, $a_0 = 2.5$, $v_{f0} = 0$, one may achieve $v_f/c = 0.13$, giving proton energy $\varepsilon_p = 2m_p v_p^2 \approx 30$ MeV.

If the double layer acquires relativistic velocity, Eq. (7.38) modifies to

$$\frac{dv_f}{dt} = \frac{2a_0^2 c^2}{\gamma_f^3 l}\frac{n_{cr} m}{n_0 m_p}\frac{1 - v_f/c}{1 + v_f/c} - \frac{n_u v_f^2}{n_0 l \gamma_f}, \tag{7.42}$$

where $\gamma_f = (1 - v_f^2/c^2)^{-1/2}$ and the energy of reflected protons is $\varepsilon_p = 2m_p v_f^2 \gamma_f^2$. Tripathi et al.[39] have solved it numerically, in conjunction with the equation $dz/dt = v_f$, for the parabolic density profile of upstream plasma, $n_u = n_{u0}(1 - z^2/L_n^2)$, and Gaussian profile of the laser pulse, $a_0^2 = a_{00}^2 \exp(-(t - z/c)^2/\tau^2)$. Such a combination of density and intensity profiles appears to be unique to ensure a fairly constant double layer velocity, leading to mono-energy proton beam with only a few percent energy spread. The optimum energy $\varepsilon_m$ at which the upstream proton energy distribution has a sharp peak, decreases with the increase in upstream plasma density (cf. Fig. 7.11).

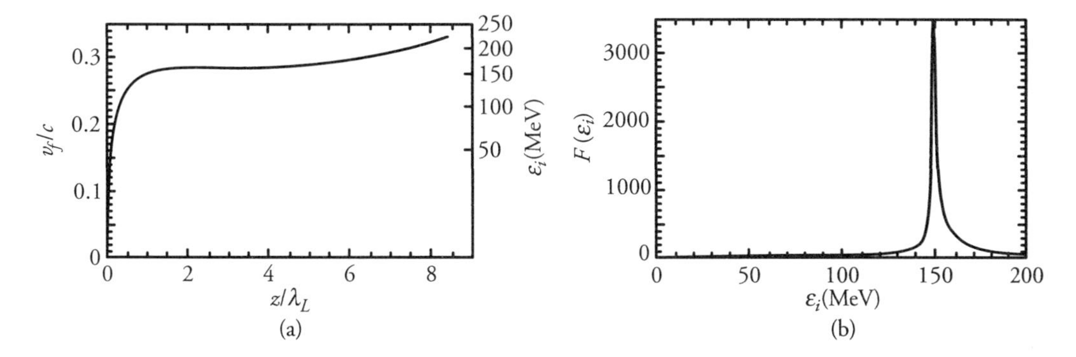


**Fig. 7.11** (a) Normalized velocity of the moving double layer irradiated by a Gaussian laser pulse and (b) energy distribution function of the accelerated ions. The parameters are: normalized laser amplitude $a_0 = 5.0$, pulse duration 30 laser periods, upstream plasma scale length 10 laser wavelengths and density 0.4 times the critical density. (From Tripathi, et al.[39])

One may mention that the reflection of upstream protons from the double layer occurs due to the space charge field in the double layer, primarily in the ion region; in the compressed electron layer, the field falls from $n_0 el/\varepsilon_0$ to zero in a shorter distance. Initially, the field in the ion layer is $E_s = n_0 ez/\varepsilon_0$. However, with time the ion layer spreads. At time $t$, it extends from $z = 0$ to $z = z_f = n_0 e^2 lt^2/2\varepsilon_0 m_p$ (in the non-relativistic limit) with ion density and space charge field

$$n_i = \frac{n_0}{1+z_f/l},\quad E_s = \frac{n_0 ez}{(1+z_f/l)\varepsilon_0}. \tag{7.43}$$

The potential difference across the layer is

$$\phi_s = \frac{n_0 ez_f^2}{2\varepsilon_0(1+z_f/l)}. \tag{7.44}$$

The upstream proton will suffer reflection from the layer when $e\phi_s > m_p v_f^2/2$, that is,

$$v_f \le \sqrt{\frac{n_0 e^2}{m_p \varepsilon_0}\frac{z_f^2}{1+z_f/l}} \approx \omega_p \sqrt{z_f l}\sqrt{\frac{m}{m_p}} \tag{7.45}$$

and maximum proton energy gain in the laboratory frame is $\varepsilon_{\rm pm} = 2m_p v_f^2 = 2m\omega_p^2\sqrt{z_f l}$. This could be a few hundred MeV. Palmer et al.[27] have observed 1 MeV proton beam with 4% energy spread using 6 ps circularly polarized $CO_2$ laser pulses of intensity $I \le 10^{16}$ W/cm$^2$ (normalized amplitude of $a_0 \le 0.5$) on a hydrogen jet with a triangular density profile of width $L_n$ = 825 μm and density peak 10 times the critical density ($n_{0\max} = 10n_{cr}$). Haberberger et al.[26] have observed 22 MeV protons with 1% energy spread.

## 7.7 Discussion

Laser-driven ion acceleration is a viable option for the production of mono-energy ions with energy in the hundreds of MeV. It carries the promise of very significantly bringing down the cost of proton therapy of cancer; in fact, it is much preferred over chemo and radiation therapy using X-rays for its minimal collateral damage to healthy tissue. The dosage imparted by a monoenergetic proton beam is concentrated at a specific stopping range so that the damage to the peripheral region is significantly reduced. Typically, a 100 MeV proton beam reaches a depth of 8 cm (c.f. Fig. 7.12) and is suitable for dealing with shallowly residing tumor cells. For other tumors one needs proton energies in the 50–250 MeV range with less than 10% energy spread. The stopping range at which the Bragg absorption peak occurs increases with beam energy. There are some recent studies on gantry design[41,42] to accommodate non-monoenergetic proton beams generated by either TNSA or RPA into human tissues.

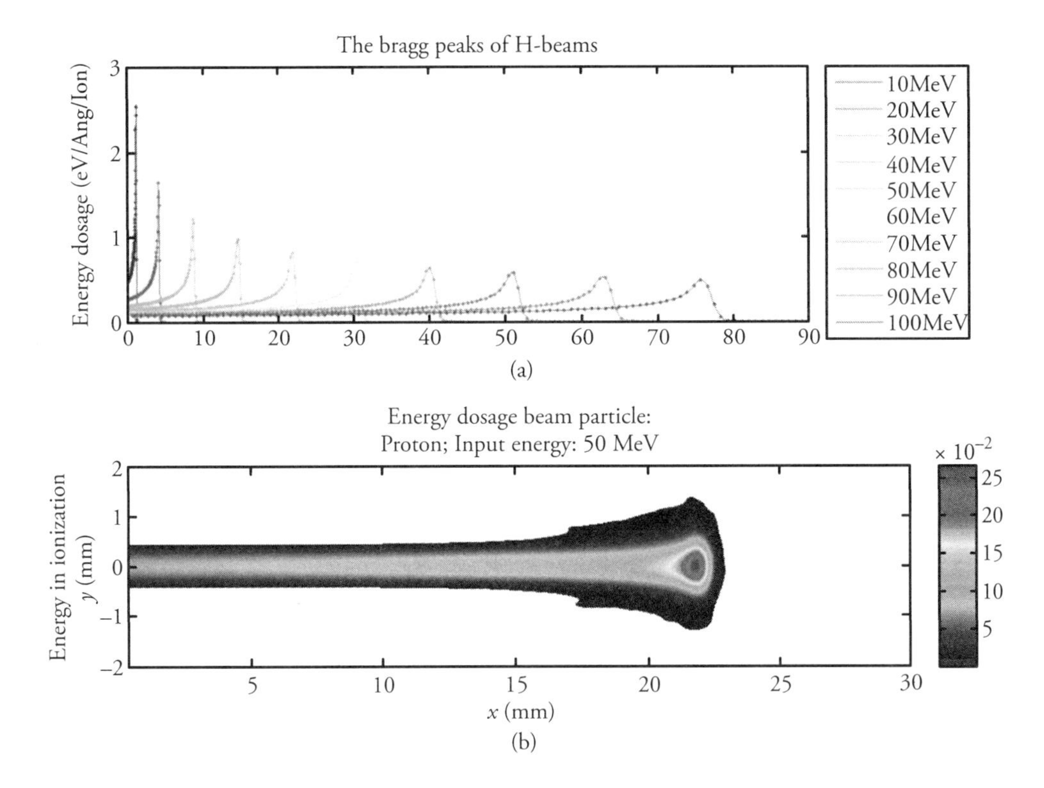


**Fig. 7.12** (Top) The Bragg peak of a monoenergetic proton beam with various input energies from 10 MeV to 100 MeV. (Bottom) Two-dimensional distribution of energy dosage due to a 50 MeV monoenergetic proton beam. (From Liu[40])

Ion beams with heavier mass, for example, carbon, can sharpen the Bragg peak and reduce the unwanted energy dosage spread. However, due to a small charge-to-mass ratio, heavier ions are more difficult to accelerate. Liu et al.[43] have carried out PIC and Monte Carlo simulations of proton energy dosage map in water employing ion beam from laser-proton accelerator. In a single-species foil, a laser beam with higher intensity is needed to reach the same energy level as with the multi-species foil. In the former, bubble structures as a sign of Rayleigh–Taylor instability, are observed to broaden the energy spectrum. The shape of the Bragg peak of the proton beam generated in the multi-species case is better focused and more favorable for medical applications.

There are many new theoretical and experimental studies[44] in the area of laser acceleration of hadron, although the threshold for laser facility requirement to have medical implication is still very challenging. The Extreme Light Infrastructure (ELI) in Europe can achieve a 10 PW laser beam for a duration of 150 fs.[45] The beam can be used for further research on ion acceleration (ELIMAIA facility) and medical applications (ELIMED group). It is an ongoing and active subject of studies, and we hope to see some practical medical applications in the near future.

**Appendix 7.1**

# Nonlinear Surface Plasma Wave

Following Liu et al.[19], we consider a vacuum $\omega_p = (n_0 e^2/m\varepsilon_0)^{1/2}$ overdense plasma interface, $z = 0$ with $z < 0$ vacuum and $z > 0$ plasma of equilibrium density $n_0$ (c.f. Fig. 7.2). A large amplitude surface plasma wave propagates along the interface with

$$E_x = F(z)e^{-i(\omega t - k_x x)}. \tag{A7.9}$$

For $z < 0$, the wave equation, in conjunction with $\nabla.\vec{E} = 0$, gives

$$F(z) = Ae^{\alpha_I z}, \quad E_z = -\frac{ik_x}{\alpha_I} Ae^{\alpha_I z} e^{-i(\omega t - k_x x)}, \tag{A7.10}$$

where $\alpha_I = \left(k_x^2 - \omega^2/c^2\right)^{1/2}$. Inside the plasma ($z > 0$), the wave equation for $E_x$, in the limit $\omega^2 << \omega_p^2$ (when $k_x \approx \omega/c$, $E_x >> E_z$ as seen in the following), leads to

$$\frac{d^2F}{dz^2} + \left(\frac{\omega^2}{c^2} - k_x^2\right)F - \frac{\omega_p^2}{c^2\gamma}\frac{n_e}{n_0}F = 0, \tag{A7.11}$$

where $\omega_p = (n_0 e^2/m\varepsilon_0)^{1/2}$, $\gamma \approx (1 + a^2/2)^{1/2}$, $a = eF/m\omega c$, $n_e$ is the electron density modified by the ponderomotive force and we have chosen $F$ to be real without loss of generality. The magnetic field of the wave can be written as

$$\vec{B} = \hat{y}(1/i\omega)(\partial F/\partial z)\exp[-i(\omega t - k_x x)]. \tag{A7.12}$$

The SPW imparts an oscillatory velocity to the electrons $v_x \cong (eE_x/im\gamma\omega)$, exerts a ponderomotive force $\vec{F}_p = -\hat{z}mc^2\partial(\gamma - 1)/\partial z, |z|$ on them, and modifies the electron density to (cf. Chapter 5)

$$n_e = n_0\left[1 + \frac{c^2}{\omega_p^2}\frac{d^2}{dz^2}(1 + a^2/2)^{1/2}\right]. \tag{A7.13}$$

Using Eq. (A7.13), Eq. (A7.11) can be written in terms of the normalized amplitude as

$$\frac{d^2a}{dz^2} + \left(\frac{\omega^2}{c^2} - k_x^2\right)a - \frac{\omega_p^2 a}{c^2(1 + a^2/2)^{1/2}} - \frac{a}{(1 + a^2/2)^{1/2}}\frac{d^2}{dz^2}(1 + a^2/2)^{1/2} = 0.$$

Multiplying both sides by $2(da/dz)dz$ and integrating over $z$, we get

$$\frac{da}{dz} = -\left(1+\frac{a^2}{2}\right)^{1/2}\left[\frac{4\omega_p^2}{c^2}\left(\left(1+a^2/2\right)^{1/2}-1\right)-\left(\frac{\omega^2}{c^2}-k_x^2\right)a^2\right]^{1/2} \tag{A7.14}$$

where we have employed the boundary conditions $a \to 0$, $da/dz \to 0$ at large $|z|$, to evaluate the constant of integration. From the third and fourth Maxwell's equations one may write

$$E_z = \frac{ik_x}{\omega^2\varepsilon/c^2 - k_x^2}\frac{\partial E_x}{\partial z}, \tag{A7.15}$$

Using the continuity of $E_x$ and $\varepsilon E_x$ at $z = 0$ we obtain

$$A = F(0)$$

$$\left.\frac{\varepsilon}{\omega^2\varepsilon/c^2 - k_x^2}\frac{dF}{dz}\right|_{z=0} = -\frac{F(0)}{\alpha_I},$$

giving the nonlinear dispersion relation

$$\alpha_I = \frac{a_0(\omega^2/c^2 - k_x^2/\varepsilon)(1+a_0^2)^{1/2}}{\left[4(\omega_p^2/c^2)\left((1+a_0^2/2)^{1/2}-1\right)-\left(\omega^2/c^2-k_x^2\right)a_0^2\right]^{1/2}} \tag{A7.16}$$

where $a_0 = a(z = 0)$. The propagation constant $k_x$ is

$$k_x = \left(\omega^2/c^2 + \alpha_I^2\right)^{1/2} \cong \omega/c + \alpha_I^2 c/2\omega.$$

The constant is mildly affected by nonlinearity. However, $\alpha_I$, the field decay constant in vacuum, and the field decay in plasma are significantly modified. For $\omega_p/\omega = 5$, $\alpha_I c/\omega$ rises from 0.2 at small $a_0$ to 0.5 at $a_0 = 4$, that is, the SPW field in vacuum is more strongly localized near the surface. Inside the plasma, the field localization also increases near the interface.

## References

1. Roth, M., A. Blazevic, M. Geissel, T. Schlegel, T. E. Cowan, M. Allen, J-C. Gauthier et al. 2002. "Energetic Ions Generated by Laser Pulses: A Detailed Study on Target Properties." *Physical Review Special Topics-Accelerators and Beams* 5 (6): 061301.

2. Ledingham, K. W. D., W. Galster, and R. Sauerbrey. 2007. "Laser-driven Proton Oncology—A Unique New Cancer Therapy?." *The British Journal of Radiology* 80 (959): 855–8.
3. Lefebvre, Erik, Emmanuel d'Humières, Sven Fritzler, and Victor Malka. 2006. "Numerical Simulation of Isotope Production for Positron Emission Tomography with Laser-accelerated Ions." *Journal of Applied Physics* 100 (11): 113308.
4. Honrubia, J. J., and M. Murakami. 2015. "Ion Beam Requirements for Fast Ignition of Inertial Fusion Targets." *Physics of Plasmas* 22 (1): 012703.
5. Macchi, Andrea, Marco Borghesi, and Matteo Passoni. 2013. "Ion Acceleration by Superintense Laser-plasma Interaction." *Reviews of Modern Physics* 85 (2): 751.
6. Daido, Hiroyuki, Mamiko Nishiuchi, and Alexander S. Pirozhkov. 2012. "Review of Laser-driven Ion Sources and their Applications." *Reports on Progress in Physics* 75 (5): 056401.
7. Borghesi, M., J. Fuchs, S. V. Bulanov, A. J. Mackinnon, P. K. Patel, and M. Roth. 2006 "Fast Ion Generation by High-intensity Laser Irradiation of Solid Targets and Applications." *Fusion Science and Technology* 49 (3): 412–39.
8. Mackinnon, A. J., Y. Sentoku, P. K. Patel, D. W. Price, S. Hatchett, M. H. Key, C. Andersen, R. Snavely, and R. R. Freeman. 2002. "Enhancement of Proton Acceleration by Hot-electron Recirculation in Thin Foils Irradiated by Ultraintense Laser Pulses." *Physical Review Letters* 88 (21): 215006.
9. Mora, Patrick. 2003. "Plasma Expansion into a Vacuum." *Physical Review Letters* 90 (18): 185002.
10. Ceccotti, T., V. Floquet, A. Sgattoni, A. Bigongiari, O. Klimo, M. Raynaud, C. Riconda et al. 2013. "Evidence of Resonant Surface-wave Excitation in the Relativistic Regime through Measurements of Proton Acceleration from Grating Targets." *Physical Review Letters* 111 (18): 185001.
11. Wagner, F., O. Deppert, C. Brabetz, P. Fiala, A. Kleinschmidt, P. Poth, V. A. Schanz et al. 2016. "Maximum Proton Energy above 85 meV from the Relativistic Interaction of Laser Pulses with Micrometer Thick ch 2 Targets." *Physical Review Letters* 116 (20): 205002.
12. Poole, P. L., L. Obst, G. E. Cochran, J. Metzkes, H. P. Schlenvoigt, I. Prencipe, T. Kluge et al. 2018. "Laser-driven Ion Acceleration via Target Normal Sheath Acceleration in the Relativistic Transparency Regime." *New Journal of Physics* 20 (1): 013019.
13. Nakatsutsumi, M., Y. Sentoku, A. Korzhimanov, S. N. Chen, S. Buffechoux, A. Kon, B. Atherton et al. 2018. "Self-generated Surface Magnetic Fields Inhibit Laser-driven Sheath Acceleration of High-energy Protons." *Nature Communications* 9 (1): 280.
14. Yan, X. Q., C. Lin, Zheng-Ming Sheng, Z. Y. Guo, B. C. Liu, Y. R. Lu, J. X. Fang, and J. E. Chen. 2008. "Generating High-current Monoenergetic Proton Beams by a Circularly Polarized Laser Pulse in the Phase-stable Acceleration Regime." *Physical Review Letters* 100 (13): 135003.
15. Qiao, B., M. Zepf, M. Borghesi, and M. Geissler. 2009. "Stable GeV Ion-beam Acceleration from Thin Foils by Circularly Polarized Laser Pulses." *Physical Review Letters* 102 (14): 145002.
16. Tripathi, V. K., Chuan-Sheng Liu, Xi Shao, Bengt Eliasson, and Roald Z. Sagdeev. 2009. "Laser Acceleration of Monoenergetic Protons in a Self-organized Double Layer from Thin Foil." *Plasma Physics and Controlled Fusion* 51 (2): 024014.
17. Eliasson, Bengt, Chuan S. Liu, Xi Shao, Roald Z. Sagdeev, and Padma K. Shukla. 2009. "Laser Acceleration of Monoenergetic Protons via a Double Layer Emerging from an Ultra-thin Foil." *New Journal of Physics* 11 (7): 073006.

18. Kar, S., K. F. Kakolee, B. Qiao, A. Macchi, M. Cerchez, D. Doria, M. Geissler et al. 2012. "Ion Acceleration in Multispecies Targets Driven by Intense Laser Radiation Pressure." *Physical Review Letters* 109 (18): 185006.

19. Henig, Andreas, S. Steinke, M. Schnürer, T. Sokollik, Rainer Hörlein, D. Kiefer, D. Jung et al. 2009. "Radiation-pressure Acceleration of Ion Beams Driven by Circularly Polarized Laser Pulses." *Physical Review Letters* 103 (24): 245003.

20. Bin, J. H., W. J. Ma, H. Y. Wang, M. J. V. Streeter, C. Kreuzer, D. Kiefer, M. Yeung et al. 2015. "Ion Acceleration Using Relativistic Pulse Shaping in Near-critical-density Plasmas." *Physical Review Letters* 115 (6): 064801.

21. Palaniyappan, Sasi, Chengkun Huang, Donald C. Gautier, Christopher E. Hamilton, Miguel A. Santiago, Christian Kreuzer, Adam B. Sefkow, Rahul C. Shah, and Juan C. Fernández. 2015. "Efficient Quasi-monoenergetic Ion Beams from Laser-driven Relativistic Plasmas." *Nature Communications* 6: 10170.

22. Scullion, C., D. Doria, L. Romagnani, A. Sgattoni, K. Naughton, D. R. Symes, P. McKenna et al. 2017. "Polarization Dependence of Bulk Ion Acceleration from Ultrathin Foils Irradiated by High-intensity Ultrashort Laser Pulses." *Physical Review Letters* 119 (5): 054801.

23. Liu, Tung-Chang, Xi Shao, Chuan-Sheng Liu, Jao-Jang Su, Bengt Eliasson, Vipin Tripathi, Galina Dudnikova, and Roald Z. Sagdeev. 2011. "Energetics and Energy Scaling of Quasi-monoenergetic Protons in Laser Radiation Pressure Acceleration." *Physics of Plasmas* 18 (12): 123105.

24. Dahiya, Deepak, Ashok Kumar, and V. K. Tripathi. 2015. "Influence of Target Curvature on Ion Acceleration in Radiation Pressure Acceleration Regime." *Laser and Particle Beams* 33 (2): 143–9.

25. Palmer, C. A. J., J. Schreiber, S. R. Nagel, N. P. Dover, C. Bellei, F. N. Beg, S. Bott et al. 2012. "Rayleigh-Taylor Instability of an Ultrathin Foil Accelerated by the Radiation Pressure of an Intense Laser." *Physical Review Letters* 108 (22): 225002.

26. Haberberger, Dan, Sergei Tochitsky, Frederico Fiuza, Chao Gong, Ricardo A. Fonseca, Luis O. Silva, Warren B. Mori, and Chan Joshi. 2012. "Collisionless Shocks in Laser-Produced Plasma Generate Monoenergetic High-Energy Proton Beams." *Nature Physics* 8 (1): 95.

27. Palmer, Charlotte A. J, N. P. Dover, I. Pogorelsky, M. Babzien, G. I. Dudnikova, M. Ispiriyan, M. N. Polyanskiy et al. 2011 "Monoenergetic Proton Beams Accelerated by a Radiation Pressure Driven Shock." *Physical Review Letters* 106 (1): 014801.

Najmudin, Z., C. A. J. Palmer, N. P. Dover, I. Pogorelsky, M. Babzien, A. E. Dangor, G. I. Dudnikova et al. 2011 "Observation of Impurity Free Monoenergetic Proton Beams from the Interaction of a $CO_2$ Laser with a Gaseous Target." *Physics of Plasmas* 18 (5): 056705.

28. Liu, C. S., V. K. Tripathi, Xi Shao, and T. C. Liu. 2015. "Nonlinear Surface Plasma Wave Induced Target Normal Sheath Acceleration of Protons." *Physics of Plasmas* 22 (2): 023105.

29. Einstein, Albert. 1905. "On the Electrodynamics of Moving Bodies." *Annalen der Physik* 17 (891): 50.

30. Robinson, A. P. L., M. Zepf, S. Kar, R. G. Evans, and C. Bellei. 2008. "Radiation Pressure Acceleration of Thin Foils with Circularly Polarized Laser Pulses." *New Journal of Physics* 10 (1): 013021.

31. Ott, Edward. 1972. "Nonlinear Evolution of the Rayleigh-Taylor Instability of a Thin Layer." *Physical Review Letters* 29 (21): 1429.

32. Pegoraro, Francesco, and S. V. Bulanov. 2007 "Photon Bubbles and Ion Acceleration in a Plasma Dominated by the Radiation Pressure of an Electromagnetic Pulse." *Physical Review Letters* 99 (6): 065002.

33. Eliasson, Bengt. 2015. "Instability of a Thin Conducting Foil Accelerated by a Finite Wavelength Intense Laser." *New Journal of Physics* 17 (3): 033026.

34. Sgattoni, Andrea, Stefano Sinigardi, Luca Fedeli, Francesco Pegoraro, and Andrea Macchi. 2015. "Laser-driven Rayleigh-Taylor Instability: Plasmonic Effects and Three-dimensional Structures." *Physical Review E* 91 (1): 013106.

35. Liu, Tung-Chang, Xi Shao, Chuan-Sheng Liu, Minqing He, Bengt Eliasson, Vipin Tripathi, Jao-Jang Su, Jyhpyng Wang, and Shih-Hung Chen. 2013. "Generation of Quasi-monoenergetic Protons from Thin Multi-ion Foils by a Combination of Laser Radiation Pressure Acceleration and Shielded Coulomb Repulsion." *New Journal of Physics* 15 (2): 025026.

36. Liu, Tung-Chang, Xi Shao, Chuan-Sheng Liu, Bengt Eliasson, Wendell T. Hill III, Jyhpyng Wang, and Shih-Hung Chen. 2015. "Laser Acceleration of Protons Using Multi-ion Plasma Gaseous Targets." *New Journal of Physics* 17 (2): 023018.

37. Macchi, Andrea, Federica Cattani, Tatiana V. Liseykina, and Fulvio Cornolti. 2005. "Laser Acceleration of Ion Bunches at the Front Surface of Overdense Plasmas." *Physical Review Letters* 94 (16): 165003.

38. He, Min-Qing, Xi Shao, Chuan-Sheng Liu, Tung-Chang Liu, Jao-Jang Su, Galina Dudnikova, Roald Z. Sagdeev, and Zheng-Ming Sheng. 2012. "Quasi-monoenergetic Protons Accelerated by Laser Radiation Pressure and Shocks in Thin Gaseous Targets." *Physics of Plasmas* 19 (7): 073116.

39. Tripathi, V. K., Tung-Chang Liu, and Xi Shao. 2017. "Laser Radiation Pressure Proton Acceleration in Gaseous Target." *Matter and Radiation at Extremes* 2 (5): 256–62.

40. Liu, Tung-Chang. 2013. "Intense Laser Acceleration of Quasi-Monoenergetic Protons." PhD diss. University of Maryland at College Park, Maryland.

41. Hofmann, Kerstin M., Umar Masood, Joerg Pawelke, and Jan J. Wilkens. 2015. "A Treatment Planning Study to Assess the Feasibility of Laser-driven Proton Therapy Using a Compact Gantry Design." *Medical Physics* 42 (9): 5120–9.

42. Masood, U., T. E. Cowan, W. Enghardt, K. M. Hofmann, L. Karsch, F. Kroll, U. Schramm, J. J. Wilkens, and J. Pawelke. 2017. "A Light-weight Compact Proton Gantry Design with a Novel Dose Delivery System For Broad-energetic Laser-accelerated Beams." *Physics in Medicine & Biology* 62 (13): 5531.

43. Liu, C. S., X. Shao, T.C. Liu, J. J. Su, M. Q. He, B. Eliasson, V. K. Tripathi, et al. 2012. "Laser Radiation Pressure Accelerator for Quasi-Monoenergetic Proton Generation and its Medical Implications." In K. Yamanouchi et al., Progress in Ultrafast Intense Laser Science, vol. VIII, pp. 177–195. Berlin Heidelberg: Springer.

44. Weber, S., S. Bechet, S. Borneis, L. Brabec, M. Bučka, E. Chacon-Golcher, M. Ciappina et al. 2017. "P3: An Installation for High-energy Density Plasma Physics and Ultra-high Intensity Laser–Matter Interaction at ELI-Beamlines." *Matter and Radiation at Extremes* 2 (4): 149–76.

45. Schreiber, Jörg, P. R. Bolton, and K. Parodi. 2016. "Invited Review Article: "Hands-on" Laser-driven Ion Acceleration: A Primer for Laser-driven Source Development and Potential Applications." *Review of Scientific Instruments* 87 (7): 071101.

# Coherent Radiation Emission

## Free Electron Laser

## 8.1 Introduction

The production of relativistic electron beams with low energy spread in laser–plasma interactions have given impetus to research in spontaneous and coherent emission of short wavelength radiation up to X-rays. Two key elements of the coherent emission process are as follows: (i) phase synchronism between the electrons and the wave and (ii) bunching of the electrons in the retarding phases of the wave so that they suffer net deceleration by the wave's field, $(\vec{J}\cdot\vec{E}<0)$.

In a Cherenkov free electron laser[1,2] (CFEL), a relativistic electron beam propagates through a dielectric loaded waveguide that supports a TM (transverse magnetic) mode with finite axial electric field, $E_z = A(r)\cos(\omega t - k_z z)$, having a phase velocity $v_p = \omega/k_z < c$. When the velocity of the beam $v_b \approx \omega/k_z$, the wave appears to the beam as an almost static field, and is capable of accelerating or decelerating the electrons efficiently. When $v_b$ is slightly greater than $\omega/k_z$, the electrons in the accelerating regions of the wave move faster and those in the retarding regions go slower. The former catch up with the latter, leading to the bunching of electrons in the retarding regions and a net transfer of energy from the electrons to the wave.

However, in free space and unmagnetized plasmas, $k \le \omega/c$ for an electromagnetic wave; hence the phase synchronism condition cannot be satisfied. The free electron lasers[3–6] (FELs) offer an alternative. They introduce a wiggler magnetic field (produced by a suitable arrangement of magnets or by a pair of helical coils), $\vec{B} = B_0\left(\hat{x}\cos\left(k_0 z\right) + \hat{y}\sin\left(k_0 z\right)\right)$ in the interaction region that acts as virtual photons of zero energy and momentum $\hbar k_0\hat{z}$. When an electron, moving with energy $E$ and momentum $\vec{p}$, emits a photon of $\hbar\omega$ energy, it gives $\hbar k\hat{z}$ momentum to the photon and $\hbar k_0\hat{z}$ momentum to the wiggler; the electron then moves with energy $E - \hbar\omega$ and momentum $\vec{p} - (\hbar k\hat{z} + \hbar k_0\hat{z})$. Using the energy–momentum relation,

$E = mc^2\left(1 + p^2/m^2c^2\right)^{1/2}$ before and after the emission, one obtains (in the limit $p >> \hbar(k + k_0)$) the condition for emission

$$\omega = (k + k_0)v_b. \tag{8.1}$$

Since $\omega = kc$ for the electromagnetic radiation, Eq. (8.1) gives the frequency of the radiation,

$$\omega = \frac{k_0 c v_b}{c - v_b} \approx 2\gamma_0^2 k_0 c, \tag{8.2}$$

where $\gamma_0 = \left(1 - v_b^2/c^2\right)^{-1/2}$ Kinetically, the process proceeds as follows. The wiggler $(0, k_0\hat{z})$ imparts an oscillatory velocity to the electrons. In the presence of a seed radiation signal $(\omega, k\hat{z})$, these electrons experience a $-e\vec{v} \times \vec{B}$ force, $\vec{F} = \hat{z}F_0 \cos(\omega t - (k + k_0)z)$. The phase synchronism of the beam with this force leads to the growth of the radiation. One may visualize the interaction in a frame moving with velocity $\vec{v}_p = \hat{z}\omega/(k + k_0)$. In this frame, the force is static,

$$\vec{F}' = \hat{z}F_0 \cos\left((k' + k_0')z'\right)$$

where the prime represents quantities in the moving frame. There are two distinctive regions (cf., Fig. 8.1): (I) the accelerating zones (where $F' > 0$) and (II) the decelerating zones (where $F' < 0$). Initially, the electrons are uniformly distributed at all $z'$. When $v_b = v_p$ as many electrons go from the accelerating zones to the retarding zones as those that go the other way; hence, there is no bunching. When $v_b$ is slightly greater than $v_p$, the accelerating zone electrons get faster and move more quickly to the retarding zones. The electrons in the retarding zones get slower and spend more time there. Thus, there is a net bunching in the retarding zones, leading to the growth of the radiation.

The emission process is often referred to as stimulated Compton scattering. For the beam electrons the wiggler appears as a backward propagating electromagnetic wave of frequency $\gamma_0 k_0 v_b$ and wave vector $-\gamma_0 k_0 \hat{z}$. Compton back-scattering of this wave produces a wave of frequency $\gamma_0 k_0 v_b$ and wave vector $\gamma_0 k_0 \hat{z}$ which in the laboratory frame has the frequency given by Eq. (8.2). An attractive feature of FEL is that its frequency scales as $\gamma_0^2$. It can be tuned over a wide frequency range by changing the beam energy.

Laser-produced ion channels, by virtue of their radial electric field, offer an alternative to the wigglers. An electron beam launched into an ion channel executes betatron oscillations (similar to the wiggled motion induced by a magnetic wiggler) at the frequency $\omega_b = \omega_p/(2\gamma)^{1/2}$, where $\omega_p$ is the plasma frequency (corresponding to the ion density and electron mass) and $\gamma$ is the Lorentz factor of electrons. This may produce X-rays by the FEL mechanism[7–9] at frequency

$$\omega \approx 2\gamma^2\omega_b. \tag{8.3}$$

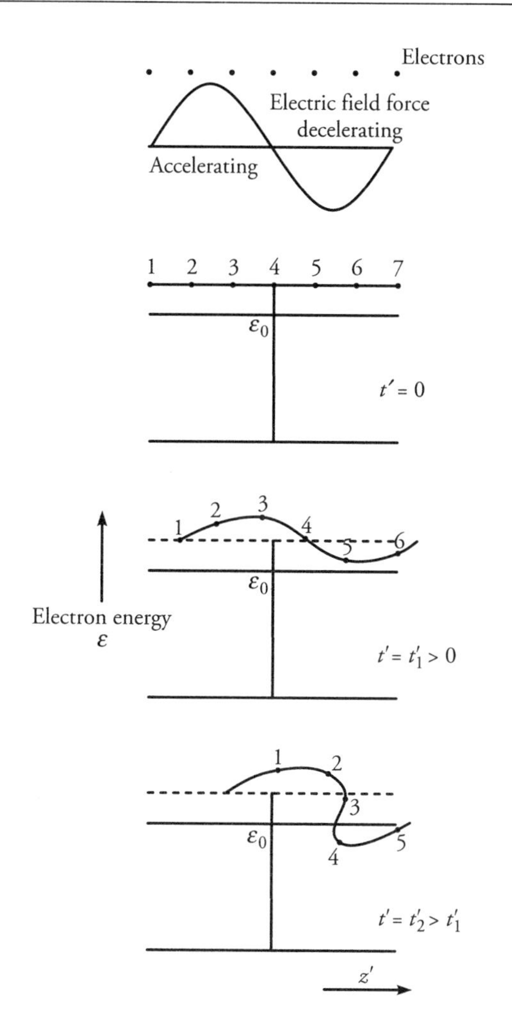


**Fig. 8.1** Energy and axial position of electrons at different instants of time. Initially all electrons have the same energy $\varepsilon_{in} > \varepsilon_0$, where $\varepsilon_0$ is the energy of a resonant electron moving with velocity $\omega/k$ with respect to the lab frame.

Rousse et al.[10] have experimentally observed X-rays of high keV energies by impinging a 0.8 μm, $3 \times 10^{18}$ W/cm$^2$, 30 fs, 18 μm spot size laser on a helium gas target with variable electron density up to $6 \times 10^{19}$ cm$^{-3}$. The X-ray emission was accompanied by the wake field generation of 20 MeV electrons with a large energy spread. Particle in cell (PIC) simulations revealed betatron oscillations of the electrons in the ion bubble just behind the laser pulse. The X-rays were incoherent. With electron beams of narrow energy spread one may realize the X-ray lasers producing coherent X-rays.

In Section 8.2, we present an analytical model of the free electron laser. In Section 8.3 we present a Vlasov formalism of coherent X-ray generation in a laser-driven ion channel by a multi-MeV electron beam using the procedure followed by Liu et al.[11] Concluding remarks are givrn in Section 8.4.

## 8.2 Free Electron Laser

A geometrical sketch of the free electron laser (FEL) is shown in Fig. 8.2. The foremost component of an FEL is an accelerator that produces an electron beam with beam voltage ≥ 500 keV (up to hundreds of MeV) and beam current ≥ 0.1 A (up to several kilo-Amperes). The beam is usually pulsed with a typical, life-time of less than 100 ns. In some experiments, one may have millisecond pulses. Since the FEL frequency sensitively depends on $\gamma_0$, the energy spread of the beam has to be kept small. The radius of an annular beam is typically 1 cm and the radial beam thickness 1–2 mm. The interaction region comprises a guide magnetic field (produced by a solenoid), a wiggler magnetic field produced by a suitable arrangement of permanent magnets or coils (cf. Fig. 8.3). The typical value of the wiggler strength is around 1 kG and the spatial wiggler period is 3 cm. Simple wigglers have $B_0$ and $k_0$ constant along

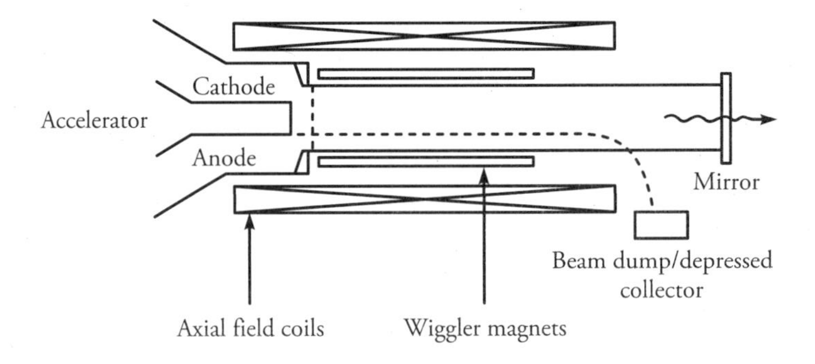


Fig. 8.2 The schematic of a free electron laser.

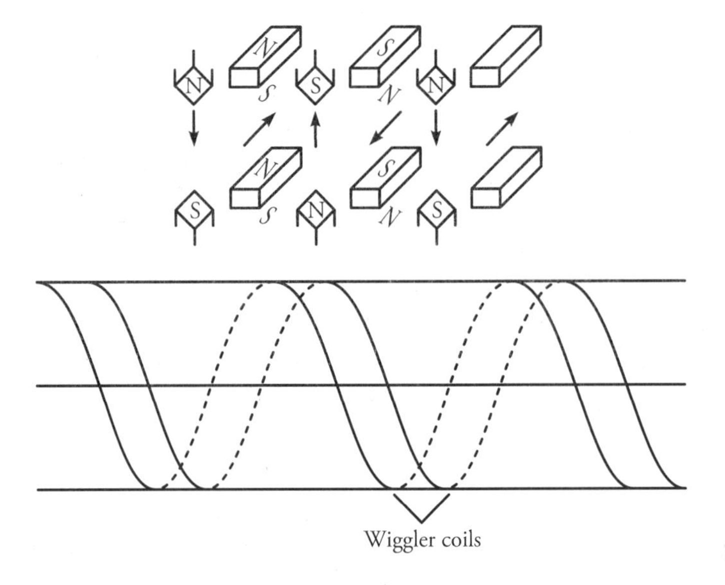


Fig. 8.3 A helical wiggler.

the length of the interaction region. One may also have a tapered wiggler whose amplitude $B_0$ and/ or wiggler period $(2\pi/k_0)$ change adiabatically along the length of the structure. The latter leads to adiabatic slowing down of the ponderomotive wave; hence, the trapped electrons also slow down leading to improved efficiency of the device. An FEL oscillator would have two reflecting mirrors with broad band high reflectance. Energy recovery from the spent beam is another important consideration. It can be accomplished by using depressed collectors. Here, however we restrict our discussion to the growth of the FEL radiation in the interaction region, ignoring boundary effects.

### 8.2.1 Growth rate

Consider the propagation of a relativistic electron beam of density $n_0$, through a circularly polarized wiggler magnetic field,

$$\vec{B}_W = B_0(\hat{x} + i\hat{y})e^{ik_0 z}, \tag{8.4}$$

where $B_0$ is the amplitude of the wiggler field.

The equilibrium velocity of the beam satisfies the equation of motion

$$\frac{d(\gamma_0 \vec{v}_0)}{dt} = -\frac{e}{m}\vec{v}_0 \times B_0(\hat{x} + i\hat{y})e^{ik_0 z}, \tag{8.5}$$

where $-e$, $m$ and $\gamma_0$ are the charge, rest mass, and gamma factor of the electrons. Since $\gamma_0$ is a constant of motion in a magnetic field, we may solve this equation by replacing $d/dt$ by $v_{0z}d/dz = ik_0 v_{0z}$, giving

$$v_{0x} = iv_{0y} = -\frac{eB_0}{m\gamma_0 k_0}e^{ik_0 z}, \tag{8.6}$$

$$\gamma_0 = \left(1 - v_0^2/c^2\right)^{-1/2} = \frac{\left(1 + e^2 B_0^2/m^2 k_0^2 c^2\right)^{1/2}}{\left(1 - v_{0z}^2/c^2\right)^{-1/2}} \approx (1 - v_{0z}^2/c^2)^{-1/2}.$$

We perturb this equilibrium by an electromagnetic wave,

$$\vec{E}_1 = A_1\left(\hat{x} + i\hat{y}\right)e^{-i(\omega t - k_1 z)},\ \vec{B}_1 = \vec{k}_1 \times \vec{E}_1/\omega. \tag{8.7}$$

The wave imparts an oscillatory velocity to the electrons,

$$\vec{v}_1 = \frac{e\vec{E}_1}{im\gamma_0\omega} \tag{8.8}$$

and exerts, in conjunction with the wiggler, a ponderomotive force on them at ($\omega$, $k$),

$$\vec{F}_p = e\nabla\phi_p = -e\left(\vec{v}_0 \times \vec{B}_1 + \vec{v}_1 \times \vec{B}_0\right),$$

with the ponderomotive potential,

$$\phi_p = \frac{eB_0A_1}{im\gamma_0k_0\omega}e^{-i(\omega t - kz)}, \tag{8.9}$$

where $k = k_0 + k_1$. The ponderomotive force produces velocity and density perturbations. Writing $v_z = v_{0z} + v_{\omega,k}$, $\gamma = \gamma_0 + \gamma_0^3 v_{0z} v_{\omega,k}$, $n = n_0 + n_{\omega,k}$ in the axial component of the momentum and continuity equation,

$$\frac{\partial(\gamma v_z)}{\partial t} + v_z\frac{\partial(\gamma v_z)}{\partial z} = F_{pz}, \tag{8.10}$$

$$\frac{\partial n}{\partial t} + \frac{\partial(nv_z)}{\partial z} = 0, \tag{8.11}$$

and linearizing them, one obtains

$$v_{zw,k} = -\frac{ek\phi_p}{m\gamma_0^3(\omega - kv_{0z})}, n_{\omega,k} = -\frac{n_0ek^2\phi_p}{m\gamma_0^3(\omega - kv_{0z})^2}. \tag{8.12}$$

The current density at $\omega$, $k_1$ is

$$\vec{J}_1 = -n_0e\vec{v}_1 - \frac{1}{2}n_{\omega,k}e\vec{v}_{0\perp}^*$$

$$= -\frac{ne^2\vec{E}_1}{mi\gamma_0\omega}\left(1 + \frac{k^2v_{0\perp}^2}{2\gamma_0^2(\omega - kv_{0z})^2}\right). \tag{8.13}$$

where $v_{0\perp} = eB_0/m\gamma_0k_0$. Using this in the wave equation

$$\nabla^2\vec{E}_1 + \frac{\omega^2}{c^2}\vec{E}_1 = -\frac{i\omega}{\varepsilon_0c^2}\vec{J}_1$$

and replacing $\nabla^2$ by $-k_1^2$, we obtain the nonlinear dispersion relation

$$(\omega^2 - k_1^2c^2)(\omega - kv_{0z})^2 = \frac{k^2v_{0\perp}^2\omega_p^2}{2\gamma_0^3}. \tag{8.14}$$

Around simultaneous zeros of the factors on the left-hand side (giving a radiation mode, and beam mode respectively), that is, for $k_1c = kv_{0z}$, we write $\omega = \omega_r + \delta$, where $\omega_r = k_1c$, $\delta << \omega_r$ to obtain the growth rate

$$\Gamma = \text{Im}\,\delta = \frac{\sqrt{3}}{2}\left(\frac{k^2 v_{0\perp}^2 \omega_p^2}{4\gamma_0^3 \omega}\right)^{1/3} \tag{8.15}$$

and $\delta_r = -\Gamma/\sqrt{3}$. The growth rate scales as $B_0^{2/3}, k_0^{-2/3}, \text{and}\, n_0^{1/3}$. For fixed $B_{0,}\ k_0$, it scales as $\omega^{-1/2}$ with radiation frequency. For typical parameters $B_0 = 1kG, 2\pi k_0^{-1} = 3\,\text{cm}$, current density≈ $10\text{A/cm}^2$, $\gamma_0 = 4$, i.e., $n_0 = 2 \times 10^9\ \text{cm}^{-3}$, $\omega_p = 2.2 \times 10^9$ rad/s, and $eB_0/mk_0c = 0.26$, we obtain $\omega$ = 320 GHz $\Gamma = 2 \times 10^9\ \text{sec}^{-1}$ or spatial growth length $c/\Gamma \approx 15$ cm.

So far we have ignored the space charge field produced by charge bunching $n_{w,k}$, which is reasonable at low beam densities. However, at a high beam density (i.e., at high beam current), when the beam plasma frequency exceeds the growth rate, the space charge field becomes significant, shifting the FEL operation from the Compton regime to the Raman regime.

## 8.2.2 Raman regime operation

Let us first discuss the space charge wave. Consider a uniform beam of cold electrons of density $n_0$ and velocity $v_{0z}\hat{z}$ subjected to an electrostatic perturbation $\vec{E} = -\nabla\phi$; $\phi = A \exp[-i(\omega t - kz)]$. Solving the momentum and continuity equations one may write the density perturbation (cf. Eq. 8.12)

$$n_{\omega,k} = -\frac{n_0 e k^2 \phi}{m\gamma_0^3(\omega - kv_{0z})^2} \tag{8.16}$$

which on using in Poisson's equation, $\nabla^2\phi = n_{w,k}\, e/\varepsilon_0$, yields the dispersion relation

$$\varepsilon = 1 + \chi_b = 0, \tag{8.17}$$

where

$$\chi_b = -\frac{\omega_p^2}{\gamma_0^3(\omega - kv_{0z})^2} \tag{8.18}$$

is the beam susceptibility. Equation (8.17) gives two space charge modes,

$$\omega = kv_{0z} \pm \frac{\omega_p}{\gamma_0^{3/2}}. \tag{8.19}$$

The one with the lower sign has $\omega\partial\varepsilon/\partial\omega < 0$, that is, it is a negative energy mode. It is the coupling of a negative energy mode ($\omega$, $k$) with the wiggler (0, $k_0$) that produces the FEL radiation ($\omega$, $k_1$) in the Raman regime. As the space charge mode feeds energy into the FEL mode, its energy becomes more negative, leading to the simultaneous growth of the beam space charge mode and the radiation mode.

In the presence of the space charge potential $\phi$ and ponderomotive potential $\phi_p$, the density perturbation can be written from Eq. (8.12) with $\phi_p$, replaced by $\phi + \phi_p$. Poisson's equation then yields

$$\varepsilon\phi = -\chi_b\phi_p. \tag{8.20}$$

One may note that around $\varepsilon \approx 0$, $\phi >> \phi_p$. The current density at ($\omega$, $k_1$) now takes the form (for $\chi_b \approx -1$, $\phi >> \phi_p$)

$$\vec{J}_1 = -n_0 e\vec{v}_1 - \frac{1}{2}n_{\omega,k}e\vec{v}_{0\perp}^{\,*} = -\frac{n_0 e^2\vec{E}_1}{mi\gamma_0\omega} - \frac{1}{2}k^2\varepsilon_0\phi\vec{v}_{0\perp}^{\,*} \tag{8.21}$$

and the wave equation Eq. (8.14) leads to

$$(\omega^2 - \omega_p^2/\gamma_0 - k_1^2c^2)\vec{E}_{1\perp} = \frac{i\omega}{2}k^2\phi\vec{v}_0^{\,*}. \tag{8.22}$$

Equations (8.20) and (8.22) yield the nonlinear dispersion relation

$$(\omega^2 - \omega_p^2/\gamma_0 - k_1^2c^2)\varepsilon = \frac{k^2 v_{0\perp}^2}{4}. \tag{8.23}$$

The simultaneous zeros of the two factors on the left-hand side, corresponding to the electromagnetic mode, $\omega = (\omega_p^2/\gamma_0 + k_1^2c^2)^{1/2}$ and beam mode $\omega = kv_b - \omega_p/\gamma_0^{3/2}$ give the frequency of operation of the FEL,

$$\omega = 2\gamma_0^2(k_0 v_b - \omega_p/\gamma_0^{3/2}). \tag{8.24}$$

Around this frequency, we expand $\omega = \omega + i\Gamma$. Equation (8.23) yields

$$\Gamma = \frac{kv_{0\perp}}{2}\left(\frac{\omega_p}{\gamma_0^{3/2}\omega}\right)^{1/2}. \tag{8.25}$$

The neglect of collective effects implies $\chi_b << 1$, that is, $\omega_p/\gamma_0^{3/2} < \Gamma$. This defines the boundary between Raman and Compton regimes of operation.

### 8.2.3 Gain estimate

As the amplitude of the ponderomotive wave (or the space charge wave) grows it tends to trap electrons. The trapped electrons, however, cannot be continuously decelerated. To understand the dynamics of trapped electrons in the ponderomotive wave, let us consider the single particle energy equation, neglecting the space charge effects,

$$mc^2 \frac{d\gamma}{dt} = -eE_p v_z \cos(\omega t - kz); \; E_p = k\left|\phi_p\right|,$$

which on writing $d/dt = v_z d/dz$ takes the form

$$\frac{d\gamma}{dz} = -\frac{eE_p}{mc^2}\cos(\omega t - kz). \tag{8.26}$$

where $\gamma = (1 - v_{0\perp}^2 / c^2 - v_z^2 / c^2)^{-1/2}$. It is useful to define the gamma factor of an electron moving with the phase velocity of the ponderomotive wave, $\gamma_r = (1 - v_{0\perp}^2 / c^2 - \omega^2 / k^2c^2)^{-1/2}$. When $\gamma$ falls to $\gamma_r$ the beam can no longer transfer energy to the wave. Let $\Delta\gamma = \gamma - \gamma_r$, $\Psi = kz - \omega t$, then one may write

$$\frac{d\Delta\gamma}{dz} = -\frac{eE_p}{mc^2}\cos(\psi)$$

$$\frac{d\psi}{dz} = k - \frac{\omega}{v_z} \approx \frac{\omega\Delta\gamma}{2c(\gamma_r - 1)^{3/2}}. \tag{8.27}$$

Introducing normalized quantities

$$\xi = z/L, \; P = \frac{L\omega\Delta\gamma}{2c(\gamma_r - 1)^{3/2}}, \; A = \frac{L^2\omega eE_p}{2mc^3(\gamma_r^2 - 1)^{3/2}},$$

where $L$ is the length of the interaction region, Eq. (8.27) can be written as

$$\frac{dP}{d\xi} = -A\cos\psi, \tag{8.28}$$

$$\frac{d\psi}{d\xi} = P. \tag{8.29}$$

One may note that an electron can lose energy to the wave as long as $\psi$ is between $-\pi/2$ and $\pi/2$. As the electron moves down the interaction region, $\psi$ could increase (for $P > 0$). One would like to choose the length of the interaction region such that the change in $\psi$ at the exit point ($\xi = 1$) is $\pi$, that is, $\psi_{exit} - \psi_{entry} \approx \pi$.

Equations (8.28) and (8.29) need be solved in conjunction with the equation governing $A$, that is, the wave equation for the radiation wave. In the general case, this has to be done numerically. In this section, we will solve them in the limit A is constant, that is, when the single pass amplification of radiation is small. Dividing Eq. (8.28) by (8.29) and integrating the resulting equation we obtain

$$P^2 = -2A\sin\psi + P_{in}^2 + 2A\sin\psi_{in}, \tag{8.30}$$

where $P_{in}$ and $\psi_{in}$ are the values of $P$ and $\psi$ at the entry point. Equation (8.30) gives the phase space trajectories of the electrons (Fig. 8.4) for different values of $P_{in}$ and $\psi_{in}$. For $P_{in}^2 + 2A\sin\psi_{in} < 2A$ all values of $\psi$ are not accessible (since $P^2$ has to be $> 0$), that is, the trajectories of particles are localized, representing trapped particles. The separatrix is given by

$$P^2 = 2A(1 - \sin\Psi). \tag{8.31}$$

At $z = 0$, the electrons enter uniformly at all times, that is, in the $P$, $\psi$ plane they lie uniformly over the horizontal line $P = P_{in}$. The ones inside the separatrix are trapped and those outside are the passing particles. At $z > 0$, some lose energy, some gain energy depending on their $\psi_{in}$. To obtain a quantitative estimate of the net energy exchange, we solve Eqs. (8.28) and (8.29) by expanding $P$ and $\psi$ to different orders in $A$,

$$P = P_0 + P_1 + P_2,\ \Psi = \Psi_0 + \Psi_1 + \Psi_2. \tag{8.32}$$

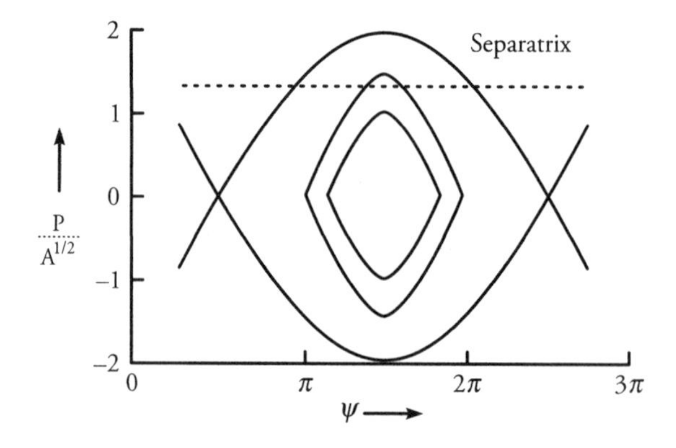


**Fig. 8.4** Phase space trajectories of trapped particles.

To the zeroth order, we have

$$P_0 = P_{in}, \Psi_0 = \Psi_{in} + P_{in}\xi. \tag{8.33}$$

To the first order,

$$\frac{dP}{d\xi} = -A\cos(\psi_{in} + P_0\xi),$$

which can be integrated as

$$P = P_0 - \frac{A}{P_0}[\sin(\psi_{in} + P_0\xi) - \sin\psi_{in}],$$

and from Eq. (8.29),

$$\frac{d\psi}{d\xi} = P_0 - \frac{A}{P_0}[\sin(\psi_{in} + P_0\xi) - \sin\psi_{in}],$$

which can be integrated as

$$\Psi = \Psi_{in} + P_0\xi + \Psi_1,$$

where

$$\psi_1 = -\frac{A}{P_0^2}[\cos(\psi_{in} + P_0\xi) - \cos\psi_{in}] + \frac{A}{P_0}\xi\sin\psi_{in}.$$

To the second order, we have

$$\frac{dP}{d\xi} = -A\cos(\psi_{in} + P_0\xi + \psi_1)$$

$$\approx -A\cos(\Psi_{in} + P_0\xi) + A\Psi_1\sin(\Psi_{in} + P_0\xi),$$

which can be integrated as

$$P = P_0 - \frac{A}{P_0}[\sin(\psi_{in} + P_0\xi) - \sin\psi_{in}] + P_2,$$

where

$$P_2 = \frac{A^2}{P_0^3}\left[\frac{1}{4}\cos 2(\psi_{\text{in}} + P_0) - \frac{1}{4}\cos 2\psi_{\text{in}}\right.$$

$$\left. + \cos\psi_{\text{in}}\cos(\psi_{\text{in}} + P_0) - \cos^2\psi_{\text{in}}\right] + \frac{A^2}{P_0^3}\sin\psi_{\text{in}}$$

$$[-P_0\cos(\Psi_{\text{in}}+P_0) + \sin(\Psi_{\text{in}} + P_0) - \sin\Psi_{\text{in}}].$$

The energy lost by an electron $\Delta P = P_{\text{in}} - P(\xi = 1)$ in passing through the interaction region is $\Delta P = -(P_1+P_2)_{\xi=1}$. Averaging this over the initial phases yields

$$\langle \Delta P \rangle = -\langle P_2 \rangle = -\frac{1}{2\pi}\int_0^{2\pi} P_2 d\psi_{\text{in}}$$

$$= \frac{A^2}{P_0^3}\left(1 - \cos P_0 - \frac{P_0}{2}\sin P_0\right) = \frac{A^2}{8}G, \tag{8.34}$$

$$G = -\frac{d}{dx}\left(\frac{\sin^2 x}{x^2}\right), \; x = P_0/2.$$

Figure 8.5 shows the variation of the gain function $G$ as a function of $P_0$ or $\gamma_0 - \gamma_r$. For $\gamma_0 > \gamma_r$, there is net transfer of energy from the particles to the wave.

At a high beam current, the amplitude of the FEL wave grows significantly during the passage of the electron through the wiggler. Hence, one must supplement the set of Eqs. (8.28) and (8.29) by an equation for $A$. Roberson and Sprangle[5] discussed this problem in considerable detail. However, we conclude this section with an estimate of efficiency. In the Compton regime of the FEL operation, the electrons give energy to waves as long as $v_b > w/k$. During the emission process, the electron velocity thus falls from $w/k$ to $(w + \delta_r)/k$, giving the efficiency

$$\eta = \frac{\gamma\left(\frac{\omega}{k}\right) - \gamma\left(\frac{\omega + \delta_r}{k}\right)}{\gamma\left(\frac{\omega}{k}\right) - 1} \approx \frac{\delta_r}{\omega}\frac{\gamma_0^3}{\gamma_0 - 1}. \tag{8.35}$$

In the Raman case, the frequency $\omega = kv_{0z} - \omega_p/\gamma_0^{3/2}$ of the space charge mode is detuned as the beam loses energy. When $k\Delta v_{0z}$ is of the order of the growth rate, the instability saturates.

Hence,

$$\eta \approx \frac{\Gamma}{\omega} \frac{\gamma_0^3}{\gamma_0 - 1}.$$

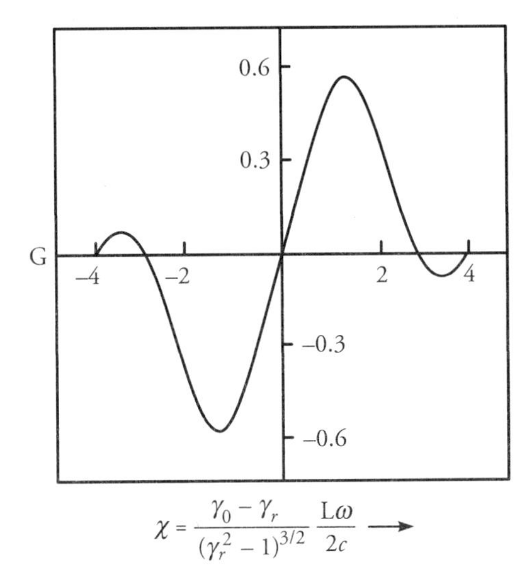


**Fig. 8.5** The gain function as a function of the initial electron energy. (cf. Ciocci et al.[1])

## 8.2.4 Tapered wiggler FEL

The efficiency of a free electron laser can be enhanced by adiabatically slowing down the ponderomotive wave so that $v_b > \omega / k(z)$ can be maintained over a long distance. This is accomplished by tapering the wiggler, for example, by making $k_0$ an increasing function of $z$. Let us examine the motion of an electron in a ponderomotive wave $E_p = A_p \cos\psi$, where $\psi = \int k(z)dz - \omega_1 t$, $k = k_1 + k_0(z)$. For the sake of simplicity, we assume $v_{0\perp}^2 << c^2$. We define a resonant gamma, $\gamma_r = (1 - \omega^2 / k^2(z)c^2)^{-1/2}$ and $\Delta\gamma = \gamma - \gamma_r$. Then, Eq. (8.27) can be written as

$$\frac{d\Delta\gamma}{dz} = -\frac{eA_p}{mc^2}\cos\psi - \frac{d\gamma_r}{dz}$$

$$\frac{d\psi}{dz} = \frac{\omega\Delta\gamma}{2c(\gamma_r - 1)^{3/2}} \tag{8.36}$$

Defining $\xi$, $P$, $A$ the same way as before and assuming $(d\gamma_r/dz)/\gamma << (d\Delta\gamma/dz)/\Delta\gamma$, the aforementioned equations can be written as

$$\frac{dP}{d\xi} = -A\cos\Psi + \alpha,$$

$$\frac{d\psi}{d\xi} = P, \tag{8.37}$$

where $\alpha = -[2c(\gamma_r^2 - 1)^{3/2} / \omega]d\gamma_r / dz$. In general, $\alpha$ is a slowly varying function of $\xi$. We assume $\alpha$ to be a constant. Writing $dP/d\xi = (dP/d\Psi)\,(d\Psi/d\xi)$, we get

$$\frac{d}{d\psi}\left(\frac{P^2}{2}\right) = -A\cos\psi + \alpha,$$

which on integration gives

$$\frac{P^2}{2} + A\sin\psi - \alpha\psi = H, \tag{8.38}$$

where $H - P_0^2 / 2 + A\sin\psi_{\text{in}} - \alpha\psi_{\text{in}}$ is a constant of integration. $H$ can be identified as a Hamiltonian, with $P^2/2$ as the kinetic energy, and $V(\psi) = A\sin\psi - \alpha\psi$ as the potential energy. In Fig. 8.6, $V$ has been plotted as a function of $\psi$ for $A > \alpha_{\text{in}}$. It has the form of a sloping corrugated roof, with maxima and minima ($dV/d\psi = 0$) at

$$\cos\Psi = \alpha/A. \tag{8.39}$$

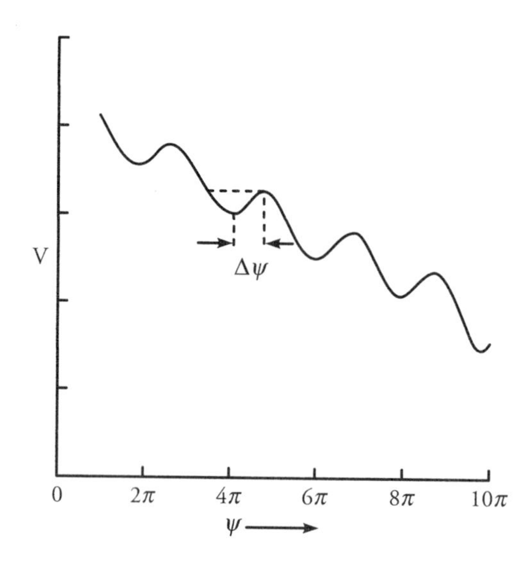


Fig. 8.6 The potential energy as a function of the phase angle $\Psi$ for $\alpha < A$.

Solutions of Eq. (8.39) can be found when $\alpha < A$. For fast tapering that is, $\alpha > A$, one would not have maxima and minima. For a minimum we have $d^2V/d\Psi^2 > 0$, that is, $A \sin\Psi_{min} < 0$. Since $\cos\psi_{min}$ is positive (cf. Eq. 8.39), $\psi_{min}$ lies in the fourth quadrant (cf. Fig. 8.7). The values of $\psi(=\psi_{max})$ corresponding to maxima in $V$ lie in the first quadrant. The separation between a minimum and the next maximum is $\Delta\psi = 2\cos^{-1}\alpha/A$, while the width of a potential energy well is $2\Delta\psi$. For an electron having $H < V_{max}$, where $V_{max} = A\sin\Psi_{max} - \alpha\Psi_{max} = A(1-\alpha^2/A^2)^{1/2} - \cos^{-1}\alpha/A$, all values of $\psi$ are not accessible because $P^2$ is a positive definite. The electron bounces back and forth inside a potential energy well. Such electrons are trapped electrons. In Fig. 8.8, phase space trajectories of electrons have been plotted with the separatrix given by

$$\frac{P^2}{2} + A\sin\psi - \alpha\psi = (A^2-\alpha^2)^{1/2} - \alpha\cos^{-1}\alpha/A. \tag{8.40}$$

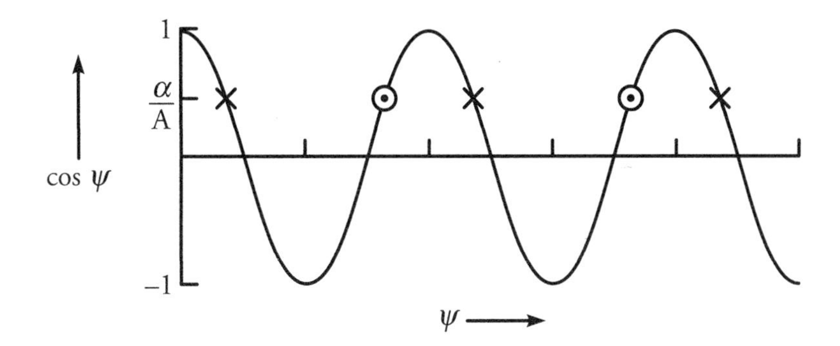


Fig. 8.7 Locations of $\Psi_{min}(0)$ and $\Psi_{max}$; $\Delta\Psi$ is the half width of the potential energy well.

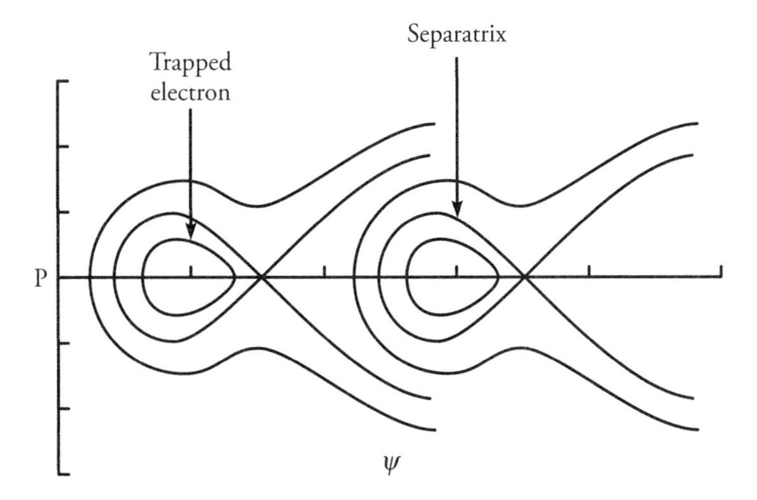


Fig. 8.8 Phase-space trajectories of electrons, with the separatrix between trapped and an-trapped electrons.

As the ponderomotive wave slows down with $z$, the trapped electrons also slow down losing energy to the FEL radiation. Let the values of $\gamma_r$ and $P$ at $\xi = 0$ and $\xi = 1$ be denoted by $\gamma_r(0)$, $P(0)$, and $\gamma_r(1)$, $P(1)$; then, the net energy radiated by the trapped electrons is

$$\Delta E = mc^2\left[\gamma_r(0)-\gamma_r(1)+\frac{L\omega\langle P(0)-P(1)\rangle}{2c(\gamma_r^2(0)-1)^{3/2}}\right]$$

where the brackets $\langle\rangle$ denote the average over the initial phases of the trapped electrons. $\langle P(0)-P(1)\rangle$ can be obtained by solving Eqs. (8.37) numerically. The efficiency of the radiation by the trapped electrons is given by

$$\eta_{tr} = \frac{\Delta E}{mc^2(\gamma(0)-1)}. \quad (8.41)$$

If we multiply this by the fraction of electrons trapped, we would get the overall efficiency of the FEL radiation. If the electrons enter the interaction region ($z = 0$) at a uniform rate, where they are distributed uniformly over $\psi$, the fraction of trapped electrons is $2\Delta\Psi/2\pi$ and the overall efficiency of the FEL radiation is

$$\eta = \frac{\Delta\psi}{2\pi}\frac{\Delta E}{mc^2(\gamma(0)-1)} \geq \frac{\Delta\psi}{2\pi}\frac{\gamma_r(0)-\gamma_r(1)}{mc^2(\gamma(0)-1)} \quad (8.42)$$

To the aforementioned expression, we must add the contribution of the un-trapped electrons; however, this contribution is relatively weak in a tapered FEL.

## 8.3 Laser-driven Ion Channel X-ray Laser

We here discuss the generation of X-ray radiation by a relativistic electron beam performing betatron oscillations in a laser generated ion channel.

Following Liu et al.[11], we consider the propagation of a high power laser in a plasma of electron density $n_0$,

$$\vec{E}_L = \hat{x}A_L e^{-r^2/r_0^2} e^{-i(\omega_L t - k_L z)}. \quad (8.43)$$

The laser imparts an oscillatory velocity to the electrons, exerts a radial ponderomotive force on them, and creates an electron-evacuated ion channel of radius $r_c$ governed by Eq. (5.54) with $a^2$ replaced by $a_L^2/2$ (due to the linear polarization), where $\gamma_L = (1 + a_L^2/2)^{1/2}$, $a_L = e|E_L|/m\omega c$. For $r_0 = c/\omega_p$, $a_L \geq 2$, one obtains $r_c \geq 0.8r_0$. In many experiments and PIC simulations, ion channels of such dimensions have been observed.

Thus, we have a cylindrical ion channel of radius $r_c$, ion density $n_0$, and radial electric field $E_{0r} = n_0 er/2\varepsilon_0$ (cf., Fig. 8.9). An electron beam of density $n_{0b}$, radius $r_b$, and Lorentz factor $\gamma >> \gamma_L$ propagates through it. The beam could be self-produced by the laser wake field or injected from outside. The laser ponderomotive force on the beam is decreased by a factor $\gamma_L/\gamma$, and hence negligible. Solving the equation of motion for a single electron, $d\vec{p}/dt = -eE_{0r}\hat{r}$ in the limit $p_z^2 >> p_x^2, p_y^2$, one obtains $d^2x/dt^2 = -\omega_b^2 x$ and a similar equation for $y$, leading to

$$x = \rho_x \sin(\omega_b t+\theta_x),\ y = \rho_y \sin(\omega_b t+\theta_y),$$

$$v_x = \rho_x \omega_b \cos(\omega_b t+\theta_x),\ v_y = \rho_y \omega_b \cos(\omega_b t+\theta_y), \tag{8.44}$$

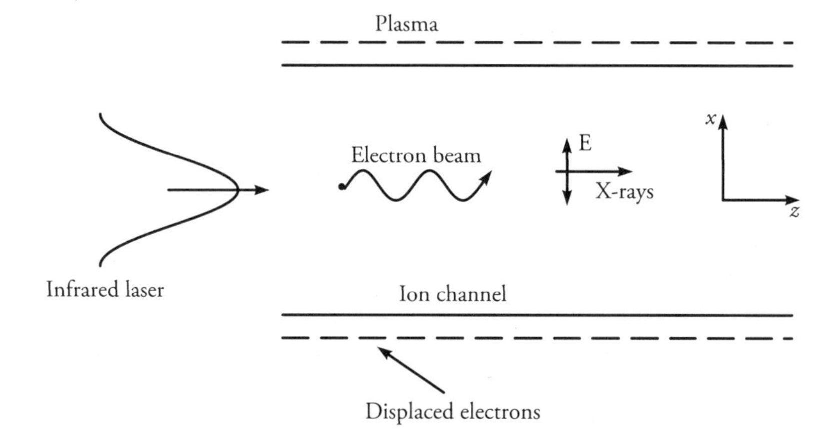


**Fig. 8.9** Betatron oscillations of an electron beam passing through an ion channel, created by an infrared laser. Coherent X-rays generated by the beam propagate along the channel axis. After Liu et al.[11].

where $\rho_x = (x^2 + v_x^2/\omega_b^2)^{1/2}$ and $\rho_y = (y^2 + v_y^2/\omega_b^2)^{1/2}$ are the constants of motion, $\theta_x$ and $\theta_y$ are the initial phases, $\omega_b = \omega_p/\sqrt{2\gamma}$ and $\gamma = \left(1 + p_z^{\,2}/m^2c^2\right)^{1/2}$. The beam electrons originating from different radial locations have different amplitudes $\rho_x$ and $\rho_y$. One may write $\rho_x = \rho\cos\phi$ and $\rho_y = \rho\sin\phi$, where $\phi$ varies from 0 to $\pi/2$ as $\rho_x$ and $\rho_y$ are positive.

The equilibrium distribution function of the beam electrons, $f_0$, may be taken to be a function of constants of motion. For a cylindrically symmetric beam, one may take

$$f_0(p_z,\rho) = A_N e^{-(p_z - p_z)^2/p_{th}^2} e^{-\rho^2/r_b^2}. \tag{8.45}$$

The normalization constant $A_N$ can be obtained by writing the number of beam electrons per unit channel length as

$$n_{0b}\pi r_b^2 = \int f_0 dp_z dp_x dp_y dx dy. \tag{8.46}$$

Writing $p_x = m\gamma v_x$, $p_y = m\gamma v_y$, $v_x = \omega_b \rho_x \cos\psi_x$, $x = \rho_x \sin\psi_x$, $y = \rho_y \sin\psi_y$, and $v_y = \omega_b \rho_y \cos\psi_y$, one may transform $dp_x dx$ by $m\gamma\omega_b \rho_x d\rho_x d\psi_x$ and $dp_y dy$ by $m\gamma\omega_b \rho_y d\rho_y d\psi_y$. Further, expressing $\rho_x = \rho\cos\phi$ and $\rho_y = \rho\sin\phi$, we may write $d\rho_x\, d\rho_y = \rho d\rho d\phi$ and solve the integral in Eq. (8.46) with limits of $\psi_x$ and $\psi_y$ from 0 to $2\pi$, $\phi$ from 0 to $\pi/2$, $\rho$ from 0 to $\infty$, and $p_z$ from $-\infty$ to $\infty$ to obtain

$$A_N = 2n_{0b}/(m^2\omega_p^2 r_b^2 \gamma_0 p_{th}\sqrt{\pi}), \tag{8.47}$$

where $\gamma_0 = \left(1 + p_{0z}^2/m^2c^2\right)^{1/2}$.

We perturb the aforementioned equilibrium by an electromagnetic (X-ray laser) perturbation,

$$\vec{E}_1 = \hat{x}A_1 e^{-i(\omega t - kz)}, \tag{8.48}$$

$\vec{B}_1 = \vec{k} \times \vec{E}_1/\omega$, where $A_1$ has very slow $x$, $y$ variations, on the scale of $r_b$. The response of beam electrons to it is governed by the Vlasov equation,

$$\frac{\partial f}{\partial t} + \vec{v}.\nabla f - e(E + \vec{v} \times \vec{B}).\frac{\partial f}{\partial \vec{p}} = 0 \tag{8.49}$$

Writing $f = f_0 + f_1$ and linearizing Eq. (8.49), we obtain

$$f_1 = eA_1\left[\frac{2(1 - kv_z/\omega)}{m\gamma\omega_b^2}\frac{\partial f_0}{\partial \rho^2} + \frac{k}{\omega}\frac{\partial f_0}{\partial p_z}\right]I_1, \tag{8.50}$$

where $I_1 = \int_{-\infty}^{t} v_x' \exp[-i(\omega t' - kz')dt']$. The prime refers to the quantities at time $t'$ and the integration is along the unperturbed trajectory of electrons determined solely by $E_{0r}\hat{r}$. Using the conditions $z' = z$ and $v_x' = v_x$ at $t' = t$, one may write $z' = z + v_z(t' - t)$ and $v_x' = \rho_x\omega_b \cos(\omega_b(t' - t) + \psi_x)$, and evaluate $I_1$, giving

$$f_1 = \frac{ieE_1\omega_b\rho_x e^{i\psi_x}}{\omega - kv_z - \omega_b}\left[\frac{2(1 - kv_z/\omega)}{m\gamma\omega_b^2}\frac{\partial f_0}{\partial \rho^2} + \frac{k}{\omega}\frac{\partial f_0}{\partial p_z}\right], \tag{8.51}$$

where the off-resonant term, having $\omega - kv_z + \omega_b$ in the denominator, has been dropped. The perturbed current density can be written as

$$\vec{J}_1 \equiv \sigma\vec{E}_1 = -e\int f_1\vec{v}dp_x dp_y dp_z. \tag{8.52}$$

The wave equation $\nabla^2 \vec{E}_1 + (\omega^2/c^2)\vec{E}_1 = -(i\omega/\varepsilon_0 c^2)\vec{J}_1$, on using Eq. (8.52) gives

$$\nabla_\perp^2 A_1 + \alpha^2 A_1 = -(i\omega/c^2\varepsilon_0)\sigma A_1, \tag{8.53}$$

where $\alpha^2 = \omega^2/c^2 - k^2$. Ignoring the right-hand side one may solve this equation with appropriate boundary conditions to obtain the mode structure $A_1 = A_{10}\Psi(x,y)$ and eigenvalues $\alpha$. Then we retain the right-hand side and presume that the mode structure remains unmodified, only the eigenvalues are modified. A reasonable mode structure could be a Gaussian, $\Psi = \exp[-(x^2+y^2)/2r_a^2]$, with $r_a^2 \geq r_b^2$. Using $A_1$ in Eq. (8.53), multiplying the resulting equation by $\Psi dxdy$, and integrating over the cross-section of the beam, and assuming $\alpha^2 \sim 1/r_a^2 << k^2$, gives

$$\omega^2 - k^2c^2 = -(i\omega/\pi r_a^2 c^2 \varepsilon_0)\int \sigma \Psi^2 dxdy$$

$$= -\omega_{pb}^2 \frac{\pi r_b^2 G}{3r_a^2}\left(\frac{r_b^2\omega_p^2}{c^2}\right)\frac{m^2c^2\gamma_0^2}{p_{\text{th}}^2}(1+\xi Z(\xi)), \tag{8.54}$$

where

$$G = \frac{2}{\pi^2}\iiint \frac{\sin\phi\cos^3\phi\cos\psi_x e^{i\psi_x} d\phi d\psi_x d\psi_y}{[1+(\cos^2\phi\sin^2\psi_x + \sin^2\phi\sin^2\psi_y)r_b^2/r_a^2]},$$

$\omega_{\text{pb}} = (n_{0b}e^2/m\varepsilon_0)^{1/2}$, $\xi = (\omega - kv_{0z} - \omega_{b0})/(3p_{\text{th}}\omega_{b0}/2mc\gamma_0)$, $v_{0z} = p_{0z}/m\gamma_0$, $\omega_{b0} = \omega_p/\sqrt{2\gamma_0}$. Here $Z(\xi)$ is the plasma dispersion function, and we have ignored the first term inside the large parenthesis of $f_1$ in Eq. (8.51) as it is $\omega_b/\omega$ times smaller than the second term. For $r_a^2 \geq r_b^2$, the geometrical factor $G \simeq 1$. Further, if $\xi > 1$, the dispersion relation takes the form,

$$(\omega^2 - k^2c^2)(\omega - kv_{0z} - \omega_{b0})^2 = RG \tag{8.55}$$

where $R = 3\pi\omega_{\text{pb}}^2\omega_p^4 r_b^4/16 r_a^2 c^2\gamma_0$. The simultaneous zeros of the two factors on the left give the operating frequency of the X-ray laser $\omega \approx 2\gamma_0^2\omega_{b0}$. Expanding $\omega$ around this frequency, $\omega = kc + \delta = kv_{0z} + \omega_{b0} + \delta$, where $\delta << kc$, one obtains $\delta = (RG/2\omega)^{1/3}(-1 + i\sqrt{3})/2$ for the unstable root, giving the growth rate

$$\Gamma = \frac{\sqrt{3}}{4}\left(3\sqrt{2}\pi\frac{r_b^4\omega_p^2}{r_a^2c^2}G\right)^{1/3}\Gamma_m, \tag{8.56}$$

where $\Gamma_m = (\omega_{\text{pb}}^2 \omega_p)^{1/3} / 2\gamma_0^{5/6}$. For a given number of beam electrons, $r_b^2 \omega_{\text{pb}}^2 = \text{constant}$, and a given radial width of the X-ray eigen-mode, $r_a$, the growth rate scales as $r_b^{2/3}$.

In the case of an electron beam with finite temperature, the expression for the growth rate is given by

$$\Gamma = \frac{\pi\omega_{\text{pb}}^2}{4\gamma_0^2\omega_{\text{b0}}} \frac{r_b^2 G}{3r_a^2} \frac{r_b^2\omega_p^2}{c^2} \frac{m^2c^2\gamma_0^2}{p_{\text{th}}^2} \xi e^{-\xi^2}. \tag{8.57}$$

In Fig. 8.10, we have plotted the growth rate, normalized by $R' = \sqrt{2}\pi^{3/2}\omega_{\text{pb}}^2\omega_p r_b^4 G / 18 r_a^2 c^2$, with the real part of the frequency of the X-ray laser, $\omega_r$, normalized by $2\gamma_0^2\omega_{b0}$ for parameters $p_{\text{th}}/mc = 0.03\gamma_0$ and $\gamma_0 = 100$. The growth rate spectrum is sharply peaked around $\omega_r = 0.95$. One may note that the finite temperature of the electron beam slightly downshifts the frequency of the X-ray radiation. This shift is larger for higher values of $p_{\text{th}}$.

The time available for the beam for the growth of X-ray instability is much larger than the length of the laser pulse as the beam and the pump laser move collinearly with very closely equal velocities. The total interaction time could be of the order of $L/c$, where $L$ is the plasma length and may be of the order of 1 ps.

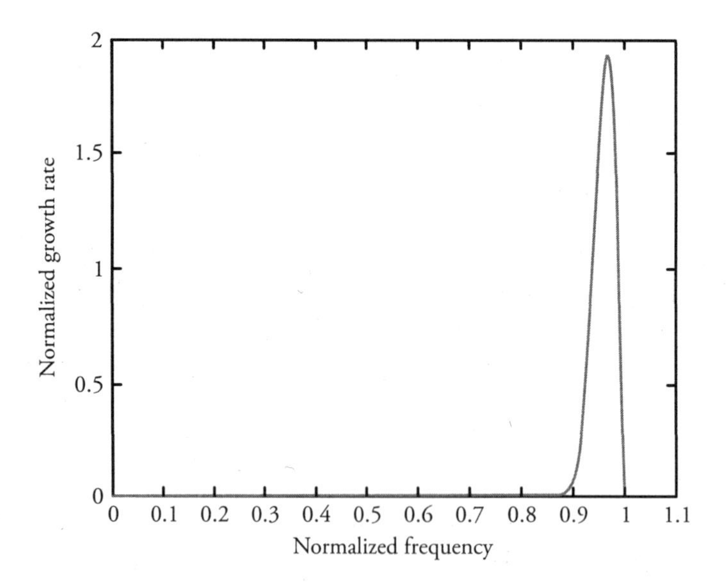


**Fig. 8.10** Normalized growth rate of an ion channel X-ray laser with normalized frequency. The parameters are $p_{\text{th}}/mc = 0.03\gamma_0$ and $\gamma_0 = 100$. (After Liu et al.[11])

## 8.4 Discussion

Free electron lasers (FELs) fill the void in the electromagnetic spectrum from millimeter waves to infrared where conventional sources are scarce. The FEL is frequency-tunable and efficient, and provides high peak powers. FELs are currently operational from submillimeter waves to UV, and X-ray lasers have also been demonstrated.[12,13] The operating frequency of a laser-driven ion channel laser scales as $n_0^{1/2}\gamma_0^{3/2}$, while the growth rate scales as $n_{0b}^{1/3}n_0^{1/6}\gamma_0^{-5/6}$. Optimum growth occurs when the radial widths of the laser pump, electron beam, and X-ray radiation are ~ $c/\omega_p$. As the instability grows, the electrons (on an average) lose energy and the betatron resonance gets detuned. At the saturation of instability, the reduction in the beam Lorentz factor is $\Delta\gamma_0 \sim \Gamma/(\partial F/\partial\gamma_0)$, where $F = \omega - kv_{0z} - \omega_{p0}/(2\gamma_0^{1/2})^{1/2}$ and $v_{0z} = c(1-1/\gamma_0^2)^{1/2}$, and the efficiency of energy conversion from the electron beam to X-rays turns out to be $\eta = \Delta\gamma_0(\gamma_0 - 1) \approx 2\Gamma/3\omega_{b0}$. It scales as $(n_{0b}/n_0\gamma_0)^{1/3}$, if one keeps $r_0 \sim c/\omega_p$. In the parameter regime given by Mangles et al.,[12] namely, a 0.8 μm, $10^{19}$ Wcm$^{-2}$, 40 fs laser impinging on a helium supersonic gas jet target with $n_0 \sim 2.3 \times 10^{19}$ cm$^{-3}$, if one takes the beam Lorentz factor $\gamma_0 \sim 50$, $n_{0b}/n_0 \sim 0.05$, the ion channel size $r_c \sim c/\omega_p$, and the channel length as 1000 μm, one obtains the growth rate $\Gamma = 2 \times 10^{12}$ sec$^{-1}$ and the spatial amplification factor $\Gamma L/c = 7$ which is substantial; hence, an X-ray laser is realizable if the beam energy spread is less than $2\Gamma/3\omega_{b0} \approx 2\%$, which is close to the energy spread of 3%, and the number of X-ray photons would be orders of magnitude higher than the number of energetic electrons. By employing a density gradient in the plasma, one may further enhance the gain of a tapered X-ray free electron laser.

## References


1. Ciocci, F., A. Doria, G. P. Gallerano, I. Giabbai, M. F. Kimmitt, G. Messina, A. Renieri, and J. E. Walsh. 1991. "Observation of Coherent Millimeter and Submillimeter Emission from a Microtron-driven Cherenkov Free-electron Laser." *Physical Review Letters* 66 (6): 699.

   Bogdankevich, L. S., Mikhail Viktorovich Kuzelev, and Anri A. Rukhadze. 1981. "Plasma Microwave Electronics." *Soviet Physics Uspekhi* 24 (1): 1.

2. Cook, A. M., R. Tikhoplav, S. Ya Tochitsky, G. Travish, O. B. Williams, and J. B. Rosenzweig. 2009. "Observation of Narrow-band Terahertz Coherent Cherenkov Radiation from a Cylindrical Dielectric-lined Waveguide." *Physical Review Letters* 103 (9): 095003.

   Alexandrov, A. F., L. S. Bogdankevich and A. A. Rukhadze. 1984. *Principles of Plasma Electrodynanics*, vol. 9. Springer series in Electrophysics, edited by G. Ecker. New York: Spinger-Verlag.

3. Motz, Hans. 1951. "Applications of the Radiation from Fast Electron Beams." *Journal of Applied Physics* 22 (5): 527–35.

   Madey, J. M. J. 1971. "Stimulated Emission of Bremsstrahlung in a Periodic Magnetic Field". *Journal of Applied Physics* 42 (5): 1906–19.

   Motz, H. 1979. "Undulators and 'Free Electron Lasers'". *Contemporary Physics* 20 (5): 547–68.

4. Marshall, T. C. 1985. *Free Electron Laser*. New York: MacMillan, p.78.

5. Roberson, C. W. and P. Sprangle. 1989. “A Review of Free-Electron Lasers”, *Physics of Fluids B: Plasma Physics* 1 (1): 3–42.

6. Freund, Henry P., and Thomas M. Antonsen. 1992. *Principles of Free-electron Lasers*. London: Chapman & Hall, 1992.

7. Chen, Kuan-Ren, J. M. Dawson, A. T. Lin, and T. Katsouleas. 1991. “Unified Theory and Comparative Study of Cyclotron Masers, Ion-channel Lasers, and Free Electron Lasers.” *Physics of Fluids B: Plasma Physics* 3 (5): 1270–8.

8. Whittum, David H., Andrew M. Sessler, and John M. Dawson. 1990. “Ion-channel Laser.” *Physical Review Letters* 64 (21): 2511.

   Whittum, David H. 1992. “Electromagnetic Instability of the Ion-focused Regime.” *Physics of Fluids B: Plasma Physics* 4 (3): 730–9.

9. Kostyukov, I., S. Kiselev, and A. Pukhov. 2003. “X-ray Generation in an Ion Channel.” *Physics of Plasmas* 10 (12): 4818–28.

10. Rousse, Antoine, Kim Ta Phuoc, Rahul Shah, Alexander Pukhov, Eric Lefebvre, Victor Malka, Sergey Kiselev et al. 2004. “Production of a keV X-ray Beam from Synchrotron Radiation in Relativistic Laser-Plasma Interaction.” *Physical Review Letters* 93 (13): 135005.

11. Liu, C. S., V. K. Tripathi, and Naveen Kumar. 2007. “Vlasov Formalism of the Laser Driven Ion Channel X-ray Laser.” *Plasma Physics and Controlled Fusion* 49 (3): 325.

12. McNeil, Brian. 2009 “Free Electron Lasers: First Light from Hard X-ray Laser.” *Nature Photonics* 3 (7): 375.

    Hand, Eric. 2009 “Science by the Femtosecond.” *Nature* 461 (7265): 708–709.

13. Pellegrini, C. 2017 “X-ray Free-electron Lasers: from Dreams to Reality.” *Physica Scripta* 2016 (T169): 014004.

# 9 Self-focusing and Filamentation

## 9.1 Introduction

The propagation of an electromagnetic wave of finite spot size or nonuniform intensity distribution is a problem of practical importance. We have seen that at low intensities, diffraction causes divergence of the wave. The situation, however, changes dramatically at higher intensity.

In the presence of a nonuniform electromagnetic wave, the electrons in a plasma experience a ponderomotive force $\vec{F}_p \sim -\nabla|E|^2$, directed opposite the intensity gradient. It causes an electron density depression in regions of higher intensity. The relativistic electron mass is also higher where the intensity is higher. Thus, the effective relative plasma permittivity $\varepsilon$ becomes an increasing function of the local intensity ($\partial\varepsilon/\partial|E|^2 > 0$) and the plasma behaves like a self-made graded index fiber. For a beam with a Gaussian distribution of intensity along its wavefront, the refractive index of the plasma is maximum on the beam axis and decreases away from it (cf. Fig. 9.1). As a consequence, the axial portion of the beam propagates with a lower phase velocity than the marginal rays and the wave front acquires a curvature, leading to self-focusing when the diffraction divergence is overpowered.[1–2] Focusing leads to an enhancement of the intensity, causing an even deeper density depression, stronger mass increase and stronger focusing. Eventually, one may obtain a completely electron-evacuated ion channel.

For electromagnetic beams of long pulse duration, the ion dynamics modifies the picture very significantly. An important parameter for a beam of spot size $r_0$, is the time for ambipolar diffusion $\tau_d = r_0/c_s$, where $c_s$ is the ion sound speed. For a pulse duration $\tau > \tau_d$, as the electrons move under the radial ponderomotive force, a radial space charge field is created that pulls the ions along, causing ambipolar diffusion of the plasma. A steady state is realized when the ponderomotive force is balanced by the thermal pressure gradient, creating a deeper density depression than when the ions were stationary.

In a low-temperature plasma, where collisions are significant, another scenario emerges. The wave causes nonuniform Ohmic heating of the electrons on the time-scale of the energy relaxation time, raising their local thermal pressure. The pressure gradient force causes ambipolar diffusion of the plasma, creating a plasma duct of even deeper density depression than the ponderomotive force does.

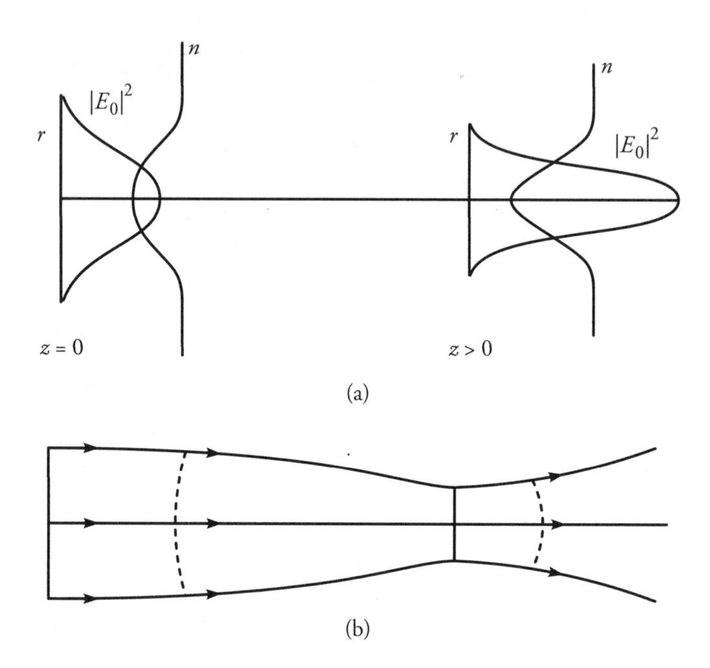


**Fig. 9.1** (a) Intensity profiles and density depressions, and (b) wave fronts at different values of $z$.

The duct continues to focus the beam in a collapsing manner until the duct is nearly depleted of plasma and diffraction divergence supersedes the nonlinear self-focusing. The beam thus acquires a minimum radius at some point beyond which it diverges. As the beam diameter increases and axial intensity decreases, the saturating effect of nonlinearity weakens; the nonlinear refraction becomes stronger and self-focusing starts. Thus, a periodic focusing is realized. There exists a specific radiation intensity profile for which the self-focusing and diffraction effects are evenly balanced and the beam propagates as a spatial soliton.

Even in an initially uniform electromagnetic wave a small perturbation in its intensity along its wave front may get amplified by the same process. This phenomenon, known as a filamentation instability,[3] causes the the beam to break into filaments.

In Section 9.2, we deduce the long time scale nonlinear plasma permittivity due to ponderomotive and Ohmic nonlinearities. We obtain the short time scale permittivity due to combined effects of relativistic mass variation and ponderomotive force in Section 9.3. In Section 9.4, we solve the wave equation in the paraxial ray approximation using the procedure reported by Akhmanov et al.[1] to study self-focusing of electromagnetic beams. We study the filamentation instability in Section 9.5. Finally some conclusions are drawn in Section 9.6.

## 9.2 Long Time Scale Nonlinear Permittivity

Consider a singly ionized plasma of equilibrium electron density $n_0^0$, electron temperature $T_e$ and ion temperature $T_i$. An electromagnetic wave propagates through it with

$$\vec{E} = \vec{A}(r, z)e^{-i(\omega t - kz)}, \tag{9.1}$$

where $\vec{A}$ is a slowly varying function of $r$ and $z$. The wave may have slow time dependence as well but the time scale is longer than the ambipolar diffusion time. At $z = 0$,

$$A^2 = A_{00}^2 e^{-r^2/r_0^2} \tag{9.2}$$

We consider the cases of collisionless and collisional plasma separately.

### 9.2.1 Collisionless plasma: ponderomotive nonlinearity

The electromagnetic wave imparts an oscillatory velocity to the electrons,

$$\vec{v} = \frac{e\vec{E}}{mi\omega} \tag{9.3}$$

and exerts a ponderomotive force, which in the non-relativistic limit (valid for long pulse experiments where the power density is usually low and $v/c << 1$) can be written as (cf. Chapter 5)

$$\vec{F}_p = e\nabla\phi_p, \text{ where } \phi_p = -\frac{e|E|^2}{4m\omega^2}, \tag{9.4}$$

where $-e$ and $m$ are the electron charge and rest mass. Under this force, as the electrons move, a space charge field $\vec{E}_s$ is created that pulls the ions along. In the steady state, the net forces on the electrons and ion fluids balance each other,

$$-e\vec{E}_s + \vec{F}_p - \frac{1}{n}\nabla(nT_e) = 0,$$

$$e\vec{E}_s - \frac{1}{n_i}\nabla(n_i T_i) = 0. \tag{9.5}$$

Under the condition of quasi-neutrality, $n = n_i$, and assuming $T_e$ and $T_i$ to be constants, these equations add to give[2]

$$(T_e + T_i)\nabla \log_e n = e\nabla\phi_p,$$

and hence

$$n = n_0^0 e^{-\alpha_p |E|^2}, \tag{9.6}$$

where $\alpha_p = e^2/[4m\omega^2(T_e + T_i)]$. It may be noted that the plasma density redistribution follows Boltzmann's law. It may also be seen that $\alpha_p |E|^2 = |v|^2 / 4v_{\text{th}}'^2$, $v_{\text{th}}' = \left[(T_e + T_i)/m\right]^{1/2}$.

The effective relative plasma permittivity can now be written as

$$\varepsilon = 1 - \frac{ne^2}{m\varepsilon_0\omega^2} = \varepsilon_L + \Phi, \tag{9.7}$$

where the linear permittivity is

$$\varepsilon_L = 1 - \frac{\omega_p^2}{\omega^2},$$

and the nonlinear contribution

$$\Phi = \frac{\omega_p^2}{\omega^2}\left(1 - e^{-\alpha_p |E|^2}\right),$$

with $\omega_p = \left(n_0^0 e^2 / m\varepsilon_0\right)^{1/2}$ being the electron plasma frequency. The nonlinear part of the permittivity, $\Phi$, increases with $|E|^2$ and saturates for $\alpha_p |E|^2 >> 1$.

### 9.2.2 Collisional plasma: Ohmic nonlinearity

When the pulse duration of the electromagnetic wave is longer than the temperature relaxation time, Ohmic heating becomes significant. The wave-induced oscillatory velocity of the electrons in a plasma with the electron–ion (plus electron–neutral particles) collision frequency $\nu$ is

$$\vec{v} = \frac{e\vec{E}}{mi(\omega + i\nu)} \tag{9.8}$$

It has a component in phase with $\vec{E}$ and causes heating of the electrons at the time average rate[2,4]

$$H = \text{Re}\left[-e\vec{E}^* \cdot \vec{v}/2\right] = e^2 |E|^2 \nu / 2m\omega^2, \tag{9.9}$$

where $\omega^2 >> \nu^2$. In the steady state, the heating rate is balanced by the power loss via collisions with ions and neutral particles, and thermal conduction

$$\frac{3}{2}\delta\nu\left(T_e + T_o\right) - \nabla.\left(\frac{\chi}{n}\nabla T_e\right) = \frac{e^2|E|^2\nu}{2m\omega^2}, \tag{9.10}$$

where $\chi = 2nv_{\text{th}}^2/\nu$ is the electron thermal conductivity, $v_{\text{th}} = (T_e/m)^{1/2}$, $\nu$ is the collision frequency, and $\delta$ is the mean fraction of excess electron energy lost in a collision. We assume that $T_e = T_i = T_0$ when the wave is not there. For elastic electron–ion collisions $\delta = 2m/m_i$, where $m_i$ is the ion mass. The characteristic times for thermal conduction and collisional energy transfer are

$$\tau_{\text{con}} \sim \nu\, r_0^2/v_{\text{th}}^2,\ \tau_{\text{coll}} \sim (\delta\nu)^{-1}.$$

For $\tau_{\text{coll}} < \tau_{\text{con}}$, that is, $\nu^2 r_0^2\delta/v_{\text{th}}^2 > 1$, one can ignore the thermal conduction term in Eq. (9.10) and obtain

$$T_e = T_o + \frac{m}{3\delta}\left(\frac{e|E|}{m\omega}\right)^2. \tag{9.11}$$

Assuming quasi-neutrality and demanding the uniformity of plasma pressure in the steady state, $n(T_e + T_i)$ = constant, one obtains the modified electron density

$$n = \frac{2\, n_0^0\, T_o}{T_e + T_o}, \tag{9.12}$$

giving the nonlinear part of the plasma permittivity[2]

$$\Phi = \frac{\omega_p^2}{\omega^2}\frac{\alpha|E|^2}{1+\alpha|E|^2}, \tag{9.13}$$

where $\alpha = e^2/[3\delta m\omega^2 T_0)] \approx 4\alpha_p/3\delta$. The nonlinearity parameter $\alpha$ in the Ohmic case can be orders of magnitude larger than the one in the case of ponderomotive nonlinearity. The nonlinear permittivity increases with $\alpha|E|^2$ and saturates for $\alpha|E|^2 >> 1$.

### 9.2.3 Ohmic nonlinearity with thermal conduction

For $\tau_{\text{con}} < \tau_{\text{coll}}$, thermal conduction is significant. Equation (9.10) can be solved analytically in the weak nonlinearity approximation. Following Sodha et al.[2], we write $T_e = T_0 + T_1$ with $T_1 << T_0$, and deduce from Eq. (9.10), in a cylindrical geometry,

$$\frac{\partial^2 T_1}{\partial r^2} + \frac{1}{r}\frac{\partial T_1}{\partial r} - k_0^2 T_1 = -\frac{e^2\nu^2|E|^2}{2m\omega^2 v_{\text{th}}^2}, \tag{9.14}$$

where $k_0^2 = 3\delta v^2 / 2v_{th}^2$. In the aforementioned equation, we have taken the radial variation of intensity to be stronger than the longitudinal variation. This equation is a modified Bessel equation with a source term. Its solution is

$$T_1 = \frac{e^2 v^2}{2m\omega^2 v_{th}^2}\left[ K_0(k_0 r)\int_0^r r' dr' I_0(k_0 r')|E|^2 + I_0(k_0 r)\int_r^\infty r' dr' K_0(k_0 r')|E|^2 \right], \tag{9.15}$$

where $I_0$ and $K_0$ are the modified Bessel functions of order 0, and $|E|^2$ is a function of $r'$. From the electron temperature perturbation, one may deduce the plasma density modification by writing $n = n_0^0 + n_1$ and employing the pressure balance, $n(T_e + T_i) = 2n_0^0 T_0$, $n_1 = -\, n_0^0 T_1 / 2T_0$.

Presuming that the wave maintains its Gaussian intensity profile, albeit with an axial intensity dependent on $z$ and a spot size of $|E|^2 = A_{00}'^2 \exp(-r^2 / r_0'^2)$ as we will see later, one may evaluate the integrals in Eq. (9.15). On the beam axis ($r = 0$), the density modification is

$$n_1(r=0) = n_{10} = in_0^0 \alpha A_{00}'^2\ I_1, \tag{9.16}$$

$$I_1 = I_0(0)\int_0^\infty t K_0(t) e^{-t^2/k_0^2 r_0'^2} dt.$$

We expand $n$ around $r = 0$ as

$$n_1 = n_{10} + c_2 r^2,\ T_1 = -\ (2T_0 / n_0^0) n_1. \tag{9.17}$$

Using this $T_1$ in Eq. (9.14) and equating $r$-independent terms on both sides, we obtain

$$c_2 = k_0^2 \alpha A_{00}'^2\ n_0^0 (1 - I_1)/4. \tag{9.18}$$

The nonlinear part of the permittivity is

$$\Phi = \Phi_0 - \Phi_2 r^2,$$

$$\Phi_0 = \omega_p^2 \alpha A_{00}'^2\ I_1/\omega^2,$$

$$\Phi_2 = \omega_p^2 \alpha A_{00}'^2 (1 - I_1) k_0^2 / 4\omega^2. \tag{9.19}$$

## 9.3 Short Time Scale Nonlinear Permittivity

In the case of short pulses that are shorter than the ion–plasma period $\left(\omega_{\text{pi}}^{-1}\right)$, the ion motion can be ignored. In this case the electron motion under the quasi-static ponderomotive force $\vec{F}_p = e\nabla\phi_p$ is inhibited by the space charge field, $\vec{E}_s = e\nabla\phi_s$, thus created. A steady state is realized when the two forces balance each other, that is, $\phi_s \approx -\phi_p$. Using this in Poisson's equation, $\nabla^2\phi_s = e\left(n - n_0^0\right)/\varepsilon_0$, one obtains the modified electron density as[5]

$$n = n_0^0\left(1 + \frac{c^2}{\omega_p^2}\nabla^2\left(1 + \frac{a^2}{2}\right)^{1/2}\right), \tag{9.20}$$

where we have employed the relativistic expression for the ponderomotive potential, $\phi_p = -(mc^2/e)(1 + a^2/2)^{1/2}$ for the linearly polarized wave of normalized amplitude from Chapter 5. Since $n$ cannot become negative, Eq. (9.20) is valid when the second term inside the parenthesis is greater than $-1$. Further, there is relativistic modification of the electron mass by the gamma factor[6] $\gamma = (1 + a^2/2)^{1/2}$. Thus, the nonlinear effective plasma permittivity is

$$\varepsilon = \varepsilon_L + \Phi,$$

$$\Phi = \frac{\omega_p^2}{\omega^2\gamma}\left(1 - \frac{n}{n_0^0}\right). \tag{9.21}$$

For a Gaussian beam, $a^2 = a'^2\exp\left(-r^2/r_0'^2\right)$, one may write, in the paraxial ray approximation,

$$\Phi = \Phi_0 - \Phi_2 r^2,$$

$$\Phi_0 = \frac{\omega_p^2}{\omega^2}\left[1 - \frac{1 + c^2 g_0/\omega_p^2}{\left(1 + a'^2/2\right)^{1/2}}\right],$$

$$\Phi_2 = \frac{\left(\omega_p^2/r_0'^2\right)a'^2}{4\,\omega^2\left(1 + a'^2/2\right)^{3/2}}\left[1 + \frac{(c^2/r_0'^2\omega_p^2)(1 - a'^2/4)}{\left(1 + a'^2/2\right)^{1/2}}\right],$$

$$g_0 = -\frac{a'^2\ /\ r_0'^2}{\left(1 + a'^2/2\right)^{1/2}}. \tag{9.22}$$

These expressions are valid as long as

$$|g_0| < \omega_p^2 / c^2, \text{or } r_0'^2 \omega_p^2 / c^2 > a'^2 / (1 + a'^2)^{1/2}$$

## 9.4 Self-focusing

In a plasma with nonlinear permittivity, $\nabla.(\varepsilon\vec{E}) = 0$ leads to $\nabla \cdot \vec{E} = -(\vec{E} \cdot \nabla\varepsilon)/\varepsilon \neq 0$. However, for $\nabla\varepsilon / \varepsilon << 1$ one may ignore the $\nabla \cdot \vec{E}$ term in the wave equation. For a wave having a rapid phase variation given by Eq. (9.11), the wave equation in the Wentzel–Kramers–Brillouin (WKB) approximation can be written as[1,2]

$$2ik \frac{\partial A}{\partial z} + \nabla_\perp^2 A + \frac{\omega^2}{c^2} \Phi A = 0 \tag{9.23}$$

where $k = \omega\varepsilon_L^{1/2} / c$. Introducing an eikonal $A = A_0(r, z) \exp[iS(r, z)]$ and separating out the real and imaginary parts of Eq. (9.23), we obtain

$$k \frac{\partial A_0^2}{\partial z} + \frac{\partial S}{\partial r} \frac{\partial A_0^2}{\partial r} + A_0^2 \nabla_\perp^2 S = 0 \tag{9.24}$$

$$2k \frac{\partial S}{\partial z} + \left(\frac{\partial S}{\partial r}\right)^2 = \frac{1}{A_0} \nabla_\perp^2 A_0 + \frac{\omega^2}{c^2} \Phi. \tag{9.25}$$

We solve this coupled set in the paraxial ray approximation by expanding $S$ as[1]

$$S = S_0 + S_2 r^2 / 2. \tag{9.26}$$

Then, Eq. (9.24) takes the form

$$k \frac{\partial A_0^2}{\partial z} + S_2 r \frac{\partial A_0^2}{\partial r} + 2A_0^2 S_2 = 0. \tag{9.27}$$

We introduce a function $f(z)$ such that

$$S_2 = \frac{1}{f} \frac{df}{dz},$$

Then, Eq. (9.27) can be written as

$$\frac{\partial G}{\partial z} = -\frac{r}{f}\frac{\partial f}{\partial z}\frac{\partial G}{\partial r}$$

when $G = A_0^2 f^2$. This equation holds when $G$ is an arbitrary function of $r/f$,

$$G = F\,(r/f)$$

or $A_0^2 = \dfrac{F(r/f)}{f^2}$.

For an initially Gaussian beam given by Eq. (9.2), one may write

$$A_0^2 = \frac{A_{00}^2}{f^2} e^{-r^2/r_0^2 f^2}. \tag{9.28}$$

where $f(z)$ is defined as a dimensionless beam width parameter. At $z = 0$, one has $f = 1$ and Eq. (9.27) reduces to Eq. (9.2). The equation of the wavefront is

$$kz + S_0 + S_2 r^2 / 2 = 0.$$

For $\partial S_0/\partial z << k$, the radius of curvature of the wavefront is

$$R = -\,k / S_2 = -f\ \left(df / dz\right)^{-1}.$$

For an initially plane wavefront $df/dz = 0$ at $z = 0$. It may be noted from Eq. (9.28) that as the beam spot size $f$ decreases, the axial intensity is correspondingly enhanced keeping the total power carried by the beam a constant, independent of $z$.

Using Eqs. (9.26) and (9.28) in Eq. (9.25) and equating different powers of r on both sides we obtain

$$2k\frac{dS_0}{dz} = -\frac{z}{r_0^2 f^2} + \frac{\omega^2}{c^2}\Phi\left(r = 0\right),$$

$$\frac{d^2 f}{dz^2} = \frac{1}{R_d^2 f^3} + \frac{f}{\varepsilon_L}\left.\frac{\partial \Phi}{\partial r^2}\right|_{r=0}, \tag{9.29}$$

where $R_d = kr_0^2$ is the Rayleigh diffraction length. The first term on the right-hand side in Eq. (9.29) is due to diffraction while the second is due to nonlinear refraction. Since $\partial\Phi/\partial r^2 < 0$, the second term causes self-focusing of the beam when it supersedes the diffraction term.

$$-\left.\frac{\partial\Phi}{\partial r^2}\right|_{r=0} \geq \frac{c^2}{\omega^2 r_0^4} \tag{9.30}$$

with $f = 1$ (at $z = 0$). The equal sign gives the threshold for self-focusing.

### 9.4.1 Long-term ponderomotive self-focusing

In this case, $\Phi$ is given by Eq. (9.7). Hence,

$$\left.\frac{\partial\Phi}{\partial r^2}\right|_{r=0} = -\frac{A_{\infty}^2}{r_0^2 f^4}\left.\frac{\partial\Phi}{\partial A_0^2}\right|_{r=0} = -\frac{\omega_p^2\ \alpha_p\ A_{00}^2}{\omega^2\ r_0^2\ f^4}\ e^{-\alpha_p A_{00}^2/f^2}. \tag{9.31}$$

At the threshold for self-focusing,

$$\alpha_p A_{00}^2 e^{-\alpha_p A_{00}^2} = \frac{c^2}{\omega_p^2 r_0^2}. \tag{9.32}$$

This equation gives the radius of the beam as a function of the axial beam intensity for self-trapping, that is, propagation without divergence or convergence. The normalized beam radius, $r_0\,\omega_p/c$, decreases with $\alpha_p\ A_{00}^2$, attains a minimum value of $r_0\omega_p/c = (2.73)^{1/2}$ at $\alpha_p\ A_{00}^2 = 1$ and then rises (cf. Fig. 9.2) due to the weakening of the self-focusing term caused by nonlinear saturation.

In the limit of $\alpha_p A_{00}^2 / f^2 << 1$ (referred to as quadratic nonlinearity), Eq. (9.29) reduces to

$$\frac{d^2 f}{dz^2} = \frac{1}{R_d^2 f^3} - \frac{1}{R_n^2 f^3} \tag{9.33}$$

giving

$$f^2 = 1 - z^2 / z_f^2, \tag{9.34}$$

where $z_f^{-2} = R_n^{-2} - R_d^{-2}$ and $R_n = \left(r_0\omega/\omega_p\right)\left(\alpha_p A_{00}^2/\varepsilon_L\right)^{-1/2}$. One may call $R_n$ as the characteristic self-focusing length. Self-focusing occurs when $R_n < R_d$. In the limit of quadratic nonlinearity, as $z$ approaches $z_f$, $f$ tends to zero, that is, the beam focuses to a point. Of course, the quadratic nonlinearity approximation breaks down there.

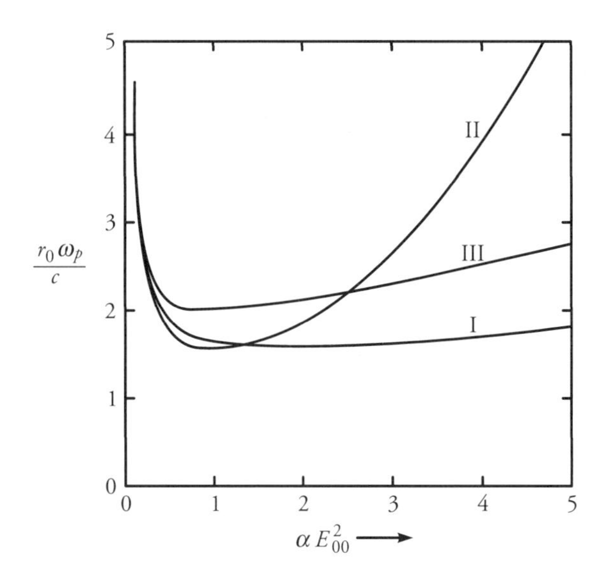


**Fig. 9.2** Radius of self-trapped radiation as a function of axial intensity. For relativistic nonlinearity (I) $\alpha = e^2/m^2\omega^2c^2$, ponderomotive nonlinearity (II) $\alpha = e^2/m^2\omega^2 v_{th}^2$, and thermal nonlinearity without thermal conduction (III) $\alpha = e^2/6\delta m^2\omega^2 v_{th}'^2$.

The threshold beam power for self-focusing (corresponding to $R_n = R_d$) is given by

$$P_{cr} = \varepsilon_L^{1/2} \frac{A_{00th}^2}{2\mu_0 c} \int_0^\infty e^{-r^2/r_0^2} 2\pi r dr = \frac{\pi c \varepsilon_L^{1/2}}{2\mu_0 \omega_p^2 \alpha_p}, \tag{9.35}$$

where $A_{00th}^2$ is the value of $A_{00}^2$ for which $R_n = R_d$.

The critical power decreases with the plasma density and increases with the plasma temperature and frequency of the wave.

When the quadratic nonlinearity approximation does not hold, one may multiply Eq. (9.29) by $2df/dz$ and integrate over z to obtain (for an initially plane wavefront, $df/dz = 0, f = 1$ at $z = 0$)

$$R_d^2\left(\frac{df}{dz}\right)^2 = 1 - \frac{1}{f^2} + \frac{r_0^2\omega_p^2}{c^2}\left[e^{-\alpha_p A_{00}^2} - e^{-\alpha_p A_{00}^2/f^2}\right]. \tag{9.36}$$

Equating the right-hand side to zero, one obtains (other than $f = 1$), the minimum spot size $f = f_{min}$ of the beam.

Figure 9.3 shows the variation of $f_{min}$ with $\alpha_p A_{00}^2$. Initially, $f_{min}$ decreases with $\alpha_p A_{00}^2$, attains a minimum around $\alpha_p A_{00}^2 \sim 2$ and then rises. For $\alpha_p A_{00}^2 > 1$,

$$f_{min}^2 \approx \frac{1}{1 + (r_0\omega_p/c)^2 \exp(-\alpha_p A_{00}^2)}.$$

The $z$-dependence of $f$ can be obtained by numerically solving Eq. (9.29), using Eq. (9.31). The spot size $f$ turns out to be a periodic function of $z$, oscillating between 1 and $f_{min}$.

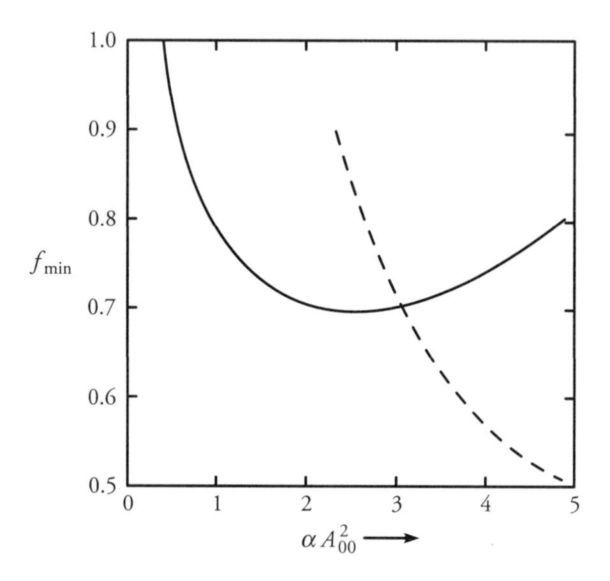


**Fig. 9.3** Size of the focal spot $r_0 f_{min}$ as a function of the power density of the beam. For thermal self-focusing with thermal conduction (dashed line) $\alpha = e^2 v^2 v_p^2 r_0^4 / 8m^2 \omega^2 v_{th}^2 c^2$. For relativistic self-focusing (solid line) $\alpha = e^2/m^2\omega^2c^2$, $r_0\omega_p/c = 1.9$. The behavior of $f_{min}$ in the other cases is similar to that of relativistic self-focusing.

## 9.4.2 Thermal self-focusing

For this case, $\Phi$ is given by Eq. (9.13) and one has

$$\left.\frac{\partial \Phi}{\partial r^2}\right|_{r=0} = -\frac{\omega_p^2\, \alpha\, A_{00}^2}{\omega^2\, r_0^2\, f^4} \frac{1}{\left(1+\alpha\, A_{00}^2 / f^2\right)^2} \tag{9.37}$$

and the threshold for self-focusing (cf. Eq. 9.30) is given by

$$\frac{\alpha A_{00}^2}{\left(1+\alpha A_{00}^2\right)^2} = \frac{c^2}{\omega_p^2 r_0^2}. \tag{9.38}$$

Figure 9.2 shows the variation of the normalized optimum spot size with $\alpha\, A_{00}^2$. The behavior is similar to the ponderomotive case. However, one must recognize that $\alpha >> \alpha_p$; hence, the threshold beam power for thermal self-focusing is orders of magnitude lower than for ponderomotive self-focusing.

Following the procedure outlined here, one obtains the equation governing $df/dz$,

$$R_d^2\left(\frac{df}{dz}\right)^2 = 1 - \frac{1}{f^2} + \frac{r_0^2\omega_p^2}{c^2}\left[\frac{1}{1+\alpha A_{00}^2} - \frac{1}{1+\alpha A_{00}^2/f^2}\right] \tag{9.39}$$

At the focus $df/dz = 0$ and the minimum spot size $f_m$ is given by

$$\frac{1}{f_m^2} = 1 + \frac{r_0^2\omega_p^2}{c^2}\left[\frac{1}{1+\alpha A_{00}^2} - \frac{1}{1+\alpha A_{00}^2/f_m^2}\right]. \tag{9.40}$$

The behavior of $f_m$ with $\alpha A_{00}^2$ (cf. Eq. 9.3) is similar to the ponderomotive case. $f$ versus $z$ can be obtained by solving Eq. (9.29) numerically. The beam spot size oscillates periodically with $z$ between $r_0$, $r_0 f_m$.

### 9.4.3 Self-focusing with thermal conduction

In the case of Ohmic nonlinearity with thermal conduction, $\Phi$ is given by Eq. (9.19); hence,

$$\left.\frac{\partial\Phi}{\partial r^2}\right|_{r=0} = -\frac{k_0^2\omega_p^2\alpha A_{00}^2}{4\,\omega^2 f^2}\left(1-I_1\right), \tag{9.41}$$

$$I_1 = I_0(0)\int_0^\infty tK_0(t)e^{-t^2/k_0^2r_0^2f^2}\,dt.$$

The threshold for self-focusing is given by

$$\alpha A_{00}^2\left(1-I_1\right) = \frac{4\,c^2}{k_0^2 r_0^4\omega_p^2}. \tag{9.42}$$

The optimum spot size for beam propagation without convergence or divergence goes inversely as one-fourth of the power of intensity. The critical power decreases with beam spot size. This is due to the fact that heat loss via thermal conduction decreases with $r_0$, hence, the nonlinearity becomes stronger.

One may solve Eq. (9.29) numerically to obtain $f$ as a function of $z$. Above the threshold power, $f$ decreases with $z$. As the spot size shrinks, thermal conduction becomes stronger and weakens the nonlinearity, causing $f$ to acquire a minimum at some value of $z = z_f$ and to rise beyond that.

### 9.4.4 Relativistic self-focusing

For high intensity short pulses, when relativistic and ponderomotive nonlinearities dominate, $\Phi$ is given by Eq. (9.22). Hence,

$$\left.\frac{\partial \Phi}{\partial r^2}\right|_{r=0} = -\Phi_2 = -\frac{\omega_p^2 a_{00}^2}{4\omega^2 r_0^2 f^4 \left(1 + a_{00}^2/2f^2\right)^{3/2}} \left[1 + \frac{4c^2\left(1 - a_{00}^2/4f^2\right)}{\omega_p^2 r_0^2 f^2 \left(1 + a_{00}^2/2f^2\right)^{1/2}}\right] \tag{9.43}$$

and the threshold for self-focusing is given by

$$\frac{a_{00}^2/4}{\left(1 + a_{00}^2/2\right)^{3/2}} \left[1 + \frac{4c^2\left(1 - a_{00}^2/4\right)}{\omega_p^2 r_0^2 \left(1 + a_{00}^2/2\right)^{1/2}}\right] = \frac{c^2}{r_0^2 \omega_p^2} \tag{9.44}$$

where $a_{00} = e A_{00}/m\omega c$. In the limit of $a_{00}^2 \ll 1$, Eq. (9.44) reduces to

$$r_0^2 \omega_p^2 a_{00}^2 / c^2 = 4/\left(1 + 4c^2/\omega_p^2 r_0^2\right) \tag{9.45}$$

giving the threshold power for self-focusing,

$$P_{cr} = \frac{\varepsilon_L^{1/2} A_{00}^2}{2\mu_0 c} \pi r_0^2 = \frac{\Pi\, \varepsilon_L^{1/2} m^2 c^3}{2\mu_0 e^2} \frac{n_{cr}}{n_0^0} \left(\frac{r_0^2 \omega_p^2 a_{00}^2}{c^2}\right)$$

$$= 4.2 \frac{n_{cr}}{n_0^0 \left(1 + 4c^2/\omega_p^2 r_0^2\right)} \mathrm{GW} \tag{9.46}$$

Above the threshold power, the beam undergoes self-focusing. As the beam focuses, the electron density depression becomes deeper and the electron mass nonlinearity becomes stronger. These in turn cause stronger focusing (in the course of propagation), and so on.

A situation may arise, as pointed out by Sun et al.[7], when a completely electron-evacuated ion channel is created in the axial region. In such a situation the paraxial theory given here does not work.

## 9.5 Filamentation Instability

Now we examine the stability of a uniform electromagnetic wave in the presence of a small-amplitude perturbation,

$$\vec{E} = \left[\vec{A}_0 + \vec{A}_1(x,z)\right] e^{-i(\omega t - kz)}, \tag{9.47}$$

where $\vec{A}_0 \| \vec{A}_1$. For simplicity, we limit ourselves to a quadratic nonlinearity and consider linear polarization $\vec{A}_0 \| \hat{y}$. Taking $A_0$ to be real, the effective permittivity of the plasma can be cast as

$$\varepsilon = \varepsilon_L + \varepsilon_2 A_0 \cdot \left(A_1 + A_1^*\right), \tag{9.48}$$

where $\varepsilon_2 = \alpha_p \omega_p^2 / \omega^2$ for long term ponderomotive nonlinearity and $\varepsilon_2 = \alpha \omega_p^2 / \omega^2$ for Ohmic nonlinearity without thermal conduction. Using Eqs. (9.7) and (9.48) in the wave Eq. (9.23), we obtain, on linearization,

$$2ik \frac{\partial A_1}{\partial z} + \frac{\partial^2 A_1}{\partial x^2} + \varepsilon_2 A_0^2 \left(A_1 + A_1^*\right) = 0. \tag{9.49}$$

Expressing $A_1 = A_{1r} + iA_{1i}$ and separating real and imaginary parts of Eq. (9.49), we get

$$2k \frac{\partial A_{1r}}{\partial z} + \frac{\partial^2 A_{1i}}{\partial x^2} = 0,$$

$$-2k \frac{\partial A_{1i}}{\partial z} + \frac{\partial^2 A_{1r}}{\partial x^2} + \frac{2k^2 \varepsilon_2}{\varepsilon_0} A_0^2 A_{1r} = 0. \tag{9.50}$$

For $A_{1r}$, $A_{1i} \sim \exp[i(q_\perp x + q_\| z)]$, one may replace $\partial/\partial x$ by $iq_\perp$ and $\partial/\partial z$ by $iq_\|$. Then these equations yield the dispersion relation

$$q_\| = -i \frac{q_\perp}{2k} \left(2k^2 \varepsilon_2 A_0^2 - q_\perp^2\right)^{1/2}. \tag{9.51}$$

An instability occurs with the spatial growth rate $\Gamma = -\text{Im}(k_\|)$ when $q_\perp^2 < 2k^2 \varepsilon_2 A_0^2$. The maximum spatial growth rate $\Gamma_{\max} = k\varepsilon_2 A_0^2 / 2$ occurs for $q_\perp = q_{\perp\text{opt}} = k\varepsilon_2^{1/2} A_0$, The growth length $\Gamma_{\max}^{-1}$ is about the same as the self-focusing length of a Gaussian beam of radius $2/q_{\perp\text{opt}}$.

In most experiments, the main beam has a finite transverse extent and undergoes self-focusing. However, smaller-scale perturbations usually grow faster than the main beam self-focusing; hence, the beam tends to break up into filaments before reaching the focus. In the nonlinear state, the filament attains a size $\sim c/\omega_p$. The inhomogeneity of the plasma does not suppress filamentation and self-focusing. Nevertheless, it is necessary for the length of the underdense region to exceed the growth length or self-focusing length for these processes to be important.

## 9.6 Discussion

High power laser–plasma experiments with gas jet targets as well as foil targets have long underdense plasma. In this kind of situation, relativistic and ponderomotive self-focusing are commonly present. At relatively lower powers, when plasma is still being formed, a strong nonlinearity in plasma permittivity arises due to tunnel ionization. This phenomenon occurs at laser intensity $I \geq 10^{14}$ W/cm$^2$ when the electric field of the laser exceeds the Coulomb field of the atom as seen by the orbiting electrons.[11] As the laser of a finite spot size (with Gaussian distribution of intensity) propagates through the gas and tunnel ionizes it, an electron density profile with the peak on the axis is created. This electron density causes defocusing of the beam. Defocusing is stronger on the axis than in the off-axis region. Hence, in certain parameters regime, a Gaussian beam acquires a ring shape intensity profile as observed experimentally by Chessa et al.[12] Liu and Tripathi[13] have developed a non-paraxial theory for this process by expanding the eikonal $S$ up to $r^4$ terms in the cylindrical polar coordinate $r$. Filamentation does not occur in cases where the nonlinearity has a tendency for defocusing.

Mishra et al.[14] have developed an elegant analytical treatment of self-focusing in multi-ion species plasmas. The study is relevant to long pulse experiments in CH plasmas and high-Z plasmas where ionization states have a distribution and plasma diffusion is involved. Ahmad et al.[15] have developed a formalism to study the evolution of an on-axis intensity spike on a Gaussian laser beam in a plasma dominated by relativistic and ponderomotive nonlinearities. A single beam width parameter characterizes the evolution of the spike. Parashar et al.[16] have studied defocusing and pulse distortion of short pulse lasers in tunnel-ionized plasmas.

Short pulse experiments are usually accompanied by self-phase modulation. Short pulses also give rise to super continuum generation[17,18] It may be visualized by writing the phase of a laser pulse $\phi = \omega t - kz$. The wave number $k$ depends on the free electron density and relativistic electron mass; hence, it is a function of time, following a pulse shape. The carrier frequency of the pulse is $\omega' = \partial\phi/\partial t = \omega - (\partial k/\partial t)z$. In case of tunnel-ionizing gases, $\partial k/\partial t < 0$; hence, $\omega' > \omega$ and a frequency upshift occurs. This effect is however, suppressed by self-defocusing of the pulse. In the case of relativistic mass nonlinearity, $\partial k/\partial t > 0$ at the front of the pulse while $\partial k/\partial t < 0$ at the rear. Thus, a frequency down-shift occurs at the front and an upshift on the rear. Self-focusing enhances these effects.[18]

## References

1. Akhmanov, Sergei Aleksandrovich, Anatolii Petrovich Sukhorukov, and R. V. Khokhlov. 1968. "Self-focusing and Diffraction of Light in a Nonlinear Medium." *Physics-Uspekhi* 10 (5): 609–36.
2. Sodha, M. S., A. K. Ghatak, and V. K. Tripathi. 1976. "V Self Focusing of Laser Beams in Plasmas and Semiconductors." In *Progress in Optics*, 13: 169–265. Elsevier.
3. Kaw, P., G. Schmidt, and T. Wilcox. 1973. "Filamentation and Trapping of Electromagnetic Radiation in Plasmas." *The Physics of Fluids* 16 (9): 1522–5.

4. Ginzburg, V. L. 1970. *The Propagation of Electromagnetic Waves in Plasma*. 2nd edition. Oxford: Pergamon.

5. Tripathi, V. K., T. Taguchi, and C. S. Liu. 2005. "Plasma Channel Charging by an Intense Short Pulse Laser and Ion Coulomb Explosion." *Physics of Plasmas* 12 (4): 043106.

6. Max, Claire Ellen, Jonathan Arons, and A. Bruce Langdon. 1974. "Self-modulation and Self-focusing of Electromagnetic Waves in Plasmas." *Physical Review Letters* 33 (4): 209.

7. Sun, Guo-Zheng, Edward Ott, Y. C. Lee, and Parvez Guzdar. 1987. "Self-focusing of Short Intense Pulses in Plasmas." *The Physics of fluids* 30 (2): 526–32.

8. Perkins, F. W., and E. J. Valeo. 1974. "Thermal Self-focusing of Electromagnetic Waves in Plasmas." *Physical Review Letters* 32 (22): 1234.

9. Esarey, Eric, Phillip Sprangle, Jonathan Krall, and Antonio Ting. 1997. "Self-focusing and Guiding of Short Laser Pulses in Ionizing Gases and Plasmas." *IEEE Journal of Quantum Electronics* 33 (11): 1879–914.

10. Kaur, Sukhdeep, and A. K. Sharma. 2009. "Self Focusing of a Laser Pulse in Plasma with Periodic Density Ripple." *Laser and Particle Beams* 27 (2): 193–9.

11. Keldysh, L. V. 1965. "Ionization in the Field of a Strong Electromagnetic Wave." *Sov. Phys. JETP* 20 (5): 1307–14.

12. Chessa, P., E. De Wispelaere, F. Dorchies, Victor Malka, J. R. Marques, G. Hamoniaux, P. Mora, and F. Amiranoff. 1999. "Temporal and Angular Resolution of the Ionization-induced Refraction of a Short Laser Pulse in Helium Gas." *Physical Review Letters* 82 (3): 552.

13. Liu, C. S., and V. K. Tripathi. 2000. "Laser Frequency Upshift, Self-defocusing, And Ring Formation in Tunnel Ionizing Gases and Plasmas." *Physics of Plasmas* 7 (11): 4360–4363.

14. Misra, Shikha, S. K. Mishra, and M. S. Sodha. 2013. "Self-focusing of a Gaussian Electromagnetic Beam in a Multi-ions Plasma."*Physics of Plasmas* 20 (10): 103105.

15. Ahmad, N., S. T. Mahmoud, and G. Purohit. 2017. "Growth of Spike in Relativistic Gaussian Laser Beam in a Plasma and its Effect on Third-harmonic Generation." *Laser and Particle Beams* 35 (1): 137–44.

16. Parashar, Jetendra, H. D. Pandey, and V. K. Tripathi. 1997. "Two-dimensional Effects in a Tunnel Ionized Plasma." *Physics of Plasmas* 4 (8): 3040–2.

17. Chin, S. L., A. Brodeur, S. Petit, O. G. Kosareva, and V. P. Kandidov. 1999. "Filamentation and Supercontinuum Generation During the Propagation of Powerful Ultrashort Laser Pulses in Optical Media (White Light Laser)." *Journal of Nonlinear Optical Physics & Materials* 8 (01): 121–46.

18. Liu, Chuan Sheng, and Vipin K. Tripathi. 2001. "Self-focusing and Frequency Broadening of an Intense Short-pulse Laser in Plasmas." *JOSA A* 18 (7): 1714–8.

# 10 Parametric Instabilities

## 10.1 Introduction

Parametric instabilities, where electromagnetic pump waves decay into other waves, is one of the most fascinating subjects of study in laboratory and space plasmas for the last six decades. As the field of laser driven fusion advanced, the issue of parametric instabilities became more and more serious as parametric coupling between the pump laser and the backward scattered electromagnetic wave, mediated by a Langmuir wave or by an ion-acoustic wave could reflect back as much as 30–50% of the laser light energy from the underdense plasmas. The plasma wave accelerates the electrons to 50 keV energy and pre-heats the core, putting in jeopardy the D–T (deuterium–tritium) pellet compression mandatory to achieve fusion. Recent experiments on indirect drive fusion failed to ignite mainly due to the deleterious Raman back scattering parametric instabilities. Thus, we treat this topic at length, starting from the basics.

Lord Rayleigh[1] made the first systematic investigation of parametric instability in 1883. He studied a stretched string attached to a prong of a tuning fork vibrating in the direction of the string. When the frequency of the fork was tuned to twice the natural frequency of the transverse vibration of the string, this vibration was observed to be amplified.

In fact, many mechanical systems, for example, the pendulums and cantilevers, execute oscillations. The frequency of oscillation depends on parameters like the length, mass, Young's modulus, and so on. A conventional way to excite oscillations is to exert an impulse or a periodic force or torque on to the system. When the frequency of the force equals the natural frequency of oscillation, the oscillator acquires a large amplitude.

There is yet another way to excite oscillations – by modulating one of the parameters characterizing the frequency. For instance, the frequency of a simple pendulum depends on its length. If we change the length periodically, we will notice that when the frequency of modulation equals twice the frequency of the pendulum, the oscillation amplitude rises exponentially in time until a nonlinear shift in frequency saturates it. This is parametric instability. In this way a child sitting on a swing makes the swing oscillate. He moves his body up and down resulting in a periodic modulation of the effective length of the swing. When the frequency of the modulation equals twice the natural frequency of the swing, the oscillation amplitude rises with time. In an LC (inductor–capacitor) circuit, one may excite growing current oscillations by modulating the capacitance.

Plasma supports many modes, each characterized by its frequency and wave number. These depend on macroscopic plasma parameters like the density and temperature of the electrons and ions. A large amplitude wave may itself modulate any of these parameters and lead to the excitation of other waves. There is yet another macroscopic parameter of plasma, the drift (average) velocity of the particles. It may not directly appear in the frequency or wave number of the waves but its modulation can excite or amplify waves. For instance, a large amplitude electromagnetic wave propagating through a plasma imparts oscillatory velocity to the electrons, and this velocity modulation is known to excite waves.

In parametric excitation of waves, besides frequency matching, one has to take into account both frequency and wave vector matching. In a three-wave coupling, a wave of frequency $\omega_0$ and wave vector $\vec{k}_0$ can excite a pair of waves of frequency and wave vector $\omega, \vec{k}$ and $\omega_1, \vec{k}_1$ when

$$\omega_0 = \omega + \omega_1, \tag{10.1a}$$

$$\vec{k}_0 = \vec{k} + \vec{k}_1. \tag{10.1b}$$

Quantum mechanically, one may view the parametric instability as a process where a pump wave photon of energy and momentum $\left(\hbar\omega_0, \hbar\vec{k}_0\right)$ produces a pair of photons/plasmons of energy and momentum $\left(\hbar\omega, \hbar\vec{k}\right)$ and $\left(\hbar\omega_1, \hbar\vec{k}_1\right)$. Then the aforementioned equations represent conservation of energy and momentum. These conditions rule out some channels of parametric excitation, for example, an electromagnetic pump wave in an unmagnetized plasma cannot excite a pair of two electromagnetic waves as $\vec{k}$ matching conditions cannot be satisfied due to dispersion. However, it can excite other pairs of waves, such as a plasma wave and an electromagnetic wave (stimulated Raman scattering) or an ion-acoustic wave and an electromagnetic wave (stimulated Brillouin scattering).

The kinetics of parametric instability in plasma is also fascinating. The density perturbation associated with the electrostatic wave beats with the oscillatory velocity of the electrons due to the large amplitude pump wave to produce a nonlinear current that drives the electromagnetic decay wave or the scattered wave. The pump and decay electromagnetic waves exert a beat frequency ponderomotive force on the electrons that drives the electrostatic wave. By this feedback mechanism, amplitudes of both of the decay waves grow exponentially with time till they saturate either by pump depletion or some other nonlinear mechanism.

There also exist another possibility. The low-frequency perturbation, instead of being a resonant eigen-mode, could be a quasi-mode, strongly Landau damped on the electrons or ions. It could be driven as a beat wave by the ponderomotive force due to the pump and the sideband (generated wave). The electron density perturbation associated with this mode beats with the oscillatory velocity of the electrons due to the pump, producing a nonlinear current driving the electromagnetic sideband. The sideband and the quasi-mode grow with time while strongly heating the particles. This process is called collective or stimulated Compton scattering; it is sometimes termed as nonlinear Landau damping. Compton scattering can be viewed as the decay of a pump photon into a lower frequency photon while the balance of energy and momentum is shared by the particle:

$$\hbar\omega_0 = \hbar\omega_1 + \Delta\varepsilon \tag{10.2a}$$

$$\hbar\vec{k}_0 = \hbar\vec{k}_1 + \Delta\vec{p}, \tag{10.2b}$$

where $(\omega_0, \vec{k}_0)$ and $(\omega_1, \vec{k}_1)$ refer to the pump and the daughter wave (sideband) and $\Delta\varepsilon$ and $\Delta\vec{p}$ are the energy and momentum imparted to the particles. For a nonrelativistic particle of mass $m$ and initial velocity $\vec{v}$, $\Delta\varepsilon \approx m\vec{v}\cdot\Delta\vec{v}$, $\Delta\vec{p} = m\,\Delta\vec{v}$, where $\Delta\vec{v}$ is the change in the velocity in the process, Eqs. (10.2a, b) yield

$$\omega_0 - \omega_1 = \left(\vec{k}_0 - \vec{k}_1\right)\cdot\vec{v} \tag{10.2c}$$

Parametric instabilities occupy a special place in plasma research where large amplitude waves are used or encountered.[3–18] Theoretical studies on parametric instabilities in laser–plasma interaction were initiated in the early 1970s; this was followed by a series of experimental studies and simulations in the 1980's. Since then the scales of the experiments have progressed from a single laser beam with kilojoules of energy and nanosecond pulses on plasma with sub-keV temperature and 100 micron scale length to 192 NIF (near infrared) beams of 10 kilojoule lasers with a total energy of 2 megajoule on a millimeter scale plasma of 4 keV. Equally impressive as these scales are the improvements in diagnostic capabilities of Thomson scattering etc. Valuable and highly reproducible experiments were carried out on the trident laser system with nearly homogeneous plasma in the laser hotspot. These developments have improved our understanding of the rich physics of parametric instabilities, and form the underlying foundation for the eventual success of laser fusion.

In Section 10.2, we will study the parametric instability of a simple pendulum, a mechanical harmonic oscillator. We will discuss the electrical LC circuit parametric amplifier in Section 10.3. In Section 10.4, we will study the parametric instabilities driven by large-amplitude electromagnetic and Langmuir waves in unmagnetized plasma and study the three-wave and four-wave processes they give rise to. We will derive a general dispersion relation for three-wave parametric instabilities of an electromagnetic pump wave in Section 10.5. In Section 10.6, we will study resonant scattering processes and build the criterion for the convective and absolute nature of a parametric instability. We will obtain the thresholds and growth rates for stimulated Raman scattering and stimulated Brillouin scattering. In Section 10.7, the non-resonant process of stimulated Compton scattering will be studied. Section 10.8 deals with four-wave parametric processes of the modulational instability, filamentation instability, and oscillating two-stream instability. In Section 10.9, we will study the decay instability of electromagnetic and electrostatic pump waves. Two-plasmon decay is discussed in Section 10.10. Section 10.11 presents the results from experiments on parametric instabilities. A general discussion is provided in Section 10.12.

## 10.2 Parametric Instability of a Pendulum

We begin by describing parametric instabilities occuring in a simple mechanical system, the non-linear pendulum.

Consider a simple pendulum of length $l$ and suspended mass $m$ (Fig. 10.1). At equilibrium, the mass (bob) is located at $x = 0$. It is given an angular displacement $\theta$ in the $x$-$y$ plane. The mass experiences two forces: the tension $T$ along the length toward the point of suspension and $mg$ vertically downward, where $g$ is the acceleration due to gravity. The net force along the (horizontal) $x$-axis is $F_x = -T\sin\theta$. Since $T \approx mg\cos\theta$, one has $F_x = -mgx/l$ for small $\theta$, and the equation of motion for the pendulum is

$$\frac{d^2x}{dt^2} + \frac{g}{l}x = 0. \tag{10.3}$$

For constant $l = l_0$, a solution of this equation is

$$x(t) = Ae^{-i\omega_r t} \tag{10.4}$$

with $\omega_r = \sqrt{g/l_0}$, that is, the pendulum oscillates with the natural frequency $\omega_r$.

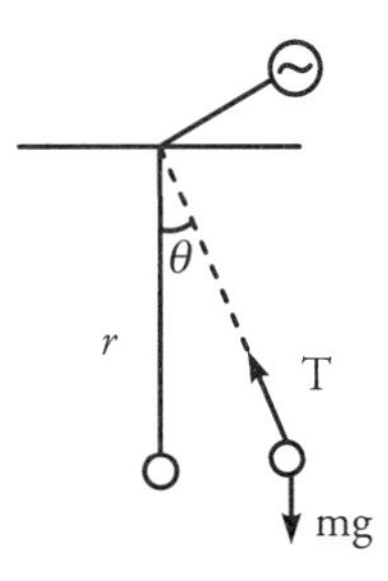


Fig. 10.1 Simple pendulum with modulated length $l = l_0 + l_1\cos(\omega_0 t)$.

If we subject the pendulum to an external force $F = F_0\exp(-i\omega t)$, Eq. (10.3) is modified to

$$\frac{d^2x}{dt^2} + \omega_r^2 x = \frac{F_0}{m}e^{-i\omega t}. \tag{10.5}$$

In the quasi-steady state, $x$ has the same time dependence as the driver, $x = a\exp(-i\omega t)$. Using this in Eq. (10.5), we obtain

$$x = -\frac{Fe^{-i\omega t}}{m\left(\omega^2 - \omega_r^2\right)}. \tag{10.6}$$

The response becomes resonantly large at $\omega = \omega_r$, that is, when the frequency of the applied force equals the natural frequency of the oscillator.

Now we remove the external force and instead modulate the length of the pendulum, $l = l_0 + l_1\cos(\omega_0 t)$ with $l_1/l_0 << 1$. Then Eq. (10.5) is modified as

$$\frac{d^2x}{dt^2} + \omega_r^2\left(1 - \mu\cos\omega_0 t\right)x = 0, \tag{10.7}$$

where $\mu = l_1/l_0$ is the index of modulation. This is Mathieu's equation with unstable solutions in certain regions of the parameter space. We solve Eq. (10.7) by perturbation theory.[2] One expects a solution of the form $x = a\exp(-i\omega t)$, with $\omega$ close to $\omega_r$. However, with this $x$, the last term in Eq. (10.7) has the time dependence proportional to $\exp(-i(\omega \mp \omega_0)t)$. One may view these terms as drivers for the oscillator. The response would be strong when the frequency of the driver is close to $\omega_r$.

### 10.2.1 Parametric instability for $\omega_0 \approx 2\omega_r$

For $\omega_0 \sim 2\omega_r$, $|\omega - \omega_0| \sim \omega_r$; hence, the $\omega - \omega_0$ frequency component will be resonantly excited while the $\omega + \omega_0$ component will be off-resonant and can be ignored. We can then write the solution of Eq. (10.7) as

$$x = a_\omega e^{-i\omega t} + a_{\omega_1}e^{-i\omega_1 t}, \tag{10.8}$$

where $\omega_1 = \omega - \omega_0$. Using this in Eq. (10.7) and separating out the $\omega$ and $\omega_1$ frequency components, we obtain the system

$$\left(\omega^2 - \omega_r^2\right)a_\omega = -\frac{\mu\omega_r^2}{2}a_{\omega_1}, \tag{10.9a}$$

$$\left(\omega_1^2 - \omega_r^2\right)a_{\omega_1} = -\frac{\mu\,\omega_r^2}{2}a_\omega, \tag{10.9b}$$

which has a non-trivial solutions for $a_\omega$ and $a_{\omega 1}$ only when the nonlinear dispersion relation

$$\left(\omega^2 - \omega_r^2\right)\left(\omega_1^2 - \omega_r^2\right) = \frac{\mu^2\omega_r^4}{4} \tag{10.10}$$

is fulfilled. Let us allow a small pump frequency mismatch $\Delta = \omega_0 - 2\,\omega_r$ with $\Delta << \omega_r$. We then solve Eq. (10.8) by writing $\omega = \omega_r + \delta, \omega_1 = \omega_r - \omega_0 + \delta = -\omega_r + \delta - \Delta$ with $\delta << \omega_r$,

$$\delta(\delta - \Delta) = -\frac{\mu^2 \omega_r^2}{16}. \tag{10.11}$$

Equation (10.11) gives

$$\delta = \frac{1}{2}\left[\Delta \pm \sqrt{\Delta^2 - \mu^2 \omega_r^2 / 4}\right]. \tag{10.12}$$

For $\mu\, \omega_r/2 > \Delta$, one obtains an unstable solution with the growth rate

$$\Gamma = \text{Im } \delta = \frac{1}{2}\left(\frac{\mu^2 \omega_r^2}{4} - \Delta^2\right)^{1/2} \tag{10.13}$$

whereas the real frequency is $\omega_0/2$, that is, it is a frequency-locked oscillation, the frequency depending only on $\omega_0$ and not on the natural frequency of the oscillation. The growth rate maximizes to $\Gamma_{max} = \mu\omega_r/4$ for $\Delta = 0$.

If one incorporates linear damping (e.g., due to the viscous drag by the air), Eq. (10.7) modifies to

$$\frac{d^2x}{dt^2} + 2\,\Gamma_d \frac{dx}{dt} + \omega_r^2 \left(1 - \mu \cos \omega_0 t\right) x = 0, \tag{10.14}$$

where $\Gamma_d$ is the linear damping rate of the oscillations. Following the same procedure as earlier, one may solve Eq. (10.14) to obtain the nonlinear dispersion relation,

$$\left(\omega^2 + 2i\,\Gamma_d \omega - \omega_r^2\right)\left(\omega_1^2 + 2i\,\Gamma_d \omega_1 - \omega_r^2\right) = \frac{\mu^2 \omega_r^2}{4} \tag{10.15}$$

Giving $\omega = \omega_r + \delta$,

$$\delta = \frac{1}{2}\left[\Delta + \sqrt{\Delta^2 - \mu^2 \omega_r^2 / 4}\right] - i\Gamma_d, \tag{10.16}$$

$$\Gamma = \text{Im}\, \delta = \frac{1}{2}\left((\mu^2 \omega_r^2 / 4) - \Delta^2\right)^{1/2} - \Gamma_d. \tag{10.17}$$

The growth rate is reduced by the linear damping rate $\Gamma_d$ and also by the pump frequency mismatch $\Delta$. The threshold for the parametric instability is given by

$$\mu^2 \geq 4\left(4\,\Gamma_d^2 + \Delta^2\right) / \omega_r^2. \tag{10.18}$$

If the damping and frequency mismatch are negligible, the instability can be driven by an infinitesimally small modulation.

### 10.2.2 Parametric instability for $\omega_0 \approx \omega_r$

If $\omega_0 \sim \omega_r$, one would excite oscillations around the natural frequency ($\pm\omega_r$). The beat of those oscillations with $\omega_0$ produces a low-frequency oscillation with $\omega << \omega_0$. A second harmonic is also generated but its amplitude is down by a factor 1/3 compared to that of the low-frequency $\omega$ mode; hence, it is neglected. We write

$$x = a_\omega e^{-i\omega t} + a_{\omega 1} e^{-i\omega_1 t} + a_{\omega 2} e^{-i\omega_2 t}, \tag{10.19}$$

where $\omega_1 = \omega - \omega_0$, $\omega_2 = \omega + \omega_0$, and separate out the different frequency components of Eq. (10.7),

$$\left(\omega^2 - \omega_r^2\right) a_\omega = -\frac{\mu\omega_r^2}{2}\left(a_{\omega 1} + a_{\omega 2}\right), \tag{10.20a}$$

$$\left(\omega_1^2 - \omega_r^2\right) a_{\omega 1} = -\frac{\mu\omega_r^2}{2} a_\omega, \tag{10.20b}$$

$$\left(\omega_2^2 - \omega_r^2\right) a_{\omega 2} = -\frac{\mu\omega_r^2}{2} a_\omega. \tag{10.20c}$$

Non-trivial solutions are possible only when fulfilling the nonlinear dispersion relation

$$\omega^2 - \omega_r^2 = -\frac{\mu^2\omega_r^4}{4}\left(\frac{1}{\omega_1^2 - \omega_r^2} + \frac{1}{\omega_2^2 - \omega_r^2}\right). \tag{10.21}$$

Defining $\Delta = \omega_0 - \omega_r$ and taking $\Delta << \omega_r$, we have

$$\omega^2 = \Delta^2 + \mu^2\omega_r\Delta/4 \tag{10.22}$$

with the growth rate $\Gamma = \text{Im}\ \omega = [-\Delta(\Delta + \mu^2\omega_r/4)]^{1/2}$. A growing solution occurs when $\Delta < 0$ and $|\Delta| < \mu^2\omega_r/4$. The growth rate has a maximum

$$\Gamma_{\max} = \mu^2\omega_r/8 \tag{10.23}$$

when $\Delta = -\mu^2\omega_r/8$. Inclusion of the damping yields

$$\Gamma = \left[-\Delta\left(\frac{\mu^2\omega_r}{4} + \Delta\right)\right]^{1/2} - \Gamma_d. \tag{10.24}$$

## 10.3 Parametric Amplifier

Next, we consider a parametric amplifier comprising an LC circuit with nonlinear capacitor. Normally a voltage source of frequency $\omega_0$, when applied to an electrical circuit, produces an electrical current of frequency $\omega_0$. At times, one needs to produce current at a different frequency. For this, one may employ an LC circuit (cf. Fig. 10.2) with one of the capacitors having a capacitance that can be modulated (for example, by applying a voltage of frequency $\omega_0$ across it) to $C = C_{00} + C_{02}\cos(\omega_0 t)$. Such a circuit is called a parametric amplifier.

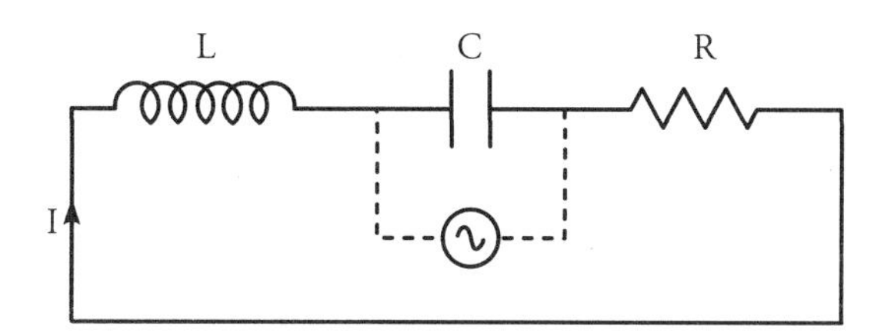


Fig. 10.2 LCR circuit with $C = C_0 + C_{02}e^{-i\omega_0 t}$.

An LC circuit is characterized by a resonant eigen-frequency of the current oscillation. A simple LC circuit, with inductance $L$ and capacitance $C$ has the eigen frequency $\omega_r = 1/\sqrt{LC}$

One may also have an LC circuit with two eigen frequencies, say $\omega_1$ and $\omega_2$. In such a case, one may excite currents at $\omega_1$ and $\omega_2$ at the expense of the voltage source at frequency $\omega_0$ if $\omega_1 + \omega_2 = \omega_0$.

### 10.3.1 Parametric excitation in an LC circuit

Consider an LC circuit (cf. Fig. 10.2) with the resistance $R$ and nonlinear capacitor $C = C_{00} + C_{02}\cos(\omega_0 t)$, where $C_{02}/C_{00} << 1$. Using Kirchoff's law for the circuit, we obtain the equation governing the current $I = dQ/dt$,

$$L\frac{dI}{dt} + RI + \frac{Q}{C_{00} + C_{02}\cos\,\omega_0 t} = 0 \tag{10.25}$$

or

$$\frac{d^2Q}{dt^2} + 2\Gamma_d\frac{dQ}{dt} + \omega_r^2\left(1 - \mu\cos\omega_0 t\right)Q = 0, \tag{10.26}$$

where $Q$ is the electric charge accumulated in the capacitor, $\Gamma_d = R/2L$, $\omega_r = 1/\sqrt{LC_{00}}$, and $\mu = C_{02}/C_{00}$ If we ignore the modulation term, then Eq. (10.26) represents a weakly damped current (or charge) oscillation with frequency $\omega_r$. In the presence of the modulation term, Eq. (10.26) is of the same form as Eq. (10.14). Writing

$$Q = a_\omega e^{-i\omega t} + a_{\omega_1} e^{-i\omega_1 t}, \tag{10.27}$$

where $\omega_1 = \omega - \omega_0$, one deduces the growth rate as Eq. (10.17) for the case of $\omega_0 = 2\omega_r - \Delta$. The current in the circuit can be written as

$$I = \frac{dQ}{dt} = -i\omega e^{\Gamma t}\left(a_\omega e^{i\omega_0 t} - a_{\omega 1} e^{-i\omega_0 t}\right). \tag{10.28}$$

$$a_{\omega_1} = -(4\delta/\mu\omega)a_\omega,$$

The actual current is the real part of this expression. The amplitude of the current grows exponentially with time. The growth rate increases with the index of modulation $\mu$. It requires a threshold value of $\mu = \mu_{th}$. For zero mismatch of the pump ($\omega_0 = 2\omega_r$), we have $\mu_{th} = 4\Gamma_d/\omega_r$.

### 10.3.2 Parametric excitation in two mode LC circuit

Consider a network with two meshes coupled through a huge capacitor $C_0 >> C_1, C_2$ (cf. Fig. 10.3). Denoting the mesh currents $I_1 = dQ_1/dt$ and $I_2 = dQ_2/dt$, the mesh equations can be written as

$$L_1 \frac{d^2 Q_1}{dt^2} + \frac{Q_1}{C_1} + \frac{Q_1 + Q_2}{C_0} = 0, \tag{10.29a}$$

$$L_2 \frac{d^2 Q_2}{dt^2} + \frac{Q_2}{C_2} + \frac{Q_2 + Q_1}{C_0} = 0, \tag{10.29a}$$

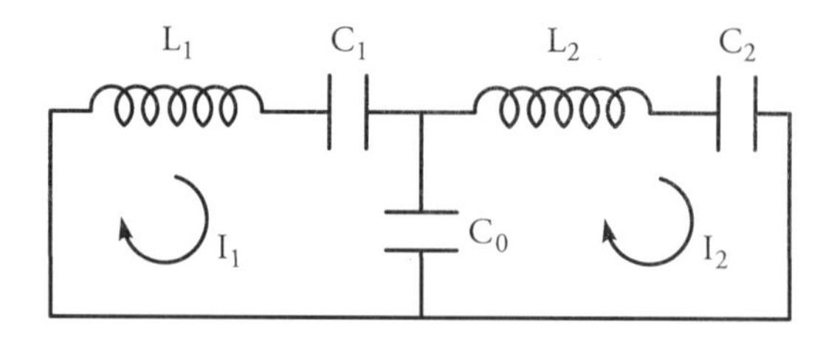


**Fig. 10.3** A network with two meshes coupled through a large capacitor $C_0$ with a modulated capacitance $C_0 = C_{00} + C_{02}\cos \omega_0 t$.

Dividing them by $L_1$ and $L_2$ and taking $C_0 = C_{00} + C_{02}\cos(\omega_0 t)$, we obtain

$$\frac{d^2Q_1}{dt^2}+\frac{C_0+C_1}{L_1C_0C_1}Q_1=\frac{\left[C_{00}-C_{02}\cos(\omega_0 t)\right]}{L_1C_{00}^2}Q_2,$$

$$\frac{d^2Q_2}{dt^2}+\frac{C_0+C_2}{L_2C_0C_2}Q_2=\frac{\left[C_{00}-C_{02}\cos\left(\omega_0 t\right)\right]}{L_2C_{00}^2}Q_1.$$

The last terms on the right-hand side of the equation couple the eigen modes to the modulation source and drive them unstable. We keep them and ignore the other terms that contain factors such as $C_1/C_0$ and $C_2/C_0$. Then the aforementioned equations can be written as

$$\frac{d^2Q_1}{dt^2}+\omega_{R1}^2Q_1=-\frac{C_{02}}{C_{00}^2L_1}\,Q_2\cos\,\left(\omega_0 t\right),\tag{10.30a}$$

$$\frac{d^2Q_2}{dt^2}+\omega_{R2}^2Q_2=-\frac{C_{02}}{C_{00}^2L_2}Q_1\cos\left(\omega_0 t\right),\tag{10.30b}$$

where $\omega_{R1}=1/\sqrt{L_1C_1}$, $\omega_{R2}=1/\sqrt{L_2C_2}$. We note that $Q_1$ and $Q_2$ have the natural frequencies $\omega_{R1}$ and $\omega_{R2}$, respectively. The driver drives them resonantly when $\omega_0 = \omega_{R1} + \omega_{R2}$. We write the solutions of Eqs. (10.29a,b) as

$$Q_1=a_{\omega 1}e^{-i\omega_1 t}$$

$$Q_2=a_{\omega 2}e^{-i\omega_2 t}$$

where $\omega_2 = \omega_1 - \omega_0$. Substituting $Q_1$ and $Q_2$ in Eqs. (10.30a,b) and collecting the coefficients of $e^{-i\omega_1 t}$ and $e^{-i\omega_2 t}$ gives, respectively,

$$\left(\omega_1^2-\omega_{R1}^2\right)\,a_{\omega 1}+\frac{C_{02}}{2\,C_{00}^2\,L_1}\,a_{\omega 2}.\tag{10.31a}$$

$$\left(\omega_2^2-\omega_{R2}^2\right)a_{\omega 2}+\frac{C_{02}}{2\,C_{00}^2\,L_2}a_{\omega 1}.\tag{10.31b}$$

They give the dispersion relation as

$$\left(\omega_1^2-\omega_{R1}^2\right)\left(\omega_2^2-\omega_{R2}^2\right)=\frac{C_{02}^2}{4\,C_{00}^4L_1L_2}.$$

Writing $\omega_1 = \omega_{R1} + \delta$ and $\omega_2 = \omega_1 - \omega_0 = -\omega_{R2} + \delta$, and taking $\delta << \omega_1, \omega_2,$ we obtain

$$\delta^2 = -\frac{C_{02}^2}{16C_{00}^4 L_1 L_2 \omega_{R1}\omega_{R2}} \tag{10.32}$$

with the growth rate

$$\Gamma = I_m(\delta) = \frac{C_{02}}{4\ C_{00}} \frac{\sqrt{C_1 C_2}}{C_{00}} . (\omega_{R1}\omega_{R2})^{1/2}. \tag{10.33}$$

Had we included resistors in the meshes of the electrical circuits, there would be a linear damping of the modes. The instability grows when

$$\Gamma > \Gamma_{th} = (\Gamma_1 \Gamma_2)^{1/2}. \tag{10.34}$$

Here, $\Gamma_1$ is the damping rate of the circuit at frequency $\omega_1$. $\Gamma_2$ is the damping rate at frequency $\omega_2$.

## 10.4 Parametric Instabilities in Plasmas

An unmagnetized plasma supports three modes of wave propagation, viz., the electromagnetic waves, the Langmuir waves, and the ion-acoustic waves. In many experiments, a large amplitude electromagnetic wave is launched from free space and is allowed to propagate into the plasma. The channels of three-wave resonant decay of electromagnetic pump wave that are kinematically allowed are as follows (cf., Fig. 10.4):

*(i)* ***Stimulated Raman scattering (SRS),*** producing a Langmuir wave and a scattered electromagnetic wave (called a sideband) at densities below one-fourth the critical density $n_0^0 \leq n_{cr}/4$ ($n_{cr}$ is the critical density corresponding to where the plasma frequency is equal to the pump frequency).

*(ii)* ***Stimulated Brillouin scattering (SBS),*** producing an ion-acoustic wave and an electromagnetic wave at densities up to the critical density, $n_0^0 \leq n_{cr}$.

*(iii)* ***Stimulated Compton scattering (SCS),*** driving a heavily Landau damped space charge mode (quasi-mode) and a scattered electromagnetic wave.

*(iv)* ***Decay instability,*** producing an ion-acoustic wave and a Langmuir wave sideband near the critical layer.

*(v)* ***Two-plasmon decay (TPD),*** producing two Langmuir waves near one-fourth the critical density $n_0^0 \sim n_{cr}/4$.

Decay into two electromagnetic waves is not allowed because the pump wave number is always greater than the sum of the wave numbers of the decay waves, $|k_0| > |k| + |k_1|$; hence, the wave vector matching condition cannot be satisfied.

For a large amplitude Langmuir wave pump, the three-wave decay channels are as follows:

(i) **Resonant decay into an ion-acoustic wave and a *Langmuir wave*,** producing a longer wavelength Langmuir sideband (inverse cascade to smaller wave numbers).

(ii) ***Nonlinear Landau damping on ions***, driving a heavily Landau damped ion space charge quasi-mode and electrostatic sideband.

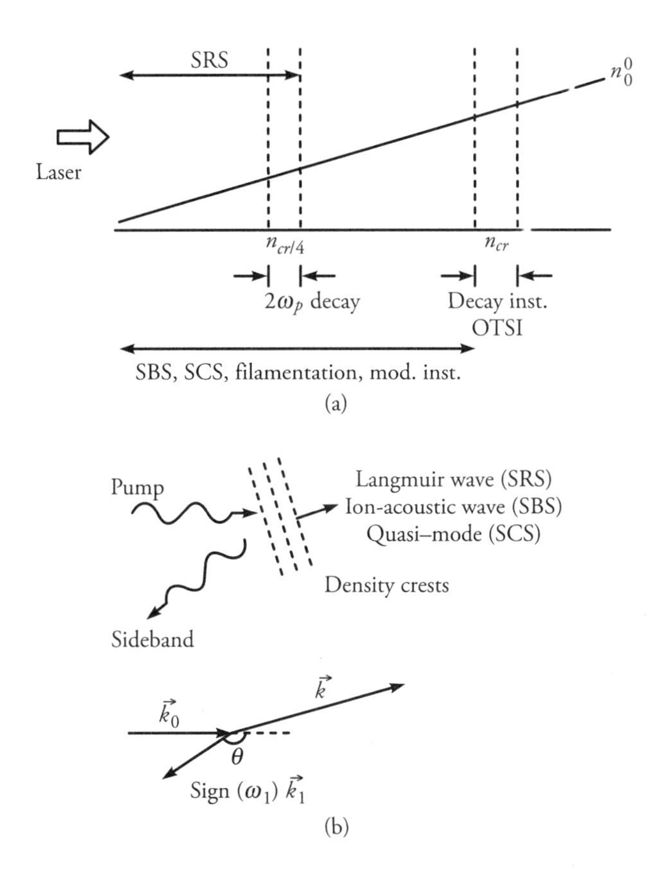


Fig. 10.4 (a) Location of various parametric processes in an inhomogeneous plasma. (b) Schematic of the scattering process.

Decay channels involving electromagnetic sidebands are relatively less important for Langmuir wave pumps because they suffer from larger convection losses.

In the aforementioned analysis, we have ignored the upper sideband $\omega_2 = \omega_0 + \omega$, $\vec{k}_2 = \vec{k}_0 + \vec{k}$, which is justified as long as it is off-resonant, that is, $\omega_2$ differs significantly from the natural frequency of the electromagnetic wave of wave vector $\vec{k}_2$, $\omega_{2\text{res}} = (\omega_p^2 + |\vec{k}_0 + \vec{k}|^2 c^2)^{1/2}$ by more than the growth rate of the parametric instability. The possibility of an upper sideband being resonant and the lower sideband off-resonant is discarded here as it is not kinematically allowed in a plasma without dc drifts. A photon cannot produce a photon of higher energy unless the low frequency mode is a negative energy mode.

However, there are processes where both the sidebands are important, leading to four-wave processes:

*(i)* ***Modulational Instability***

When the upper and lower sidebands are close to the pump frequency and both almost lie on the dispersion curve of the high-frequency wave, the perturbation (not a normal mode) propagates with phase velocity close to the group velocity of the pump wave. A small modulation in the amplitude of the pump wave builds up as the pump propagates. The instability involves the pump wave $(\omega_0, \vec{k}_0)$, two sidebands $(\omega_0 \pm \omega, \vec{k}_0 \pm \vec{k})$, and a low-frequency, non-resonant, driven mode $(\omega, \vec{k})$

*(ii)* ***Filamentation Instability***

When the wave vector of the density perturbation $\vec{k} \perp \vec{k}_0$ and $k << k_{0,}$ the density perturbation tends to converge the electromagnetic energy around the density minima where the refractive index has maxima. The ponderomotive force of the enhanced wave intensity causes an increased density depression leading to a purely growing density perturbation in space and breakup of the pump wave into filaments.

*(iii)* ***Oscillating Two-Stream Instability***

In this process, a long wavelength pump wave near the critical layer excites a short wavelength standing Langmuir wave and a purely growing density perturbation. In the regions where the electric fields of the pump and Langmuir waves are parallel, the plasma is pushed away to the regions where the fields are antiparallel. The depressed density regions attract electric field energy from the neighborhood leading to deeper density depressions and enhancement of the short wavelength Langmuir mode. The result is strong Langmuir turbulence involving collapse of the density depressions and efficient acceleration of electrons.

The parametric instabilities, for example, SRS and TPD, that excite Langmuir waves, also produce hot electrons[19–22] and are a matter of concern for laser driven fusion. The SRS instability imposes serious constraints on ignition designs in indirect drive fusion because of the large volume of quasi-homogeneous low density plasma in gas filled hohlraums.[26] Direct drive implosions, involving shorter density scale lengths are effected by TPD.[27] Rosenberg et al.[28] have recently reported that there are up to 2.9% laser energy conversions to hot electrons through the SRS and TPD processes in ignition scale direct drive fusion experiments. Oscillating two-stream instability (OTSI) is a potential source of hot electrons near the critical density.

## 10.5 Dispersion Relation for the Parametric Instability with Electromagnetic Pump Wave

Consider a plasma of equilibrium electron density $n_0^0$, electron temperature $T_e$, and ion temperature $T_i$. An electromagnetic pump wave propagates through the plasma with electric and magnetic fields

$$\vec{E}_0 = \vec{A}_0 \; e^{-i(\omega_0 t - k_0 z)} \tag{10.35a}$$

$$\vec{B}_0 = \vec{k}_0 \times \vec{E}_0 \,/\, \omega_0. \tag{10.35b}$$

The electric field imparts an oscillatory velocity to the electrons,

$$\vec{v}_0 = \frac{e\,\vec{E}_0}{mi\omega_0}. \tag{10.36}$$

It also excites a pair of waves – a low-frequency electrostatic wave of potential

$$\phi = Ae^{-i\left(\omega t - \vec{k}\cdot\vec{r}\right)} \tag{10.37}$$

and an electromagnetic sideband wave,

$$\vec{E}_1 = \vec{A}_1 e^{-i\left(\omega_1 t - \vec{k}_1\cdot\vec{r}\right)} \tag{10.38a}$$

$$\vec{B}_1 = \vec{k}_1 \times \vec{E}_1 \,/\, \omega_1, \tag{10.38b}$$

where $\vec{k}_1 = \vec{k}_0 - \vec{k}$ $\omega_1 = \omega_0 - \omega^*$ and the asterisk denotes the complex conjugate. The sideband is a resonant mode, approximately satisfying the dispersion relation $\omega_1^2 = \omega_p^2 + k_1^2 c^2$.

As a linear response, the sideband imparts an oscillatory velocity to electrons,

$$\vec{v}_1 = \frac{e\,\vec{E}_1}{mi\omega_1} \tag{10.39}$$

and the electrostatic wave gives rise to electron and ion density perturbations (cf. Eq. 2.54)

$$n^L = \frac{k^2\varepsilon_0}{e}\chi_e\phi \tag{10.40a}$$

$$n_i^L = -\frac{k^2\varepsilon_0}{e}\chi_i\phi, \tag{10.40b}$$

where $\chi_e$ and $\chi_I$ are the electron and ion susceptibilities at $\omega$, $\vec{k}$, and we have assumed the ions to be singly ionized.

The pump and sideband waves exert a low-frequency ponderomotive force on the electrons,

$$\vec{F}_{p\omega} = -\frac{m}{2}\left(\vec{v}_0 \cdot \nabla \vec{v}_1^* + \vec{v}_1^* \cdot \nabla \vec{v}_0\right) - \frac{e}{2}\left(\vec{v}_0 \times \vec{B}_1^* + \vec{v}_1 \cdot \vec{B}_0\right),$$

which on using the identity $\nabla(\vec{A}\cdot\vec{B}) \equiv \vec{A}\cdot\nabla\vec{B} + \vec{B}\cdot\nabla\vec{A} + \vec{A}\times(\nabla\times\vec{B}) + \vec{B}\times(\nabla\times\vec{A})$, can be written as $\vec{F}_{p\omega} = e\nabla\phi_{p\omega}$, with

$$\phi_{p\omega} = -\frac{m}{2e}\vec{v}_0.\vec{v}_1^* = -\frac{e\ \vec{E}_0.\ \vec{E}_1^*}{2\ m\ \omega_0\omega_1^*}. \tag{10.41}$$

The ponderomotive potential $\phi_{p\omega}$ gives rise to a nonlinear electron density perturbation $n^{NL}$, obtained from Eq. (10.40a) with $\phi$ replaced by $\phi_{p\omega}$. The net electron density perturbation at $\omega$, $k$ is

$$n = \frac{k^2\varepsilon_0}{e}\chi_e\left(\phi + \phi_{p\omega}\right). \tag{10.42}$$

The ponderomotive force on the ions is smaller than that on the electrons by a factor of the ion–electron mass ratio; and hence, their contribution to the nonlinear density perturbation can be ignored, that is, $n_i \approx n_i^L$. Using the electron and ion density perturbations in Poisson's equation, $\nabla^2\phi = e(n - n_i)/\varepsilon_0$, and replacing $\nabla^2$ by $-k^2$, we get

$$\varepsilon\phi = -\chi_e\phi_{p\omega} = \chi_e\frac{e\vec{E}_0.\vec{E}_1^*}{2m\omega_0\omega_1^*}, \tag{10.43}$$

where $\varepsilon = 1 + \chi_e + \chi_i$ and the ponderomotive force appears as the driver for the potential $\phi$ of the electrostatic wave. Equation (10.43) relates $\phi$ to the electric field of the scattered wave.

The current density at the sideband frequency can be written as $\vec{J}_1 = \vec{J}_1^L + \vec{J}_1^{NL}$, where

$$\vec{J}_1^L = -en_0^0\vec{v}_1 = -\frac{n_0^0 e^2}{mi\omega_1}\vec{E}_1, \tag{10.44}$$

$$\vec{J}_1^{NL} = -e\ n^*\vec{v}_0 / 2 = k^2\varepsilon_0\left(1+\chi_i^*\right)\frac{e\ \vec{E}_0\phi^*}{2\ mi\omega_0}, \tag{10.45}$$

and where we have used Eq. (10.42) and written $\chi_e(\phi + \phi_{p\omega}) = -(1 + \chi_i)\phi$ according to Eq. (10.43). Using $\vec{J}_1$ in the wave equation,

$$\nabla^2 \vec{E}_1 - \nabla\left(\nabla.\vec{E}_1\right) + \frac{\omega_1^2}{c^2}\vec{E}_1 = -\frac{i\omega_1}{c^2\varepsilon_0}\vec{J}_1 \tag{10.46}$$

and replacing $\nabla$ by $i\vec{k}_1$, we obtain

$$D_1\vec{E}_1 + c^2\vec{k}_1\left(\vec{k}_1.\vec{E}_1\right) = -\frac{k^2\left(1+\chi_i^*\right)\omega_1}{2m\omega_0}\, e\vec{E}_0\phi^* \tag{10.47}$$

where $D_1 = \omega_1^2 - \omega_p^2 - k_1^2c^2$. Taking the dot product of Eq. (10.47) with $\vec{k}_1$, we obtain the electrostatic contribution to the nonlinear term of the electromagnetic wave as

$$\vec{k}_1.\vec{E}_1 = -\frac{k^2\left(1+\chi_i^*\right)\omega_1 e\phi^*}{2\, m\omega_0\left(\omega_1^2 - \omega_p^2\right)}\vec{k}_1.\vec{E}_0. \tag{10.48}$$

Using this in Eq. (10.47) and taking $\omega_1^2 - \omega_p^2 \approx k_1^2c^2$, we obtain

$$D_1\vec{E}_1 = -\frac{k^2\left(1+\chi_i^*\right)\omega_1}{2m\omega_0}\left(\vec{E}_0 - \frac{c^2\vec{k}_1}{\omega_1^2 - \omega_p^2}\vec{k}_1.\vec{E}_0\right)e\phi^*, \tag{10.49}$$

relating the field of the scattered electromagnetic wave to the potential of the electrostatic wave. Except near the quarter critical density (which corresponds to the pump frequency), $\omega_1^2 - \omega_p^2$, the denominator of the last term, can be replaced by $k_1^2c^2$. Equations (10.43) and (10.49) are the coupled equations for $\phi$ and $\vec{E}_1$. Substituting for $\vec{E}_1$ from Eq. (10.49) into Eq. (10.43), we obtain the nonlinear dispersion relation, coupling the electrostatic mode ($\varepsilon \approx 0$) and the electromagnetic mode ($D_1 \approx 0$),

$$\varepsilon D_i^* = \mu_c, \tag{10.50}$$

where $\varepsilon$ is the electrostatic dielectric function and $D_1$ is the electromagnetic dispersion function. The coupling coefficient is

$$\mu_c = -\chi_e\left(1+\chi_i\right)\frac{k^2\left|v_0\right|^2}{4}\left(1 - \frac{c^2k_1^2}{\omega_1^2 - \omega_p^2}\cos^2\delta_1\right), \tag{10.51}$$

and $\delta_1$ is the angle between $\vec{E}_0$ and $\vec{k}_1$. The coupling coefficient $\mu_c$ is maximum for $\delta_1 = \pi/2$, that is, when the scattered (sideband) wave propagates perpendicular to the electric field of the pump wave. Equation (10.50) is the general dispersion relation for stimulated Raman, Compton, and Brillouin scattering processes. A dispersion relation of the same form is recovered from kinetic theory, where the kinetic effects are contained in the electron and ion susceptibilities. The sideband electromagnetic wave is not affected by kinetic effects. However, it may suffer linear damping due to collisions. In this case, $D_1$ is modified to

$$D_1 = \omega_1^2 - \omega_p^2\left(1 - i\nu / \omega_1\right) - k_1^2 c^2, \tag{10.52}$$

where $\nu$ is the electron–ion collision frequency.

## 10.6 Resonant Scattering

Equation (10.50), in the absence of the pump ($v_0 \rightarrow 0$), yields $D_1 = 0$ and/or $\varepsilon = 0$, giving the dispersion relation of linear eigen modes of the plasma. In a resonant three-wave parametric process, we look for a solution to the nonlinear dispersion relation near the simultaneous zeros of $\varepsilon$ and $D_1$. Let $\varepsilon = \varepsilon_r + i\varepsilon_i$ and $\omega = \omega_r - i\Gamma_d$ be the root of $\varepsilon(\omega) = 0$ with $\varepsilon_r(\omega_r) = 0, \Gamma_d = \varepsilon_i/(\partial\varepsilon_r/\partial\omega_r)$. Similarly, $\omega_1 = \omega_{1r} - i\Gamma_{d1}$ is the root of $D_1(\omega_1) = 0$ with $\omega_{1r} = \left(\omega_p^2 + k_1^2 c^2\right)^{1/2}$, $\Gamma_{d1} = \nu\, \omega_p^2 / 2\omega_1^2$. In the presence of the pump, we express $\omega = \omega_r + i\Gamma$ and $\omega_1^* = \omega_0 - \omega = \omega_{1r} - i\Gamma$. Then, $\varepsilon = i(\Gamma + \Gamma_d)\partial\varepsilon_r/\partial\omega_r$, $D_1^* = -2i\left(\Gamma + \Gamma_{d1}\right)\omega_{1r}$ and Eq. (10.50) yields

$$\left(\Gamma + \Gamma_d\right)\left(\Gamma + \Gamma_{d1}\right) = \Gamma_M^2 \equiv \frac{\chi_e^2 k^2 v_0^2 \sin^2\delta_1}{8\left(\omega_0 - \omega\right)\partial\varepsilon / \partial\omega}, \tag{10.53}$$

with the solution

$$\Gamma = \frac{1}{2}\left[-\Gamma_d - \Gamma_{d1} + \sqrt{4\Gamma_M^2 + (\Gamma_d - \Gamma_{d1})^2}\,\right], \tag{10.54}$$

where we have dropped the subscript $r$ from the frequency for the sake of brevity. Setting $\Gamma = 0$, one obtains the threshold for the parametric instability

$$\Gamma_M^2 = \Gamma_d \Gamma_{d1}. \tag{10.55}$$

Above the threshold, the growth rate $\Gamma$ is proportional to $v_0^2$ at modest pump power (when $\Gamma_M$ lies between $\Gamma_d$ and $\Gamma_{1d}$); whereas at large powers ($\Gamma_M^2 >> \Gamma_d, \Gamma_{1d}$), $\Gamma = \Gamma_M$ and it is proportional to $v_0$.

### 10.6.1 Absolute and convective nature of parametric instability

The region of parametric interaction is usually limited either by an inhomogeneity (cf. Chapter 11), the finite size of the plasma, or a finite extent of the pump. In these cases, the parametric instability could be convective, where decay waves get spatially amplified as they pass through the interaction region, or absolute, where the waves grow exponentially with time at a fixed point in space. We may give an initial value to the amplitude of one of the decay waves at some point in space, say $z = 0$, and examine its long-time behavior. If the amplitude at $z = 0$ grows exponentially with time, the instability is absolute, otherwise it is convective.

To examine this issue, we will follow the treatment of Nishikawa and Liu,[2] limiting ourselves to one dimension where the interaction region is localized in $z$ and the wave fields vary with $z$. We will allow the amplitude of the decay waves to possess a slow space–time dependence, $A(z, t)$, $A_1(z, t)$. From Eq. (10.43) one may deduce the equation governing $A(z, t)$ by writing $\omega = \omega + i\partial/\partial t$, $k = k - i\partial/\partial z$ in $\varepsilon = \varepsilon_r + i\varepsilon_i$, expanding $\varepsilon_r$ around its zero,

$$\varepsilon(\omega,k) = i(\partial\varepsilon_r/\partial\omega)\partial/\partial t - i(\partial\varepsilon_r/\partial k)\partial/\partial z + i\varepsilon_i ,$$

and operating $\partial/\partial t$, $\partial/\partial z$ on the amplitude of the wave,

$$\frac{\partial A}{\partial t} + v_g \frac{\partial A}{\partial z} + \Gamma_d A = -i\chi_e \frac{|v_0|}{2(\partial\varepsilon/\partial\omega)\omega_1} A_1^*, \tag{10.56}$$

where $v_g = \partial\omega/\partial k$ is the group velocity of the electrostatic wave. Similarly, one may write $\omega_1 = \omega_1 + i\partial/\partial t$, $k_1 = k_1 - i\partial/\partial z$ and expand $D_1(\omega_1, k_1)$ in Eq. (10.49) to obtain

$$\frac{\partial A_1}{\partial t} + v_{g1} \frac{\partial A_1}{\partial z} + \Gamma_{d1} A_1 = -i\chi_e \frac{k^2}{4}|v_0| A^*, \tag{10.57}$$

where $v_{g1} = c^2k_1/\omega_1$ is the group velocity of the scattered wave. Introducing the normalized amplitudes, $a = kA(\varepsilon_0(\partial\varepsilon/\partial\omega)/4\hbar)^{1/2}$, $a_1 = A_1(\varepsilon_0/2\hbar\omega_1)^{1/2}$ (such that $|a|^2$ represents the number density of plasmons (phonons), called action, and $|a_1|^2$ the density of scattered photons), these equations can be written as

$$\frac{\partial a}{\partial t} + v_g \frac{\partial a}{\partial z} + \Gamma_d a = i\Gamma_M a_1^*, \tag{10.58a}$$

$$\frac{\partial a_1}{\partial t} + v_{g1} \frac{\partial a_1}{\partial z} + \Gamma_{d1} a_1 = i\Gamma_M a^*, \tag{10.58b}$$

We take the initial amplitude of the scattered wave as

$$a_1 = a_{10}\delta(z) \text{ at } t = 0$$

and look at the asymptotic behavior of $a_1$ at $z = 0$ with time. Taking the Laplace-Fourier transforms of Eqs. (10.58a), (10.58b), with

$$a^*_{p,q} = \int_0^\infty dt \int_{-\infty}^\infty dz a^* e^{-pt-iqz},$$

$$a_{1p,q} = \int_0^\infty dt \int_{-\infty}^\infty dz a_1 e^{-pt-iqz},$$

we obtain

$$(p + iqv_g + \Gamma_d)a^*_{p,q} = -i\Gamma_M a_{1p,q},$$

$$(p + iqv_{g1} + \Gamma_{d1})a_{1p,q} = i\Gamma_M a^*_{p,q} + a_{10}$$

leading to

$$a_1 = \frac{a_{10}}{(2\pi)^2}\int_{-\infty}^{\infty} dq \int_{\sigma-i\infty}^{\sigma+i\infty} dp a^* \frac{p + iqv_g + \Gamma_d}{D(p,q)} e^{pt+iqz}, \tag{10.59}$$

where

$$D(p,q) = (p + iqv_g + \Gamma_d)(p + iqv_{g1} + \Gamma_{d1}) - \Gamma_M^2 \tag{10.60}$$

$\sigma$ is so chosen that all the poles of the integrand lie on the left of the $p$ integration path. The long time behavior of $a_1$ is governed by the contribution of the poles, determined by

$$p(q) = \frac{1}{2}[-(iq(v_g + v_{g1}) + \Gamma_d + \Gamma_{1d}) \pm \{4\Gamma_M^2 + (\Gamma_d - \Gamma_{d1} + iq(v_g - v_{g1}))^2\}^{1/2}]$$

For $q|v_g - v_{g1}| >> \Gamma_d, \Gamma_{d1}, \Gamma_M$, these roots are approximately

$$p(q) = -iqv_g, -iqv_{g1}. \tag{10.61}$$

For $v_g v_{g1} > 0$ (forward scattering), one can deform the $q$ integration path in such a way as to bring all the poles to the left of the imaginary $p$-axis, that is, Re $p(q) < 0$ for all the poles. Then the contribution of all poles in the $p$ integration vanishes at large $t$. Thus, in this case, the parametric instability is convective.

This procedure does not work when $v_g v_{g1} < 0$ (back scatter). In this case, we first carry out the $p$ integration by picking the contributions of the poles. Of the two poles, the one lying on the right in the $p$-plane governs the long time behavior. Denoting this pole as $p = p'(q)$, we obtain

$$a_1(t,0) = \frac{a_{10}}{2\pi}\int_{-\infty}^{\infty} dq \frac{p'(q) + iqv_g + \Gamma_d}{(\partial D / \partial p)_{p=p'(q)}} e^{p'(q)t}. \tag{10.62}$$

The $q$ integral can be evaluated using the the saddle point method. The saddle point at $t \to \infty$ is determined by $\partial p'(q)/\partial q = 0$, which yields the saddle point location at

$$q_s = i\left[\frac{\Gamma_d - \Gamma_{d1}}{v_g - v_{g1}} - \frac{v_g - v_{g1}}{|v_g - v_{g1}|}\frac{\Gamma_M}{\sqrt{|v_g v_{g1}|}}\right],$$

$$p'(q_s) = 0 \text{ when } \Gamma_m = \frac{(\Gamma_d v_g - \Gamma_{d1} v_{g1})|v_g - v_{g1}|}{2(v_g - v_{g1})\sqrt{|v_g v_{g1}|}} = \frac{1}{2}\sqrt{|v_g v_{g1}|}\left(\frac{\Gamma_d}{v_{g1}} - \frac{\Gamma_{d1}}{v_g}\right)\operatorname{sgn}(v_g v_{g1}). \tag{10.63}$$

The amplitude at $z = 0$ then is proportional to $e^{\Gamma t}$. The growth occurs when

$$\Gamma_M > \frac{1}{2}\sqrt{|v_g v_{g1}|}\left(\frac{\Gamma_d}{|v_{g1}|} + \frac{\Gamma_{d1}}{|v_g|}\right). \tag{10.64}$$

This gives the condition for absolute instability. When the condition is not satisfied, the instability is convective as $\Gamma$ is –ve and $a_1(t, 0)$ damps out with time as $t \to \infty$.

## 10.6.2 Convective amplification

For convective instability, Eqs. (10.58a) and (10.58b), on taking $\partial/\partial t = 0$, $\Gamma_d > v_g \partial/\partial z$, lead to

$$v_{g1}\frac{\partial a_1}{\partial z} = \left(\frac{\Gamma_M^2}{\Gamma_d} - \Gamma_{d1}\right)a_1,$$

giving the amplitude of the scattered wave after passing through the interaction region of length $L$, in terms of its amplitude prior to entering the region, $a_{10}$ (noise level), as

$$a_1 = a_{10} e^G \tag{10.65}$$

with the amplification factor

$$G = \left( \frac{\Gamma_M^2}{\Gamma_d} - \Gamma_{d1} \right) L / |v_{g1}| \approx \frac{\Gamma_M^2 L}{\Gamma_d |v_{g1}|}. \tag{10.66}$$

Amplification demands that $\Gamma_M^2 > \Gamma_d \Gamma_{d1}$. Further, the threshold for convective amplification may be defined as $G \geq 1$, giving

$$\Gamma_{M\text{th}}^2 = (\Gamma_{d1} + |v_{g1}| / L) \Gamma_d. \tag{10.67}$$

### 10.6.3 Growth of the absolute instability

When the threshold for absolute instability is exceeded, we may deduce the equation for scattered wave amplitude from Eqs. (10.58a) and (10.58b), taking $v_{g1}$ to be –ve with $|v_{g1}| >> v_g$ (valid for stimulated Raman and Brillouin scattering), linear damping of the scattered wave to be negligible, and time variation of $a^*$, $a_1$ as $e^{-i\Omega t}$,

$$\frac{\partial^2 a_1}{\partial z^2} + 2\alpha \frac{\partial a_1}{\partial z} + \frac{\Gamma_M^2 + i\Omega(\Gamma_d - i\Omega)}{v_g |v_{g1}|} a_1 = 0 \tag{10.68}$$

where $\alpha = (\Gamma_d - i\Omega)/2v_g + i\Omega/2|v_{g1}|$. Writing $a_1 = \psi e^{-\alpha z}$, the aforementioned equation reduces to

$$\frac{\partial^2 \psi_1}{\partial z^2} + q^2 \psi_1 = 0, \tag{10.69}$$

where $q^2 = [\Gamma_M^2 + i\Omega(\Gamma_d - i\Omega)] / v_g |v_{g1}|] - \alpha^2$. We can obtain a similar equation for the amplitude of the electrostatic wave. For the interaction region localized between $z = 0$ and $L$, the quantization condition can be taken as $qL = \pi$. Writing $\Omega = i\Gamma$ in the expression for $q^2$, this condition gives the growth rate as

$$\Gamma = (\Gamma_M^2 v_g / |v_{g1}| - \pi^2 v_g^2 / L^2)^{1/2} - \Gamma_d. \tag{10.70}$$

For a parabolic profile of the pump,

$$\Gamma_M^2 = \Gamma_{M0}^2(1 - z^2 / L^2), \tag{10.71}$$

Equation (10. 69) takes the form

$$\frac{\partial^2 \psi_1}{\partial \xi^2} + (\lambda - \xi^2)\psi_1 = 0,$$

giving the eigenfunctions and eigenvalues

$$\psi_1 = C_1 H_n(\xi) e^{-\xi^2/2}, \lambda = 2n + 1, n = 0, 1, 2, \ldots,$$

where

$$l = \left[ L\sqrt{v_g |v_{g1}|} / \Gamma_{M0} \right]^{1/2}, \ \xi = z / l, \ \lambda = \left( \frac{\Gamma_{M0}^2}{v_g |v_{g1}|} - \frac{(\Gamma_d - i\Omega)^2}{4v_g^2} \right) l^2$$

and $H_n(\xi)$ are Hermite polynomials. For the fundamental mode, the eigenvalue condition gives the growth-rate as

$$\Gamma = 2\Gamma_{M0} \left( \frac{v_g}{|v_{g1}|} \right)^{1/2} \left( 1 - \frac{\sqrt{v_g |v_{g1}|}}{L\Gamma_{M0}} \right)^{1/2} - \Gamma_d \tag{10.72}$$

with the threshold value of $\Gamma_{M0}$,

$$\Gamma_{M0\text{th}} = \frac{\sqrt{v_g |v_{g1}|}}{2L} [1 + (1 + \Gamma_d^2 L^2 / v_g^2)^{1/2}].$$

The threshold increases with the reduction of size of the interaction region and rise in linear damping rate of the electrostatic wave. Above the threshold, the growth rate is less than $\Gamma_{M0}$ (the peak value of the uniform medium growth rate with uniform pump amplitude) by a factor two times the square root of the group velocity ratio of the electrostatic wave and the scattered wave. For SRS, in a plasma of 1 keV electron temperature, $\Gamma \approx \Gamma_{M0}/2$.

It may be worthwhile to mention here that one important scenario of the saturation of parametric instabilities is by pump depletion. As the decay waves grow at the expense of pump energy, the pump amplitude must fall to maintain energy conservation. The decay waves couple with each other to produce a nonlinear current at the pump frequency. Using this current in the wave equation, one obtains the equation governing the normalized pump amplitude $a_0 = A_0(\varepsilon_0 / 2\hbar\omega_0)^{1/2}$, similar to Eq. (10.58b),

$$\frac{\partial a_0}{\partial t} + v_{g0}\frac{\partial a_0}{\partial z} + \Gamma_{d0}a_0 = -i\Gamma'_M a a_1$$

where $\Gamma'_M \equiv [\chi_e^2 k^2 \hbar e^2 / (4m^2\omega_1\omega_0\varepsilon_0\partial\varepsilon / \partial\omega)]^{1/2}$ and $\Gamma_{d0}$ and $v_{g0}$ are the linear damping rate and the group velocity of the pump wave. One may recognize that $\Gamma_M = \Gamma'_M a_0$. Equations (10.58a) and (10.58b), in conjunction with the aforementioned equation, form a symmetric coupled set. If one ignores the convection terms (arising through the $z$ derivative of amplitudes) and linear damping of waves, these equations reveal that

$$\frac{d}{dt}|a|^2 = \frac{d}{dt}|a_1|^2 = -\frac{d}{dt}|a_0|^2, \tag{10.73}$$

representing conservation of action. For each pump wave photon lost, a scattered wave photon and a decay wave plasmon are produced. The set gives periodic temporal variation of the decay wave and pump wave actions. However, parametric instabilities usually saturate by different nonlinear processes.

### 10.6.4 Stimulated Raman scattering

In this process, the pump electromagnetic wave is scattered off a Langmuir wave. Since the frequencies of the Langmuir wave and the scattered sideband electromagnetic wave are each greater than $\omega_p$, stimulated Raman scattering (SRS) occurs when $\omega_p < \omega_0/2$, that is, below the quarter critical density. The wave number of the Langmuir wave is given by

$$k = \left(k_0^2 + k_1^2 - 2k_0k_1\cos\theta_1\right)^{1/2},$$

where $\theta_1$ is the scattering angle between $\vec{k}_1$ and $\vec{k}_0$. The wave number $k$ is largest for back-scattering ($\theta_1 = \pi$). In the far underdense region ($\omega_p << \omega_0$), $k \sim 2k_0 \sin(\theta_1/2)$. To ensure weak damping of the Langmuir wave ($kv_{th}/\omega_p < 0.35$), the region of SRS is limited to

$$(n_{cr} / n_0)^{1/2} v_{th} / c \le 0.2.$$

At lower densities, the SRS becomes stimulated Compton scattering.

From Eq. (10.54), the growth rate for SRS, taking $\chi_e \sim -1$, $\partial\varepsilon/\partial\omega = 2/\omega$ and ignoring damping, can be written as

$$\Gamma = \Gamma_M = \frac{k\,|v_0|}{4}\left(\frac{\omega}{\omega_0 - \omega}\right)^{1/2} |\sin\,\delta_1|\,. \tag{10.74}$$

It maximizes for $\delta_1 = \pi/2$ and $\theta_1 = \pi$ (back-scattering). In the underdense region, we have

$$\Gamma_M = \frac{|v_0|}{2c}\left(\frac{n_0^0}{n_{cr}}\right)^{1/4} \omega_0. \tag{10.75}$$

The back-scattered SRS is stabilized when $\Gamma_d \Gamma_{d1} \geq \Gamma_M^2$. On including the linear damping of decay waves, the growth rate is given by Eq. (10.54) with $\Gamma_M$ given by Eq. (10.75). In Fig. 10.5, the normalized growth-rate of the stimulated Raman back-scattering has been plotted as a function of the normalized density, including the effect of linear damping of the Langmuir wave. The parameters are: $T_e$ = 3 keV, $|v_0|/c$ = 0.01 corresponding to a laser intensity of $3 \times 10^{15}$ W/cm$^2$ at 0.351 μm wavelength. At low densities ($n_0^0 / n_{cr} \leq 0.05$), the Langmuir wave is strongly Landau damped on the electrons ($\Gamma_d > \Gamma_M$) and the parametric instability becomes stimulated Compton scattering with a smaller growth rate. As the density rises, the linear Landau damping is small and the process returns to being SRS with a swift rise in growth rate.

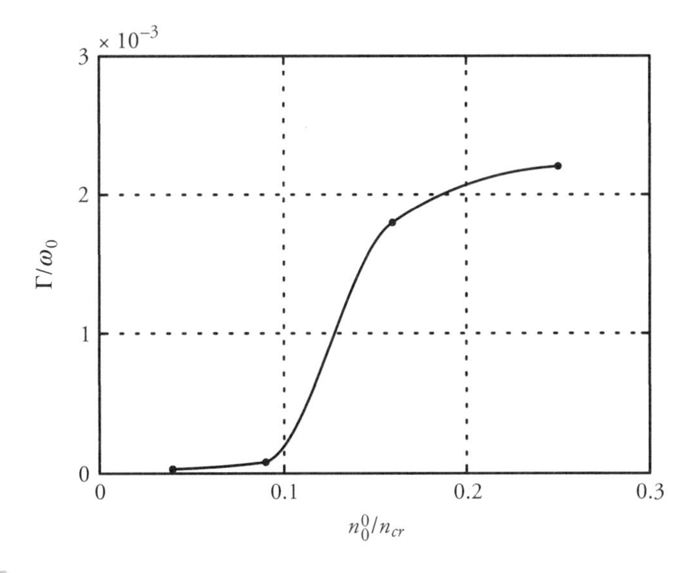


**Fig. 10.5** Normalized growth rate of the SRS (back-scattering) instability as a function of normalized density for $T_e$ = 3 keV, laser intensity $3 \times 10^{15}$ W/cm$^2$ at 0.351 μm. At low densities, the SRS instability goes over to stimulated Compton scattering.

In plasmas ionized to high charge states and irradiated with short wavelength laser, the collisional damping of both the daughter waves is strong and the threshold condition for SRS is

$$\Gamma_M \geq \frac{\nu\omega_p}{2\omega_1}. \tag{10.76}$$

Several experiments have demonstrated collisional suppression of SRS.[29] For gold targets irradiated with a 0.26 μm laser, collisional stabilization occurs for intensities less than $10^{15}$ W/cm$^2$. In cases where low frequency density fluctuations exist in the plasma, the effective collision frequency (resistivity) is enhanced and the threshold power for SRS is increased.

In the experiments, the quantity of interest is Raman reflectivity, defined as the ratio of intensities of the Raman scattered electromagnetic wave and the incident pump wave,

$$R_{SRS} = \frac{\eta_1 |E_1|^2}{\eta_0 |E_0|^2}$$

where $\eta_{0,1} = (1 - \omega_p^2 / \omega_{0,1}^2)^{1//2}$. Using Eq. (10.43), with $\chi_e = -1, \varepsilon = \varepsilon_r + i\varepsilon_i = i\Gamma \partial\varepsilon_r / \partial\omega$ $+ 2i\Gamma_d / \omega = 2i(\Gamma + \Gamma_d)/\omega$, $E_0 = iv_0 m\omega_0 / e$, and $n = k^2 \varepsilon_0 \chi_e \phi / e$ giving $\phi = -ne/(k^2 \varepsilon_0)$, one obtains

$$E_1^* = \frac{2\omega_1^* \varepsilon}{iv_0 \chi_e} \phi = \frac{4\omega_1^* (\Gamma + \Gamma_d)}{\omega v_0} \frac{ne}{k^2 \varepsilon_0}.$$

We may express $R_{SRS}$ in terms of the density fluctuation level of the Langmuir wave, which can be measured by Thomson scattering,

$$R_{SRS} = 16 \frac{\eta_1}{\eta_0} \frac{(\Gamma + \Gamma_d)^2}{\omega^2} \left| \frac{n}{n_0^0} \right|^2 \left| \frac{\omega_p}{kv_0} \right|^4. \qquad (10.77)$$

One may also express SRS reflectivity in a different way by recognizing that for each scattered wave photon produced one plasmon is produced, $|E|^2 \approx |E_1|^2\, \omega_p / \omega_1$. In conjunction with $n = n_0^0 keE / mi\omega^2$, it leads to

$$R_{SRS} = \frac{\omega_1}{\omega} \left| \frac{n}{n_0^0} \right|^2 \frac{n_0^0 m(\omega_p / k)^2}{\varepsilon_0 |E_0|^2}$$

in the underdense region ($\omega_p^2 << \omega_1^2$). In the limit $\Gamma_d = 0$, i.e. when the linear damping of the Langmuir wave is neglected, Eq. (10.77) reduces to the above expression. However, one needs to understand how the SRS saturates. Kline et al.[29] have reported very interesting results on SRS (cf. Fig. 10.6). At high plasma density (corresponding to $\omega_p/kv_{th} > 4$), Thomson scattering measurements (cf. Fig. 10.6) indicate that the SRS saturates by down-cascading of the Langmuir wave to a longer wavelength through the Langmuir decay instability (LDI), in which the primary Langmuir wave (generated in the SRS process) excites a secondary Langmuir wave and an ion-acoustic wave. For $\omega_p/kv_{th} < 3.5$, no secondary Langmuir wave was observed; however, a large level of broadband primary Langmuir waves were observed with high-level Raman reflectivity. Drake et al.[30] carried out experiments with 0.35 μm, 0.015–4 × $10^{15}$ W/cm$^2$ laser in one-tenth critical density CH plasma and demonstrated that high SRS reflectivity was due to an absolute Raman instability, orders of magnitude above the level of convective amplification (cf. Fig. 10.7). Fernandez et al.[31] and Montgomery et al.[32] observed orders of magnitude rise (inflation) in Raman reflectivity with a small increase in the laser intensity at plasma density corresponding to $kv_{th}/\omega \sim 0.35$ (cf., Fig. 10.8), where the Langmuir wave, according to linear theory, is strongly Landau damped and one would not have expected a high level of SRS. Wang et al.[33] have explained these results as a nonlinear transition from convective to absolute instability. At intensities higher than the convective instability threshold value, the resonant electrons get trapped in the plasma wave and flatten the distribution function around $v_z \approx \omega/k$ within a bounce period $2\pi\omega_b^{-1}$ ($\omega_b = (ek^2|\phi|m)^{1/2}$), drastically reducing the Landau damping and the threshold for the absolute instability of the Raman back-scattering and upstaging the SRS growth rate. The instability saturates only by wave-breaking of the Langmuir wave, leading to an increase of the reflectivity by several orders of magnitude (inflation). The SRS is observed to have a strong correlation with the number of hot electrons with temperature $T_h \approx m(\omega/k)^2/2$. The theory and simulation is so far only for homogeneous Plasmas. However, NIF experiment[28] showed that an SRS instability occurs in the boundary region between the gold plasma from the hohlraum and the helium plasma in the gas. This is an inhomogeneous plasma where the 1/4 critical

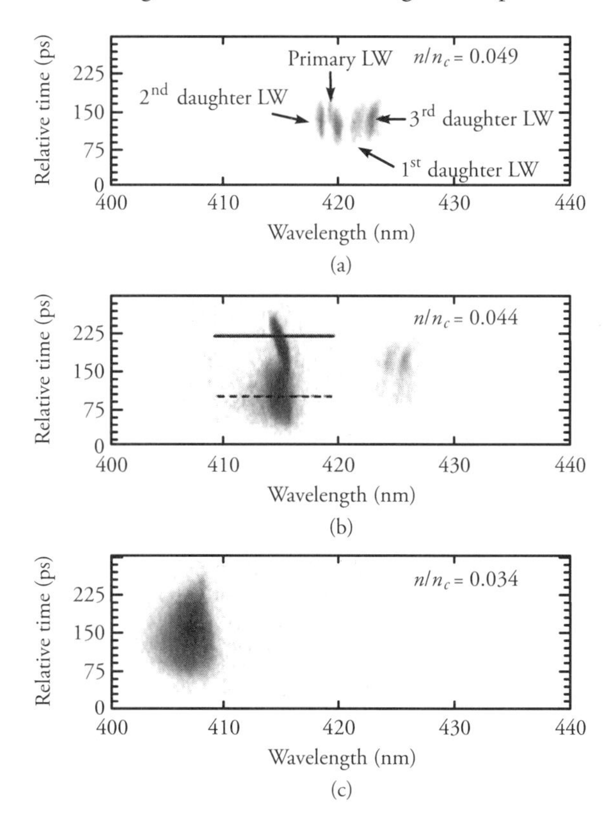


**Fig. 10.6** Thomson streak measurements of Langmuir waves in SRS, for $T_e$ = 600 eV and different $kv_{th}/\omega_p$ (= $k\lambda_D$) and laser intensity values (a) 0.27, 4.4 × $10^{15}$ W/cm$^2$, (b) 0.29, 4.4 × $10^{15}$ W/cm$^2$, (c) 0.343, 1.1 × $10^{16}$ W/cm$^2$, where $n_c$ is the critical density. (After Kline et al.[29])

absolute Raman instability (cf. Section 11.3) dominates the reflected signal, and weaker signals from the underdense regime with a gap that is not yet understood.

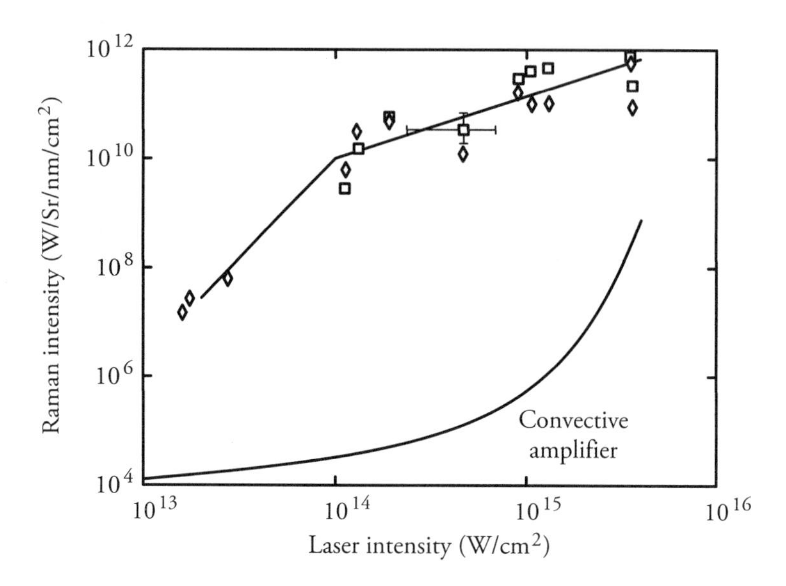


**Fig. 10.7** Intensity of Raman back-scatter of 540, 500 nm, corresponding to $n/n_{cr} = 0.11$, 0.075 when the maximum density is $0.15n_{cr}$, $0.11n_{cr}$ and the density scale length is 370, 450 μm. The convective amplification curve is based on the formula derived in Chapter 11 with a thermal noise source. (After Drake et al.[30])

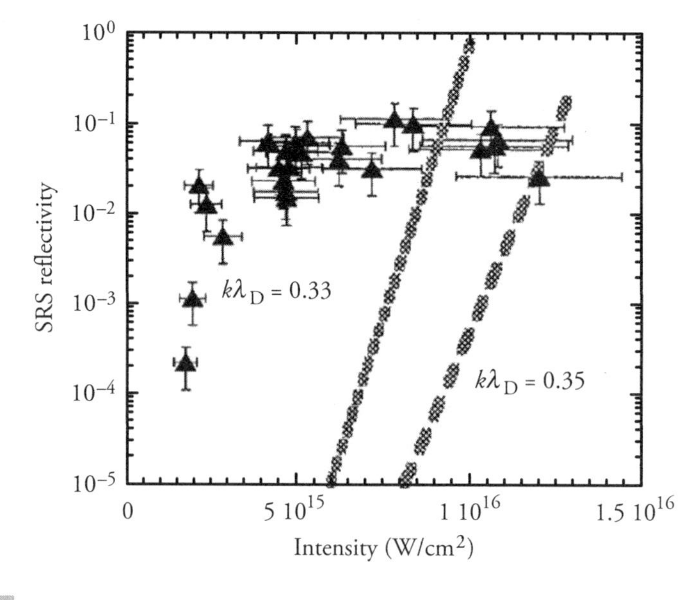


**Fig. 10.8** Raman reflectivity versus laser intensity for plasma conditions corresponding to $kv_{th}/\omega_p (= k\lambda_D) = 0.33$–0.35. The dashed curves are from a theoretical model. (After Montgomery et al.[32])

The indirect drive inertial confinement fusion experiments[34] employ a large number of beams crossing each other. To a plasma with finite flow velocity, the frequencies of these lasers appear different; cross beam energy transfer leads to the creation of sound waves and density fluctuations. These fluctuations greatly enhance the high frequency anomalous resistivity, raising the threshold for SRS and lowering the growth rate. Experiments with a single laser beam noted a different scenario with the addition of a small percentage of hydrogen in the plasma where the SRS instability saturates via the excitation of a Langumir wave driven by the decay instability (LDI). The hydrogen enhances the Landau damping of the ion-acoustic wave, suppressing the LDI and consequently raising the level of SRS.[35]

Drake et al.[24] carried experiments on a high-$Z$ gold foil using 0.53 μm, $10^{14}$–$3 \times 10^{16}$ W/cm$^2$ laser of 150–1880 μm spot size and observed Raman side scatter. The back-scattering was strong. Equally strong was the side scattering at 60° to the back normal. As the SRS reflectivity rose to 10%, the hot electron fraction also rose to about 10%.

## 10.6.5 Stimulated Brillouin scattering

In a non-isothermal plasma with $T_e >> T_i$, the ion-acoustic wave is weakly damped. Stimulated Brillouin scattering (SBS) of the electromagnetic pump wave occurs off the ion-acoustic mode. Since $\omega < \omega_{pi}$, $\omega_1 \approx \omega_0$, $|k_1| \approx k_0$, $k \approx 2k_0\sin(\theta_1/2)$. Taking $\omega \approx kc_s$, $\partial\varepsilon/\partial\omega \approx 2\,\omega_{pi}^2/\left(k^2c_s^2\omega\right)$, and $\chi_e = \omega_{pi}^2/k^2c_s^2$, the growth rate, in the absence of linear damping of the decay waves, can be written as

$$\Gamma_M = \omega_{pi}\frac{|v_0|}{4\,c_s}\left(\frac{c_s}{c}\right)^{1/2}|\sin\delta_1|\left(2\sin\left(\theta_1/2\right)\right)^{1/2}$$

$$= \omega_{pi}\frac{|v_0|}{4c_s}\left(\frac{2c_s}{c}\right)^{1/2} \tag{10.78}$$

for the back-scatter. This expression is valid when $\Gamma_M < \omega$. The growth rate, including the linear damping of decay waves, is given by Eq. (10.54) with $\Gamma_M$ given by Eq. (10.78).

At large pump powers, when $\Gamma_M \sim \omega$, $\varepsilon(\omega, k)$ is no longer close to zero and the low frequency mode goes over to a reactive quasi-mode. In this case, Eq. (10.50) with $\varepsilon \approx \left(\omega^2 - k^2c_s^2\right)\omega_{pi}^2/\left(\omega^2k^2c_s^2\right)$ takes the form of a cubic equation for $\omega$,

$$\left(\omega^2 - k^2c_s^2\right)\left(\omega - \left(k_0^2 - k_1^2\right)c^2/2\omega_0\right) = -k^2|v_0|^2\,\omega_{pi}^2/8\omega_0$$

for $\delta_1 = \pi/2$, this equation can be solved numerically. For $\omega^2 >> k^2c_s^2$ and $\omega >> \left(k_0^2 - k_1^2\right)c^2/2\omega_0$, it yields

$$\omega = \left( \frac{\omega_{pi}^2}{8\,\omega_0} k^2 |v_0|^2 \right)^{1/3} \frac{1 + i\sqrt{3}}{2} . \tag{10.79}$$

The growth rate and oscillation frequency of the low-frequency reactive quasi-mode are comparable to each other. In Fig. 10.9, the normalized growth rate of the stimulated Brillouin back-scattering has been plotted as a function of normalized density, by solving Eq. (10.50) for the following parameters: $T_e$ = 3 keV, $T_e/T_i$ = 9, ion charge state $Z_i$ = 1, $|v_0|/c$ = 0.01 (corresponding to a laser intensity of 3 × $10^{15}$ W/cm$^2$ at 0.351 μm wavelength and ignoring the linear damping of the sideband). At low density, the growth rate is suppressed by the ion Landau damping of the acoustic wave. At higher density, the linear damping is weak and the growth rate rises gradually. At higher intensity, the ion wave goes over to the reactive quasi-mode and the growth rate is large.

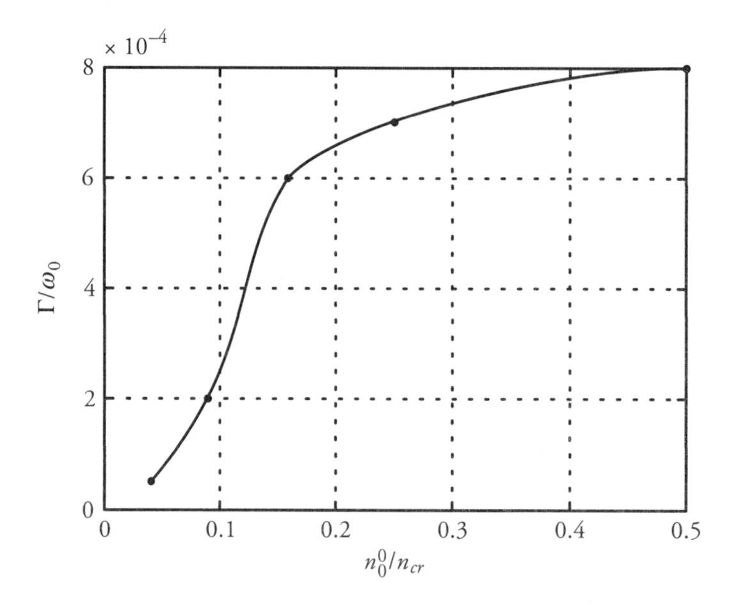


**Fig. 10.9** Normalized growth rate of the SBS (back-scattering) instability as a function of normalized density for $T_e$ = 3 keV, $T_e/T_i$ = 9, laser intensity 3 × $10^{15}$ W/cm$^2$ at 0.351 μm. At low densities the SBS intability is in the heavily ion Landau damped (stimulated Compton scattering) regime.

The addition of hydrogen in a high-Z plasma increases the linear damping of the ion-acoustic wave and suppresses the instability. In some laser–plasma experiments, the SRS initially dominates, since weakly damped ion-acoustic waves cannot exist in a plasma with an electron-to-ion temperature ratio not much greater than one. Later, as the electron temperature rises, the ion damping decreases, and SBS sets in.[36,37] The ion-acoustic wave produced in the SBS process may even beat with the Langmuir wave of the SRS process, to produce a beat wave strongly Landau damped on the electrons, thus diminishing the growth rate of the SRS instability. Another scenario is the decay of the Langmuir wave generated in the SRS process into a secondary Langmuir wave and an ion-acoustic wave (see Section 10.9.2).

The SBS spectra show that SBS occurs over a wide density range, starting from one-thirtieth of the critical density to the critical density. Lasers coupling to the ion-acoustic waves play an important role in crossed beam energy transfer. Kirkwood et al.[38] have observed efficient energy transfer between crossed laser beams through the mediation of ion-acoustic waves.

## 10.7 Non-resonant Scattering: Stimulated Compton Scattering

When the space charge mode has a phase velocity close to the thermal velocity of the electrons or ions ($\omega \approx kv_{\text{th}}$ or $\omega \approx kv_{\text{thi}}$), that is, $\varepsilon$ has a large imaginary part and the linear damping of the mode is strong, the parametric instability is termed stimulated Compton scattering or nonlinear Landau damping. Writing $\omega = \omega_r + i\Gamma, D_1^* = -2i(\Gamma + \Gamma_{d1})\omega_{1r}$ in Eq. (10.50), one obtains the growth rate

$$\Gamma = \frac{k^2 |v_0^2| \sin^2 \delta_1}{8(\omega_0 - \omega)} \text{Im}\left[\frac{\chi_e(1+\chi_i)}{\varepsilon}\right] - \Gamma_{1d}. \tag{10.80}$$

Above the threshold, $\Gamma$ varies proportional to $|v_0|^2$.

### 10.7.1 Compton scattering off electrons

The electrostatic mode in the stimulated Compton scattering process having $\omega \simeq kv_{\text{th}}$ is a driven mode or quasi-mode, strongly Landau damped on the electrons. Since $k < 2k_0$, the frequency of this quasi-mode $\omega < 2\omega_0 v_{\text{th}}/c << \omega_0$. It could be close to $\omega_p$ or far from it. From Eq. (10.80), on writing $\chi_e = \varepsilon - 1 - \chi_i$ and taking $\Gamma_{1d} = 0$, $\delta_1 = \pi/2$, the growth rate of the stimulated Compton scattering (SCS) can be written as

$$\Gamma = \frac{k^2 |v_0|^2 (1+\chi_i)^2}{8\omega_0} \frac{\chi_{\text{ei}}}{(1+\chi_i+\chi_{\text{er}})^2 + \chi_{\text{ei}}^2}, \tag{10.81}$$

where $\chi_{\text{er}}$ and $\chi_{\text{ei}}$ are the real and imaginary parts of the electron susceptibility $\chi_e$. For $\omega \approx kv_{\text{th}} > \omega_{\text{pi}}$.

$$\Gamma = \frac{|v_0|^2 / v_{\text{th}}^2}{8\omega_0} \frac{\omega_p^2}{1 + \omega_p^2/\omega^2 + 2\omega_p^4/\omega^4}. \tag{10.82}$$

The growth rate goes linearly with the pump intensity.

### 10.7.2 Compton scattering off ions

The electromagnetic pump wave can also undergo stimulated Compton scattering off ions with $\omega \sim kv_{\text{thi}} << kv_{\text{th}}$. In this case, $\chi_e \sim \omega_{\text{pi}}^2 / k^2 c_s^2$, $k_1 \approx k_0$, $k \approx 2k_0 \cos(\theta/2)$ and Eq. (10.80), for $\Gamma_1 = 0$, $\delta_1 = \pi/2$, $\omega < \omega_{\text{pi}}$ gives the growth rate

$$\Gamma \equiv \frac{|v_0|^2 \omega_p^2}{8v_{\text{th}}^2 \omega_0} \frac{T_e / T_i}{(1 + 0.3\, T_e / T_i)^2 + 0.3T_e^2 / T_i^2}. \tag{10.83}$$

The growth rate of this process is comparable to or more than that of the Compton scattering off electrons, but the real frequency is smaller. It occurs in isothermal underdense plasmas.

## 10.8 Four-wave Parametric Processes

In the case of modulational and filamentation instabilities, one has to consider both the lower sideband $\left(\omega_0 - \omega^*, \vec{k}_0 - \vec{k}\right)$ and the upper sideband $\left(\omega_0 + \omega, \vec{k}_0 + \vec{k}\right)$. Taking the upper sideband fields as

$$\vec{E}_2 = \vec{A}_2 e^{-i\left(\omega_2 t - \vec{k}_2 \cdot \vec{r}\right)},$$

$$\vec{B}_2 = \vec{k}_2 \times \vec{E}_2 / \omega_2, \tag{10.84}$$

and the electron quiver velocity at $\omega_2$ as $\vec{v}_2 = e\vec{E}_2 / (mi\omega_2)$, the low-frequency ponderomotive potential turns out to be

$$\varphi_{p\omega} = -\frac{m}{2}(\vec{v}_0 \cdot \vec{v}_1^* + \vec{v}_0^* \cdot \vec{v}_2)$$

$$= -\frac{e}{2m\omega_0}\left(\frac{\vec{E}_0 \cdot \vec{E}_1^*}{\omega_1} + \frac{\vec{E}_0^* \cdot \vec{E}_2}{\omega_2}\right). \tag{10.85}$$

The electron density perturbation due to $\phi$ and $\phi_{p\omega}$ is still given by Eq. (10.42). However, Eq. (10.43) governing $\phi$ now has an additional term,

$$\varepsilon\phi = \chi_e \frac{e}{2m\omega_0}\left(\frac{\vec{E}_0 \cdot \vec{E}_1^*}{\omega_1^*} + \frac{\vec{E}_0^* \cdot \vec{E}_2}{\omega_2}\right). \tag{10.86}$$

The nonlinear current density at $\omega_2$, $\vec{k}_2$ is

$$\vec{J}_2^{\rm NL} = -en\vec{v}_0 / 2 = k^2\varepsilon_0\left(1+\chi_i\right)\frac{e\,\vec{E}_0\phi}{2\,mi\omega_0}. \tag{10.87}$$

Using this in the wave equation (Eq. 2.24), we obtain

$$D_2\vec{E}_2 = -\frac{k^2\left(1+\chi_i\right)\omega_2}{2m\omega_0}\left(\vec{E}_0 - \frac{\vec{k}_2}{k_2^2}\,\vec{k}_2\cdot\vec{E}_0\right)e\phi, \tag{10.88}$$

where $D_2 = \omega_2^2 - \omega_p^2 - k_2^2c^2$. Substituting for $\vec{E}_2$ from Eq. (10.88) and $\vec{E}_1$ from Eq. (10.49) in Eq. (10.86), we obtain the nonlinear dispersion relation

$$\varepsilon = \mu_c\left(\frac{1}{D_1^*}+\frac{1}{D_2}\right), \tag{10.89}$$

where $\mu_c$ is given by Eq. (10.51).

### 10.8.1 Modulational instability

By this process, the pump wave gets amplitude modulated by a density perturbation co-propagating with the pump with $\vec{k}\,||\,\vec{k}_0$. Since an amplitude modulated wave is expressible as a superposition of a carrier wave (pump) and two sidebands, the inclusion of the upper sideband is necessary. Using that $\omega/k \sim \partial\omega_0/\partial k_0$ in an under-dence plasma, one has $\omega \approx kv_{go} = kc\left(1-\omega_p^2/\omega_0^2\right)^{1/2}$, $D_{1,2} = \left[\omega^2 - k^2c^2 \mp 2\omega_0\left(\omega - kv_{g0}\right)\right]$, with $\chi_e = -\omega_p^2/\omega^2$, and $\chi_i = -\omega_{\rm pi}^2/\omega^2$, the four-wave nonlinear dispersion relation, Eq. (10.89), can be written as

$$\frac{\omega_p^2-\omega^2}{\omega_{\rm pi}^2-\omega^2} = -\frac{\omega_p^2}{\omega^2}\frac{k^2\left|v_0\right|^2}{2}\left[\frac{\omega^2-k^2c^2}{\left(\omega^2-k^2c^2\right)^2-4\,\omega_0^2\left(\omega-kv_{go}\right)^2}\right]. \tag{10.90}$$

This equation gives an unstable root for $\omega^2 < \omega_{\rm pi}^2$. Writing $\omega = kv_{go} + i\Gamma$, we obtain

$$\Gamma^2 = \frac{\omega_{\rm pi}^2}{8\omega_0^2}\frac{\omega_p^2}{\omega_0^2}\frac{\left|v_0\right|^2}{c^2}\omega^2, \tag{10.91}$$

where we assume that $2\omega_0(\omega - kv_{go}) > \omega^2 - k^2c^2$, that is, $\Gamma > \omega_p^2\,\omega^2/2\omega_0^3$ requiring

$$\frac{|v_0|}{c} > \frac{\omega_p}{\omega_0}\frac{\omega}{\omega_{pi}}\sqrt{2}.$$

In the opposite limit, the growth rate is small.

In the case of laser beams of finite spot size (having Gaussian distribution of intensity in the transverse plane), a different kind of modulational instability arises in low density plasmas, encountered in laser electron acceleration.[39] The laser beam couples to a plasma wave co-moving with the phase velocity equal to the group velocity of the laser. The plasma wave creates density crests and troughs with their values being maximum on the laser axis and fading off with $r$. The sections of the laser moving with the density troughs self-focus while those moving with the crests diverge. The laser intensity thus acquires axial modulation, deepening the density perturbation. This further modulates the laser, leading to the growth of the parametric instability.

## 10.8.2 Filamentation instability

The filamentation instability leads to self-focusing and break-up in filament of the beam. In this case, a low frequency density perturbation has $\vec{k}$ perpendicular to $\vec{k}_0$ so that the different parts of the pump wave front experience different indices of refraction. The beamlets converge toward the refractive index maxima, that is, into density depressions, enhancing the density perturbation.[20] For $\vec{k} \perp \vec{k}_0$, we have $k_1^2 = k_2^2 = k^2 + k_0^2$ and $D_{1,2} = \omega^2 - k^2c^2 \mp 2\omega\omega_0$. Taking $kv_{th} > \omega > kv_{thi}$ $\chi_e \approx \omega_{pi}^2 / k^2c_s^2$, and $\chi_i = -\omega_{pi}^2/\omega^2$, so that the plasma electrons follow the variations in the ponderomotive potential $\phi_p$ caused by the high-frequency wave (the pump and sidebands), the dispersion relation, Eq. (10.89), in the limit $k^2c^2 > 2\omega\omega_0$, can be set as

$$\omega^2 - k^2c_s^2 = -\frac{|v_0|^2}{2c^2}\left(\omega_{pi}^2 - \omega^2\right), \tag{10.92}$$

which gives a purely growing mode ($\omega = i\Gamma$) when $|v_0|^2/2c^2 > 2k^2c_s^2/\omega_{pi}^2$, with the growth rate

$$\Gamma \approx \frac{|v_0|}{c\sqrt{2}}\omega_{pi}. \tag{10.93}$$

It is useful to obtain the spatial growth rate of the zero frequency mode. For $\omega = 0$, $\vec{k} = \vec{k}_\perp + k_z\hat{z}$ (where $\vec{k}_\perp$ is perpendicular to $\hat{z}$), Eq. (10.89) can be written as

$$1 = \frac{\omega_{pi}^2}{k^2c_s^2}\frac{1}{1+T_i/T_e}\frac{k^2|v_0|^2}{2}\frac{k^2c^2}{k^4c^4 - 4\omega_0^2k_z^2c^2}, \tag{10.94}$$

which gives the spatial growth rate $G = ik_z$,

$$G = \frac{k_\perp}{2}\left[\frac{\omega_p^2}{\omega_0^2}\frac{|v_0|^2}{(1+T_i/T_e)v_{th}^2} - \frac{k_\perp^2 c^2}{\omega_0^2}\right]^{1/2}, \tag{10.95}$$

where we have assumed $k_\perp^2 > k_z^2$ and $\omega_0^2 > \omega_p^2$. The first term inside the square root is due to nonlinear refraction; whereas the second one is due to diffraction. The diffraction gets stronger when the filament size $\left(\sim k_\perp^{-1}\right)$ is smaller, eventually overpowering the nonlinear refraction. The growth rate maximizes to

$$G_{\max} = \frac{\omega_0}{4\,c}\frac{|v_0|^2}{v_{th}^2(1+T_i/T_e)}\frac{n_0^0}{n_{cr}} \tag{10.96}$$

at

$$k_\perp = k_{\perp opt} = \frac{\omega_p}{\sqrt{2}\,c}\frac{|v_0|}{v_{th}}\frac{1}{(1+T_i/T_e)}. \tag{10.97}$$

Equation (10.96) is the same expression as that obtained in Chapter 9 for the case of the ponderomotive nonlinearity.

### 10.8.3 Oscillating two-stream instability

The oscillating two-stream instability (OTSI) is a process in which a long wavelength pump wave excites a short wavelength Langmuir wave and a purely growing density perturbation. The dynamics of the OTSI can be understood as follows. Consider a dipole ($k_0 = 0$) pump,

$$\vec{E}_0 = \hat{x}A_0\cos\omega_0 t, \tag{10.98}$$

producing an electron drift velocity $\vec{v}_0 = -\hat{x}(eA_0/m\omega_0)\sin\omega_0 t$. Let there also exist a small amplitude standing Langmuir wave (cf. Fig. 10.10)

$$\vec{E}_s = \hat{x}A_s\cos(kx)\cos\omega_0 t. \tag{10.99}$$

In the zones $(2n-½)\pi/k < x < (2n+½)\pi/k$ (where $n = 0, 1, 2, \ldots$), $\vec{E}_0$ and $\vec{E}_s$ are parallel to each other at all times. At other values of $x$, they are anti-parallel. These fields exert a force on the electrons, pushing them from the parallel to the anti-parallel regions, and creating density depressions

$$\delta n \approx (e\,\phi_p/T_e)n_0^0; \tag{10.100}$$

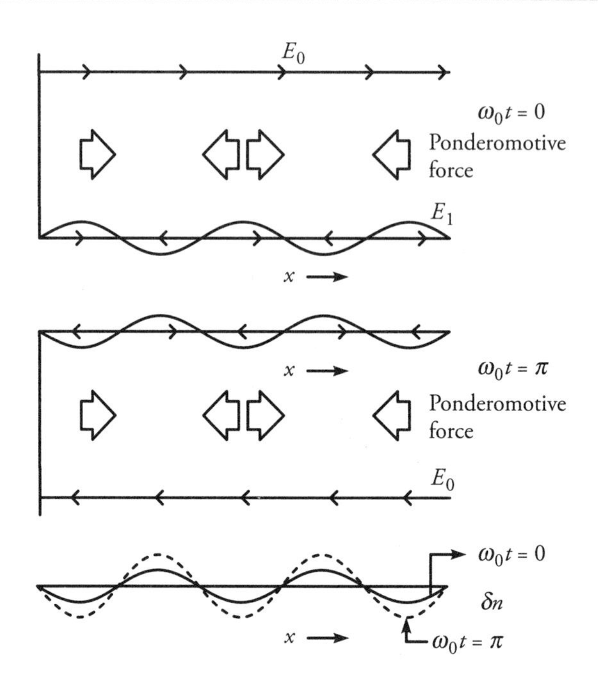


Fig. 10.10 Growth of the oscillating two-stream instability.

where $\phi_p = \left(eA_0 A_s / 2m\omega_0^2\right)\cos(kx)$. These depressions in conjunction with $\vec{v}_0$ produce nonlinear density oscillations (according to the continuity equation)

$$n_s^{\mathrm{NL}} = -\frac{n_0^0 ek\left|v_0\right|^2 A_s}{4T_e\omega_0^2}\cos\omega_0 t\sin kx. \tag{10.101}$$

and a space charge potential (according to Poisson's equation) $\phi_s' = -\,en_s^{\mathrm{NL}} / k^2\varepsilon_1$, creating the electric field

$$\vec{E}_s' = -\hat{x}\frac{\partial\varphi_s'}{\partial x} = -\hat{x}\frac{n_0^0 e^2\left|v_0\right|^2 E_s\cos(\omega_0 t)\cos(kx)}{4\varepsilon_0\omega_0^2 T_e\varepsilon_s}, \tag{10.102}$$

where $\varepsilon_s = 1 - \left(\omega_p^2 + 3\,k^2 v_{\mathrm{th}}^2\right)\omega_0^2$ is the plasma permittivity at $(\omega_0, \vec{k})$. For $\vec{E}_s'$ to be in phase with $\vec{E}_s$, so that $\vec{E}_s$ may grow with time, one must have $\varepsilon_S < 0$, that is, the frequency mismatch

$$\Delta = \omega_0 - \omega_p\left(1 + 3\,k^2 v_{\mathrm{th}}^2 / 2\omega_p^2\right) < 0 \tag{10.103}$$

Further, for $\vec{E}_s$ to have large magnitude, $\varepsilon_S$ should be as small as possible, that is, the frequency mismatch $\Delta$ should be of the order of the growth rate.

Let us consider the nonlinear coupling of the pump, given by Eq. (10.98), the low frequency potential given by Eq. (10.37), and the sideband potentials given by

$$\phi_1 = A_1 e^{-i(\omega_1 t - \vec{k}_1.\vec{r})}, \phi_2 = A_2 e^{-i(\omega_2 t - \vec{k}_2.\vec{r})}, \tag{10.104}$$

where $\omega_1 = \omega_0 - \omega^*$, $\omega_2 = \omega_0 + \omega$, $\vec{k}_1 = -\vec{k}$, $\vec{k}_2 = \vec{k}$, and where $\omega << kv_{thi}$, $kv_{th}$. The last inequality implies that the low frequency ion and electron responses are adiabatic, with density perturbations

$$n = n_0^0 e(\phi + \phi_{p\omega}) / T_e,\ n_i = -n_0^0 e\phi / T_i.$$

These density perturbations are formally given by Eqs. (10.42) and (10.40b). The low frequency ponderomotive potential due to the pump and the sideband Langmuir waves is given by Eq. (10.85), which on using $\vec{E}_1 = -i\vec{k}_1\phi_1$, and $\vec{E}_2 = -i\vec{k}_2\phi_2$, can be written as

$$\phi_{p\omega} = \frac{\vec{k}_1.\vec{v}_0}{2\omega_1^*}\phi_1^* + \frac{\vec{k}_2.\vec{v}_0^*}{2\omega_2}\phi_2. \tag{10.105}$$

Equation (10.86) governing $\phi$ can be written as

$$\varepsilon\phi = -\chi_e\phi_{p\omega} = -\frac{\chi_e}{2}\left[\frac{\vec{k}_1.\vec{v}_0}{\omega_1^*}\phi_1^* + \frac{\vec{k}_2.\vec{v}_0^*}{\omega_2}\phi_2\right]. \tag{10.106}$$

The low frequency density perturbation $n$ beats with $\vec{v}_0$ to produce nonlinear density perturbations $n_1^{NL}, n_2^{NL}$ at the sideband frequencies. For $n_1^{NL}$, the continuity equation

$$\frac{\partial n_1^{NL}}{\partial t} + \nabla.\left(\frac{1}{2}n^*\vec{v}_0\right) = 0$$

leads to

$$n_1^{NL} = n^*\frac{\vec{k}_1.\vec{v}_0}{2\omega_1} = -\frac{k^2\varepsilon_0\left(1+\chi_i^*\right)}{e}\frac{\vec{k}_1.\vec{v}_0}{2\omega_1}\phi^*, \tag{10.107}$$

where we have employed $\chi_e(\phi + \phi_p) = -(1 + \chi_i)\phi$ from Eq. (10.43). The linear density perturbation at $\omega_1$, $\vec{k}_1$ due to $\phi_1$, is

$$n_1^L = \left(k_1^2\varepsilon_0 / e\right)\chi_{1e}\phi_1, \tag{10.108}$$

where is the electron susceptibility at $\omega_1$, $\vec{k}_1$. Hence, the total electron density perturbation at $\omega_1$, $\vec{k}_1$ is $n_1 = n^L + n_1^{NL}$. One may write the density perturbation at the upper sideband $n_2$ in a similar manner. Employing $n_1$, $n_2$ in Poisson's equation, $\nabla^2\phi_{1,2} = n_{1,2}e/\varepsilon_0$, and writing $\nabla^2\phi_{1,2} = -k_{1,2}^2\phi_{1,2}$, we obtain

$$\varepsilon_1\phi_1 = -\frac{e\,n_1^{\mathrm{NL}}}{k_1^2\varepsilon_0} = \frac{k^2}{k_1^2}\frac{\vec{k}_1.\vec{v}_0}{2\omega_1}\left(1+\chi_i^*\right)\phi^*,$$

$$\varepsilon_2\phi_2 = -\frac{e n_2^{\mathrm{NL}}}{k_2^2\varepsilon_0} = \frac{k^2}{k_2^2}\frac{\vec{k}_2.\vec{v}_0}{2\omega_2}\left(1+\chi_i\right)\phi. \tag{10.109}$$

Using $\phi_{1,}$ $\phi_2$ in terms of $\phi$ from these equations in Eq. (10.106), we obtain the nonlinear dispersion relation

$$\varepsilon = \mu_d\left(\frac{1}{\varepsilon_1^*}+\frac{1}{\varepsilon_2}\right), \tag{10.110}$$

where

$$\mu_d = \frac{k^2}{k_1^2}\frac{\left|\vec{k}_1.\vec{v}_0\right|^2}{4\omega_0^2}\chi_e\left(1+\chi_i\right)$$

and $\varepsilon_j = 1-(\omega_p^2+3k^2v_{\mathrm{th}}^2/2)/\omega_j^2$, $j$ = 1, 2. Introducing the frequency mismatch $\Delta$ (cf. Eq. 10.103), we may write

$$\varepsilon_1 \approx 2\left(\Delta-\omega\right)/\omega_0,\ \varepsilon_2 \approx 2\left(\Delta+\omega\right)/\omega_0.$$

Then, Eq. (10.110) takes the form

$$\omega^2 = \Delta^2 + S\Delta, \tag{10.111}$$

where

$$S = \frac{\omega_{\mathrm{pi}}^2\left|\vec{k}\cdot\vec{v}_0\right|^2}{4k^2c_s^2\left(1+T_i/T_e\right)\omega_0} \tag{10.112}$$

The growth occurs for $\Delta < 0$, $|\Delta| < S$. In this case, the real part of $\omega$ is zero. The maximum growth occurs for $\Delta = -S/2$; the maximum growth rate is

$$\Gamma_{\max} = \frac{\left|v_0\right|^2}{8\,v_{\mathrm{th}}^2}\frac{\omega_p}{\left(1+Ti/Te\right)}, \tag{10.113}$$

where we have taken $\vec{k}\,||\,\vec{E}_0$. With the inclusion of the linear damping rate $\Gamma_{1d}$ of the sidebands, the growth rate of the OTSI turns out to be

$$\Gamma = \Gamma_{max} - \Gamma_{1d}. \tag{10.114}$$

The OTSI diverts laser energy into short wavelength Langmuir waves on the time scale longer than the ion plasma period. It occurs near the critical layer and may cause heating of the electrons through Landau damping of the short wavelength Langmuir waves.

## 10.9 Decay Instability

Near the critical layer, an electromagnetic pump wave can decay into an ion-acoustic wave and a Langmuir wave sideband. It is also probable that the incident p polarized light may undergo linear mode conversion producing a large amplitude Langmuir wave which then decays into a pair of ion-acoustic and Langmuir waves.

### 10.9.1 Electromagnetic pump

Consider the decay of an electromagnetic pump wave, given by Eq. (10.35a, b), into an ion-acoustic wave $\phi = A\exp[-i(\omega t - \vec{k}\cdot\vec{r})]$ and a Langmuir wave sideband $\phi_1 = A_1\exp[-i(\omega_1 t - \vec{k}_1\cdot\vec{r})]$, where $\omega_1 = \omega_0 - \omega^*$, $\vec{k}_1 = \vec{k}_0 - \vec{k}$. Since $\omega < \omega_{pi}$, $\omega_1$ is close to $\omega_0$. Further, $\omega_1^2 >> k_1^2 v_{th}^2$ to ensure weak damping. The decay occurs near the critical layer, where $\omega_0 \approx \omega_p + k_0^2c^2/2\omega_p$, $\omega_1 \approx \omega_p + 3k_1^2 v_{th}^2/2\omega_p$, $k_1 \approx k_0 c/v_{th}\sqrt{3}$, and $\omega \approx kc_s$. The phase matching condition $kc_s = \omega_0 - \omega_1$ gives

$$kc_s = \Big[k_0 c_s \cos\theta_1 - (m/m_i)^{1/2}\omega_{pi}/3 + \Big\{\big(k_0 c_s \cos\theta_1 - (m/m_i)^{1/2}\omega_{pi}/3\big)^2 - k_0^2 c^2 m/3m_i\Big\}^{1/2}\Big], \tag{10.115}$$

where $\theta_1$ is the angle between $\vec{k}_1$ and $\vec{k}_0$. As $k_0 \to 0$, we have $k \to 0$. The decay instability produces short wavelengths.

The low frequency mode is driven by the ponderomotive force due to the electromagnetic pump wave and the decay Langmuir wave, while the Langmuir wave is driven by the beating of the ion wave driven density perturbation and the pump driven velocity of electrons. The coupled mode equations for $\phi$, $\phi_1$, viz., Eqs. (10.106) and (10.109), lead to the dispersion relation in Eq. (10.110) with the upper sideband neglected,

$$\varepsilon\varepsilon_1^* = \mu_d, \tag{10.116}$$

where

$$\mu_d = \frac{\omega_{pi}^4}{\omega^2 k^2 c_s^2}\frac{|\vec{k}_1.\vec{v}_0|^2}{4\omega_0^2}.$$

Expressing $\omega = \omega_r + i\Gamma$, $\omega_1 = \omega_{1r} + i\Gamma$, where $\omega_r$ and $\omega_{1r}$ (= $\omega_0 - \omega_r$) are the roots of $\varepsilon_r(\omega_r) = 0$ and $\varepsilon_{ir}(\omega_{ir}) = 0$, we may write

$$\varepsilon = i(\Gamma + \Gamma_d)\frac{\partial \varepsilon_r}{\partial \omega_r} = i\ (\Gamma + \Gamma_d)\ 2\ \omega_{\text{pi}}^2 /\ \omega_r^3 \tag{10.117a}$$

$$\varepsilon_1^* = -\ i(\Gamma + \Gamma_{d1})\frac{\partial \varepsilon_{1r}}{\partial \omega_{1r}} = -i(\Gamma + \Gamma_{d1})2 / \omega_{1r}, \tag{10.117b}$$

where $\Gamma_d = \varepsilon_i/(\partial\varepsilon_r/\partial\omega_r)$, $\Gamma_{1d} = \varepsilon_{1i}/(\partial\varepsilon_{1r}/\partial\omega_{1r})$ are the linear damping rates of the decay waves, and subscripts $i$ and $r$ refer to the imaginary and real parts of the quantities. For brevity, we now suppress the subscript $r$ on $\omega_r$ and $\omega_{1r}$. Using Eqs. (10.117a, b) in Eq. (10.116), one obtains

$$(\Gamma + \Gamma_d)\ (\Gamma + \Gamma_{d1}) = \Gamma_M^2, \tag{10.118}$$

where

$$\Gamma_M^2 = \frac{\omega\omega_p}{8}\frac{\left|\vec{k}_1.\vec{v}_0\right|^2}{k_1^2 v_{\text{th}}^2}. \tag{10.119}$$

The threshold for the instability is given by $\Gamma_M^2 = \Gamma_d\Gamma_{d1}$. Much above the threshold, the growth rate is $\Gamma \approx \Gamma_M$. It maximizes for $\vec{k}\ ||\ \vec{v}_0$ and scales linearly with the pump amplitude.

When the low-frequency mode is strongly Landau damped on the ions, the decay process becomes nonlinear Landau damping, giving a growth rate

$$\Gamma = -\Gamma_{d1} + \frac{\omega_p}{2}\frac{\left|\vec{k}\ .\vec{v}_0\right|^2}{k^2 v_{\text{th}}^2}\frac{\omega_{\text{pi}}^2}{k^2 c_s^2}\text{Im}\left(-\ \frac{1}{\varepsilon}\right). \tag{10.120}$$

## 10.9.2 Electrostatic pump

Large amplitude Langmuir waves can resonantly decay into the same pair of decay waves as mentioned in Section 10.9.1. However, since $|\omega_1| < \omega_0$, we have $|k_1| < k_0$, that is, the decay produces longer wavelength modes. This can lead to an inverse cascade process where the Langmuir waves decay into progressively smaller wave number/longer wavelength. Denoting the angle between $\vec{k}_0$ and $\vec{k}_1$ as $\theta_1$, the frequency of the low-frequency mode can be written as $\omega \sim k_0c_s(1 - \cos\theta_1)$, which maximizes for the back-scatter to $\omega = 2k_0c_s$. The growth rate for this channel is given by Eq. (10.118). For nonlinear Landau damping, $\Gamma$ is given by Eq. (10.120). One must remember that in the case of electrostatic pump, $\vec{v}_0\ ||\ \vec{k}_0$, hence the maximum growth occurs when $\vec{k}\ ||\ \vec{k}_0$. Further, if the electrostatic pump wave is generated through the mode conversion of an electromagnetic wave, the amplitude of the pump wave could be larger than that of the electromagnetic wave; hence, the decay instability of the electrostatic pump is stronger.

The decay instability plays a prominent role in the saturation of the SRS instability. When the ratio of the phase velocity of the Langmuir wave produced in SRS to the electron thermal speed exceeds 3.5, the SRS is seen to saturate via the decay instability of the Langmuir wave.

## 10.10 Two-plasmon Decay

Near the quarter critical density $\left(n_0^0 \approx n_{cr}/4\right)$, the pump electromagnetic wave can excite a pair of Langmuir waves

$$\phi = Ae^{-i\left(\omega t - \vec{k}.\vec{r}\right)},\ \phi_1 = A_1 e^{-i\left(\omega_1 t - \vec{k}_1.\vec{r}\right)}, \tag{10.121}$$

where $\omega_1 = \omega_0 - \omega$, $\vec{k}_1 = \vec{k}_0 - \vec{k}$. Since all the three interacting waves are high-frequency modes, the motion of the ions can be neglected. For the electron response, one must solve the Vlasov equation since kinetic effects are important on both the decay waves. In the limit $\omega^2, \omega_1^2 >> \left(k^2, k_1^2\right) v_{th}^2$, the kinetic effects are not important in the nonlinear coupling term though very important in the dispersion and damping of the decay waves. Hence, we follow a cold fluid theory to evaluate the coupling coefficient. The linear velocity and density perturbation of the electrons due to $\phi$ are, respectively,

$$v = -\frac{ek\phi}{m\omega}, \tag{10.122}$$

and

$$n = -\frac{n_0^0 ek^2\phi}{m\omega^2}. \tag{10.123}$$

The response $(v_1, n_1)$ to $\phi_1$ can be similarly written. Retaining the nonlinear terms in the momentum and continuity equations, we obtain

$$\vec{v}^{NL} = \frac{m}{2i\omega}\left(\vec{v}_0.\nabla\vec{v}_1^* + \vec{v}_1^*.\nabla\vec{v}_0 + e\vec{v}_1^* \times \vec{B}_0/m\right) = -\frac{e\vec{k}_1.\vec{v}_0\ \phi_1^*}{2m\omega\omega_1}\vec{k}, \tag{10.124}$$

$$n^{NL} = \frac{\vec{k}}{\omega}.\left(n_0^0\vec{v}^{NL} + \frac{1}{2}n_1^*\vec{v}_0\right) = -\frac{n_0^0\ e\vec{k}.\vec{v}_0}{2m\omega\omega_1}\left(\frac{\vec{k}^2}{\omega} + \frac{\vec{k}_1^2}{\omega_1}\right)\phi_1^*. \tag{10.125}$$

The nonlinear density perturbation $n_1^{NL}$ at $\omega_1$, $\vec{k}_1$ can be obtained from $n^{NL}$ by replacing $\phi_1$ by $\phi^*$. Using $n^{NL}$ in Poisson's equation, $\nabla^2\phi = e(n^L + n^{NL})/\varepsilon_0$, and employing $n^L = \varepsilon_0 k^2 \chi_e \phi/e$, we obtain

$$\varepsilon\phi = \frac{\omega_p^2}{2\,k^2}\frac{\vec{k}\cdot\vec{v}_0}{\omega\omega_1}\left(\frac{\vec{k}^2}{\omega} + \frac{\vec{k}_1^2}{\omega_1}\right)\phi_1^*. \tag{10.126}$$

Similarly, one may write an equation for $\phi_1$. Solving the coupled set, we get

$$\varepsilon\varepsilon_1 = \frac{\left(\vec{k}\cdot\vec{v}_0\right)^2}{4k^2k_1^2}\left(\frac{k^2}{\omega} + \frac{k_1^2}{\omega_1}\right)^2, \tag{10.127}$$

where $\varepsilon$ and $\varepsilon_1$ are the dielectric functions at $(\omega, k)$ and $(\omega_1, k_1)$, respectively. Expanding $\varepsilon$ and $\varepsilon_1$ around the resonant frequencies, that is, taking

$$\omega = \omega + i\Gamma, \varepsilon = i(\Gamma + \Gamma_d)\partial\varepsilon/\partial\omega,\ \varepsilon_1 = i(\Gamma + \Gamma_{d1})\partial\varepsilon_1/\partial\omega_1,$$

one obtains

$$\left(\Gamma + \Gamma_d\right)\left(\Gamma + \Gamma_{d1}\right) = \Gamma_M^2 \equiv \frac{\left(\vec{k}.v_0\right)^2}{16\,k^2\,k_1^2}\left(\vec{k}_0.\left(\vec{k}_0 + 2\vec{k}\right)\right)^2. \tag{10.128}$$

For $\vec{k}_0 \| \hat{z}$, and $\vec{E}_0 \| \hat{y}$ (with no loss of generality), $\Gamma_M$ maximizes to $k_0|v_0|/2$ when $k_z = k_y = k_0/2$. This growth rate is the same as that of stimulated Raman scattering at the quarter critical layer. It must be noted that the growth rate for the two-plasmon decay vanishes for a dipole pump ($k_0 = 0$). The decay occurs at densities slightly below the quarter critical. As one goes to lower densities, the Landau damping of the decay waves becomes stronger, suppressing the instability for $kv_{\text{the}}/\omega_p > 0.3$, that is, $(\omega_0/2 - \omega_p)/\omega_p > 0.1$.

Tanaka et al.[40] made one of the first observations of two-plasmon decay (TDP) of 0.351 μm, $10^{13}$– $10^{15}$ W/cm$^2$ laser in CH plasma with a large density gradient, by observing half harmonic radiation as a consequence of mode conversion (inverse of resonance absorption) of the decay Langmuir waves into the electromagnetic radiation. The TPD signal was accompanied by a stronger SRS signal. Baldis et al.[41] carried direct measurements of TPD generated plasma waves in $CO_2$ laser produced plasma, while others inferred it from $3\omega_0/2$ emissions. Michel et al.[15] have reported hot electron measurements in TPD with polarization smoothed 0.351 μm 18 laser beams on planar targets and 60 beams on spherical targets. The overlapped intensity threshold for hot electron generation increases by a factor of 2 to 3 with multi-beam lasers. Above the threshold, the fraction of the laser energy converted to hot electrons increases exponentially with overlapped laser intensity. At higher intensity, the increase is much slower. Up to nearly 1% energy conversion is observed (cf. Fig. 10.11).

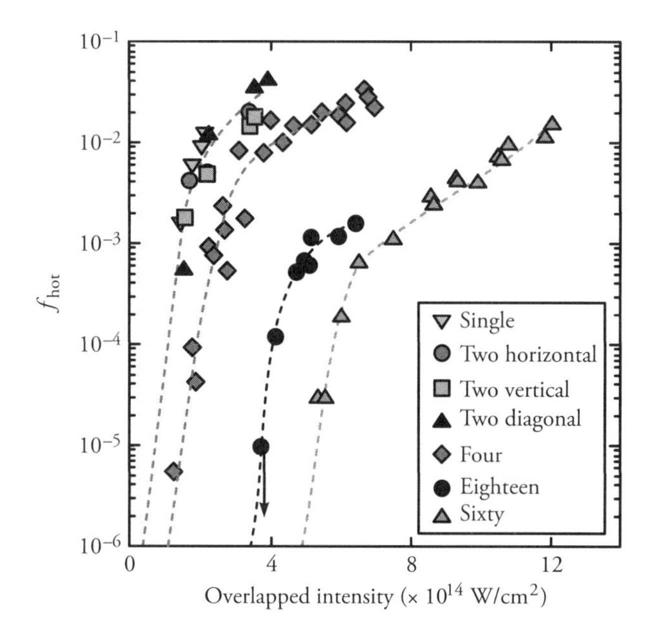


**Fig. 10.11** Fractional energy transfer to hot electrons as a function of vacuum overlapped laser intensity using multi-beam irradiation. (After Michel et al.[15])

## 10.11 Recent Experiments and Simulations

There have been extensive experimental and simulation studies of parametric instabilities. For laser fusion related studies, Montgomery[42] has recently reviewed the experiments of last two decades. It has been recognized that stimulated Raman and Brillouin scattering can be detrimental to laser fusion. Controlling these instabilities is a prerequisite for inertial confinement fusion, particularly for indirect drive fusion, where nanosecond laser beams, directed toward the gold-coated hohlraum wall, pass through a large quasi-homogeneous H-He plasma, surrounded by a highly ionized state of gold plasma. The high-Z plasma turns the laser energy into X-rays that compress the DT pellet for fusion.

Particularly significant are the results from National Ignition Facility (NIF). With the completion of the construction of 192 laser beams with 1.8 megajoules of laser power a few years ago, there was a high expectation of achieving ignition of laser fusion. Yet, the experiments have so far not been successful. One of the reasons identified was the high level of Raman scattering of the laser beam by the hohlraum plasma.

The early study with the Nova laser resulted in high level of Raman scattering, up to 50% of the laser energy, as reported by Fernandez et al.[31] in 2000. Experiments with the trident laser system with a pre-formed homogeneous plasma in the laser hotspot have been highly reproducible with about 10% Raman scattering (cf. Fig. 10.12). Particularly interesting is the observed difference in behaviors of the scattered light with respect to the ratio of the plasma wave phase velocity to the electron thermal speed (or $(k\lambda_D)^{-1}$). If the ratio exceeds 3.5, Raman scattering is observed to saturate by the parametric decay of the Langmuir wave into another Langmuir wave and an ion-acoustic wave (LDI). For lower ratio of the phase velocity to thermal speed, there are more resonant electrons with the wave causing heavy Landau damping according to the linear theory. However, nonlinearly, plasma waves can trap a significant fraction of the electrons and flatten the distribution function around the phase velocity, leading to a significant reduction of the Landau damping rate. The Raman scattering level then increases several orders of magnitude to over 10 percent with a small increment of laser power. This inflation of reflectivity appears to be a manifestation of a nonlinear transformation of the convective instability to the absolute instability of SRBS.[33] It is caused by the lowering of the threshold power for the absolute instability by the flattening of the distribution function by convective SRBS. Once the absolute instability sets in, it grows until other nonlinear effects begin to limit its growth, such as wave breaking, wave collapse, caviton or soliton formation, trapped electron parametric instability, or Brillouin backscattering whichever has the shortest time scale for growth and strongest effects.

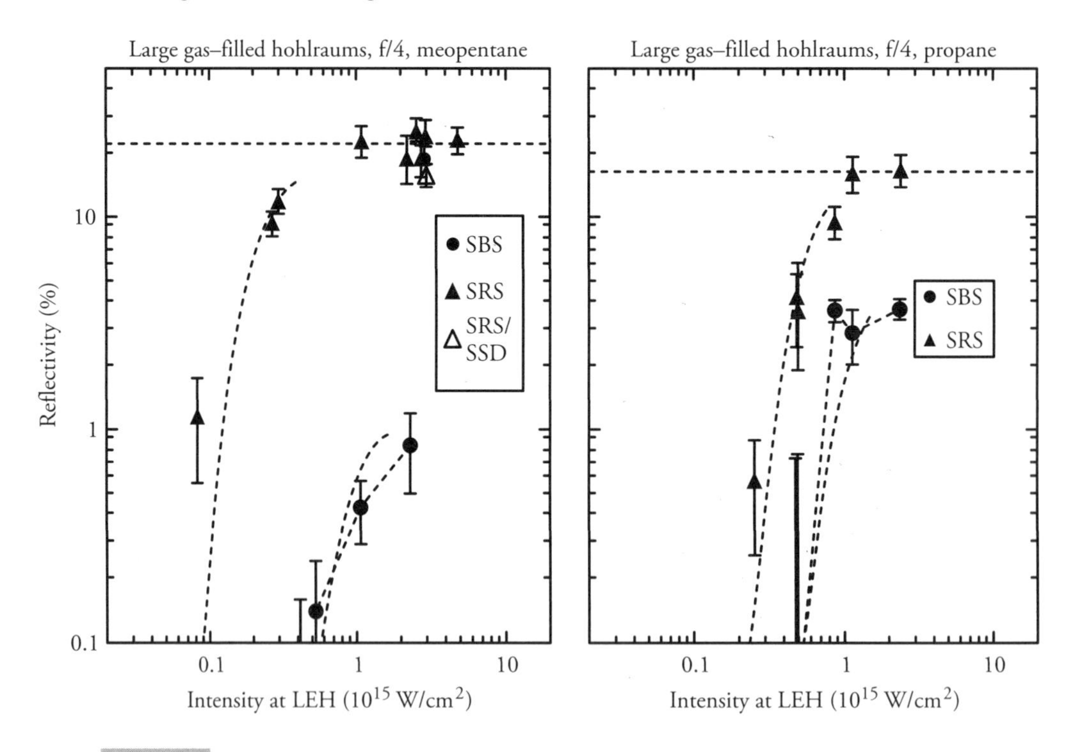


**Fig. 10.12** Raman and Brillouin reflectivities versus intensity in a gas-filled hohlraum with $n/n_{cr} = 0.11(k\lambda_D = 0.325)$ (left) and $n/n_{cr} = 0.065(k\lambda_D = 0.43)$ (right). (After Fernandez et al.[31])

Meezan et al.[43] have carried out experiments under varied conditions in a gas-filled hohlraums, using a multitude of crossed laser beams of wavelength 0.351 μm and intensity 2–10 × $10^{14}$ W/cm$^2$. With 80: 20 hydrogen: helium gas in the hohlraum and $n_e/n_{cr} \sim 0.022$, a Raman reflectivity of 20% was observed (cf. Fig. 10.13).

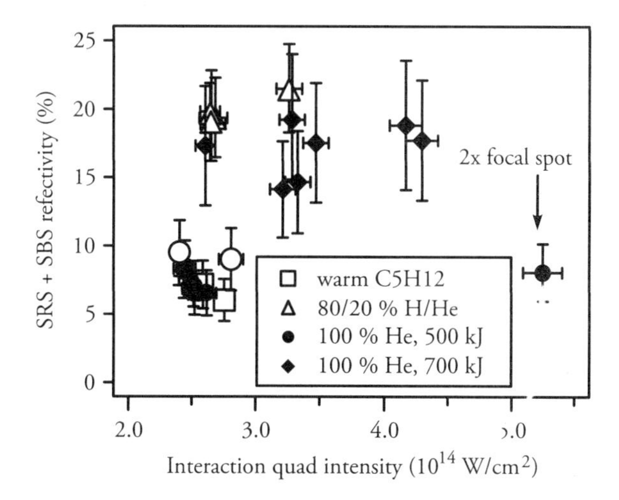


**Fig. 10.13** Total (SRS+SBS) reflectivity versus intensity in the inner cone of interaction. Shots with polarization smoothing are denoted by filled symbols. (After Meezan et al.[41])

Rosenberg et al.[28] have reported novel experimental results at NIF on SRS from direct drive ignition scale plasma. A planar CH target was simultaneously irradiated by a large number of laser beams, producing a plasma of density scale-length ~ 500–700 μm, and electron temperature $T_e$ ~ 3–5 keV with laser intensity 6–16 × $10^{14}$ W/cm$^2$ at 0.351 μm. Strong signals were observed around 0.7 μm, indicating SRS at quarter critical density. Side scattering was more prominent. There was, however, no indication of the two-plasmon decay (TPD). This was consistent with the theoretical estimates of threshold intensities. In the long density scale length plasma of the experiment, the SRS threshold was exceeded while the TPD was barely at the threshold. An important consequence of the parametric instability was the generation of hot electrons with a temperature ~45–55 keV, diverting 2–3% of the laser energy, which is very substantial and a serious threat to implosion.

Nonlinear wave processes such as Raman scattering are also important for laser electron acceleration. We refer the reader to an excellent review by Esarey, Schroder and Leeman[44] for details.

## 10.12 Discussion

High power laser–plasma interaction involves a wide variety of parametric processes that can significantly impact the efficiency of laser energy coupling to plasma. Table 10.1 summarizes the maximized growth rates of various channels of parametric instabilities.

Stimulated Raman and Brillouin scattering can cause strong reflection of laser energy from the underdense regions, long before reaching the critical layer. Compton scattering causes heating of the electrons and ions. Filamentation instability breaks up the laser beam into filaments, creating channels of higher laser intensity and depressed electron density. At the quarter critical density layer, two-plasmon decay competes with Raman scattering and leads to three-half harmonic generation. The decay instability and oscillating two-stream instability are potential contenders near the critical layer. They can significantly modify the plasma profile. The OTSI produces shorter wavelength Langmuir waves that can heat the electrons.

**Table 10.1** Maximized growth rates of various channels of parametric instability

|  |  |
|---|---|
| SRS instability | $\Gamma = \dfrac{k\lvert v_0\rvert}{4}\left(\dfrac{\omega}{\omega_0-\omega}\right)^{1/2}$ |
| SBS instability | $\Gamma = \omega_{pi}\dfrac{\lvert v_0\rvert}{4c_s}\left(\dfrac{2c_s}{c}\right)^{1/2}$ |
| SCS instability off electrons | $\Gamma = \dfrac{\lvert v_0\rvert^2\omega_p^2}{32v_{th}^2\omega_0}$ |
| SCS instability off ions | $\Gamma = \dfrac{\lvert v_0\rvert^2\omega_p^2}{8v_{th}^2\omega_0}\dfrac{T_e/T_i}{(1+0.3T_e/T_i)^2+0.3T_e^2/T_i^2}$ |
| Decay instability | $\Gamma = \left(\dfrac{\omega\omega_p}{8}\right)^{1/2}\dfrac{\lvert\vec{k}_1\cdot\vec{v}_0\rvert}{k_1v_{th}}$ |
| $2\omega_p$ Decay instability | $\Gamma = k_0\lvert v_0\rvert/2$ |
| Modulational instability | $\Gamma = \dfrac{\omega_{pi}\omega_p\omega}{2\omega_0^2}\dfrac{\lvert v_0\rvert}{c}$ |
| Filamentation instability | $\Gamma \approx \dfrac{\lvert v_0\rvert}{c\sqrt{2}}\omega_{pi}$ |
| OTSI | $\Gamma = \dfrac{\lvert v_0\rvert^2}{8v_{th}^2}\dfrac{\omega_p}{1+T_i/T_e}$ |

The parametric instability for a spherical plasma has also been studied analytically with the threshold power and growth rate obtained.[45] This is relevant for a spherical target plasma. The discussion in this chapter is restricted to coherent monochromatic pump waves. Some experiments have employed pump wave of random phase or finite frequency width $\Delta\omega_0$. As long as $\Delta\omega_0 < \Gamma$, the growth rate, the frequency width is not important as all the frequency components of the pump wave can feed energy to the same pair of daughter waves. For

$\Delta\omega_0 > \Gamma$, only a narrow spectrum of the pump wave is in resonance with a pair of daughter waves, reducing the effective intensity of the pump participating in the parametric process to $I_{\text{eff}} = I_0\Gamma/\Delta\omega_0$. Since the growth rate of a resonant parametric process goes as $I_{\text{eff}}^{1/2}$, the growth rate goes as

$$\Gamma = \frac{\Gamma_M^2}{\Delta\omega_0},$$

where $\Gamma_M$ is the growth rate when the pump is monochromatic. The frequency width of the pump thus suppresses the parametric instability, as observed in experiments[16] of Brillouin and Raman scattering.

## References


1. Rayleigh, Lord. 1883. "XXXIII. On Maintained Vibrations." *The London, Edinburgh, and Dublin Philosophical Magazine and Journal of Science* 15 (94): 229–35.
2. Nishikawa, K. and C.S. Liu. 1976. "General Formalism of Parametric Excitation" in *Advances in Plasma Physics,* vol. 6, edited by A. Simon and W.B. Thompson. New York: John Wiley. p. 3.
3. Liu, C. S. and P.K. Kaw. 1976. "Parametric Instabilities in Homogenous Unmagnetized Plasmas" in *Advances in Plasma Physics,* vol. 6, edited by A. Simon and W.B. Thompson. New York: John Wiley. p. 83.
4. Kruer, W. L. 1988. *The Physics of Laser Plasma Interactions.* Addison Wesley.
5. Drake, James F., Predhiman K. Kaw, Yee-Chun Lee, G. Schmid, Chuan S. Liu, and Marshall N. Rosenbluth. 1974. "Parametric Instabilities of Electromagnetic Waves in Plasmas." *The Physics of Fluids* 17 (4): 778–85.
6. Cohen, Bruce I., and Claire Ellen Max. 1979. "Stimulated Scattering of Light by Ion Modes in a Homogeneous Plasma: Space-Time Evolution." *The Physics of Fluids* 22 (6): 1115–32.
7. Silin, V. P. 1965. "Parametric Resonance in a Plasma." *Soviet Physics-JETP* 21: 1127.
8. Nishikawa, Kyoji. 1968. "Parametric Excitation of Coupled Waves. II. Parametric Plasmon-Photon Interaction." *Journal of the Physical Society of Japan* 24 (5): 1152–8.
9. Forslund, D. W., J. M. Kindel, and E. L. Lindman. 1975. "Theory of Stimulated Scattering Processes in Laser-irradiated Plasmas." *The Physics of Fluids* 18 (8): 1002–016.
10. Gorbunov, Leonid Mikhailovich. 1973. "Hydrodynamics of Plasma in a Strong High-frequency Field." *Physics-Uspekhi* 16 (2): 217–35.
11. Liu, C. S., and V. K. Tripathi. 1986. "Consequence of Filamentation on Stimulated Raman Scattering." *The Physics of Fluids* 29 (12): 4188–91.
12. Liu, Chuan Sheng, and V. K. Tripathi. 1986. "Parametric Instabilities in a Magnetized Plasma." *Physics Reports* 130 (3): 143–216.
13. Sagdeev, R. Z., V. D. Shapiro, and V. I. Shevchenko. 1991. "Anomalous Dissipation of Strong Electromagnetic Waves in a Laser Plasma". *Handbook of Plasma Physics.*, vol. 3, edited by M. N. Rosenbluth and R. Z. Sagdeev. Elsevier Science Publishing. p.271.

14. Gibbon, Paul, and Eckhart Förster. 1996. "Short-pulse Laser-Plasma Interactions." *Plasma Physics and Controlled Fusion* 38 (6): 769.

15. Michel, D. T., A. V. Maximov, R. W. Short, J. A. Delettrez, D. Edgell, S. X. Hu, I. V. Igumenshchev et al. 2013. "Measured Hot-electron Intensity Thresholds Quantified by a Two-plasmon-decay Resonant Common-wave Gain in Various Experimental Configurations." *Physics of Plasmas* 20 (5): 055703.

16. Rose, Harvey A., D. F. DuBois, and B. Bezzerides. 1987. "Nonlinear Coupling of Stimulated Raman and Brillouin Scattering in Laser-Plasma Interactions." *Physical Review Letters* 58(24): 2547.

17. Walsh, C. J., D. M. Villeneuve, and H. A. Baldis. 1984. "Electron Plasma-wave Production by Stimulated Raman Scattering: Competition with Stimulated Brillouin Scattering." *Physical Review Letters* 53 (15): 1445.

18. Lindl, John D., Peter Amendt, Richard L. Berger, S. Gail Glendinning, Siegfried H. Glenzer, Steven W. Haan, Robert L. Kauffman, Otto L. Landen, and Laurence J. Suter. 2004. "The Physics Basis for Ignition Using Indirect-drive Targets on the National Ignition Facility." *Physics of Plasmas* 11 (2): 339–491.

19. Phillion, D. W., D. L. Banner, E. M. Campbell, R. E. Turner, and K. G. Estabrook. 1982. "Stimulated Raman Scattering in Large Plasmas." *The Physics of Fluids* 25 (8): 1434–43.

20. Figuerova, H., C. Joshi, H. Azechi, N.A. Ebrahim, and K. Estabrook. 1987. "Stimulated Raman Scattering, Two-plasmon Decay, and Hot Electron Generation from Underdense Plasmas at 0.35 μm." *The Physics of Fluids* 27: 1887.

21. Seka, W., E. A. Williams, R. S. Craxton, L. M. Goldman, R. W. Short, and K. Tanaka. 1984. "Convective Stimulated Raman Scattering Instability in UV Laser Plasmas." *The Physics of Fluids* 27 (8): 2181–6.

22. Simon, A., R. W. Short, E. A. Williams, and T. Dewandre. 1983. "On the Inhomogeneous Two-plasmon Instability." *The Physics of fluids* 26 (10): 3107–18.

23. Ebrahim, N. A., H. A. Baldis, C. Joshi, and R. Benesch. 1980. "Hot Electron Generation by the Two-plasmon Decay Instability in the Laser-Plasma Interaction at 10.6 μm." *Physical Review Letters* 45 (14): 1179.

24. Drake, R. P., R. E. Turner, B. F. Lasinski, K. G. Estabrook, E. M. Campbell, C. L. Wang, D. W. Phillion, E. A. Williams, and W. L. Kruer. 1984. "Efficient Raman Sidescatter and Hot-electron Production in Laser-Plasma Interaction Experiments." *Physical Review Letters* 53 (18): 1739.

25. Yaakobi, B., A. A. Solodov, J. F. Myatt, J. A. Delettrez, C. Stoeckl, and D. H. Froula. 2013. "Measurements of the Divergence of Fast Electrons in Laser-Irradiated Spherical Targets." *Physics of Plasmas* 20 (9): 092706.

26. Lindl, John. 1995. "Development of the Indirect-drive Approach to Inertial Confinement Fusion and the Target Physics Basis for Ignition and Gain." *Physics of Plasmas* 2 (11): 3933–4024.

27. Myatt, J. F., J. Zhang, R. W. Short, A. V. Maximov, W. Seka, D. H. Froula, D. H. Edgell et al. 2014. "Multiple-beam Laser–Plasma Interactions in Inertial Confinement Fusion." *Physics of Plasmas* 21 (5): 055501.

28. Rosenberg, M. J., A. A. Solodov, J. F. Myatt, W. Seka, P. Michel, M. Hohenberger, R. W. Short et al. 2018. "Origins and Scaling of Hot-Electron Preheat in Ignition-Scale Direct-Drive Inertial Confinement Fusion Experiments." *Physical Review Letters* 120 (5): 055001.

29. Kline, J. L., D. S. Montgomery, L. Yin, D. F. DuBois, B. J. Albright, B. Bezzerides, J. A. Cobble et al. 2006. "Different $k\lambda_D$ Regimes for Nonlinear Effects on Langmuir Waves." *Physics of Plasmas* 13 (5): 055906.

30. Drake, R. P., E. A. Williams, P. E. Young, Kent Estabrook, W. L. Kruer, H. A. Baldis, and T. W. Johnston. 1988. "Evidence that Stimulated Raman Scattering in Laser-produced Plasmas is an Absolute Instability." *Physical Review Letters* 60 (11): 1018.
31. Fernández, Juan C., J. A. Cobble, D. S. Montgomery, M. D. Wilke, and B. B. Afeyan. 2000. "Observed Insensitivity of Stimulated Raman Scattering on Electron Density." *Physics of Plasmas* 7 (9): 3743–50.
32. Montgomery, D. S., J. A. Cobble, J. C. Fernandez, R. J. Focia, R. P. Johnson, N. Renard-LeGalloudec, H. A. Rose, and D. A. Russell. "Recent Trident Single Hot Spot Experiments: Evidence for Kinetic Effects, and Observation of Langmuir decay Instability Cascade." *Physics of Plasmas* 9 (5): 2311–20.
33. Wang, Y. X., Q. Wang, C. Y. Zheng, Z. J. Liu, C. S. Liu, and X. T. He. 2018. "Nonlinear Transition from Convective to Absolute Raman Instability with Trapped Electrons and Inflationary Growth of Reflectivity." *Physics of Plasmas* 25: 100702
34. Hopkins, LF Berzak, N. B. Meezan, S. Le Pape, L. Divol, A. J. Mackinnon, D. D. Ho, M. Hohenberger et al. 2015. "First High-convergence Cryogenic Implosion in a Near-vacuum Hohlraum."*Physical Review Letters* 114 (17): 175001.
35. Kirkwood, R. K., B. J. MacGowan, D. S. Montgomery, B. B. Afeyan, W. L. Kruer, J. D. Moody, K. G. Estabrook et al. 1996. "Effect of Ion-wave Damping on Stimulated Raman Scattering in High-Z Laser-produced Plasmas." *Physical Review Letters* 77 (13): 2706.
36. Walsh, C. J., D. M. Villeneuve, and H. A. Baldis. 1984. "Electron Plasma-wave Production by Stimulated Raman Scattering: Competition with Stimulated Brillouin Scattering." *Physical Review Letters* 53 (15): 1445.
37. Rose, Harvey A., D. F. DuBois, and B. Bezzerides. 1987. "Nonlinear Coupling of Stimulated Raman and Brillouin Scattering in Laser-Plasma Interactions." *Physical Review Letters* 58 (24): 2547.
38. Kirkwood, R. K., B. B. Afeyan, W. L. Kruer, B. J. MacGowan, J. D. Moody, D. S. Montgomery, D. M. Pennington, T. L. Weiland, and S. C. Wilks. 1996. "Observation of Energy Transfer between Frequency-mismatched Laser Beams in a Large-scale Plasma." *Physical Review Letters* 76 (12): 2065.
39. Esarey, Eric, Jonathan Krall, and Phillip Sprangle. 1994. "Envelope Analysis of Intense Laser Pulse Self-modulation in Plasmas."*Physical Review Letters* 72 (18): 2887.
40. Tanaka, K., L. M. Goldman, W. Seka, M. C. Richardson, J. M. Soures, and E. A. Williams. 1982. "Stimulated Raman Scattering from UV-laser-produced Plasmas." *Physical Review Letters* 48 (17): 1179.
41. Baldis, H. A., and C. J. Walsh. 1981. "Experimental Observations of Nonlinear Saturation of the Two-plasmon Decay Instability."*Physical Review Letters* 47 (23): 1658.
42. Montgomery, David S. 2016. "Two Decades of Progress in Understanding and Control of Laser Plasma Instabilities in Indirect Drive Inertial Fusion." *Physics of Plasmas* 23 (5): 055601.
43. Meezan, N. B., L. J. Atherton, D. A. Callahan, E. L. Dewald, S. Dixit, E. G. Dzenitis, M. J. Edwards et al. 2010. "National Ignition Campaign Hohlraum Energetics." *Physics of Plasmas* 17 (5): 056304.
44. Esarey, E., C. B. Schroeder, and W. P. Leemans. 2009. "Physics of Laser-driven Plasma-based Electron Accelerators." *Reviews of Modern Physics* 81 (3): 1229.
45. Liu C. S., M. N. Rosenbruth, and R. B. White. 1975 "Parametric Instabilities in Spherical Inhomogeneous Plasmas," *Proceedings of 5th International Conference on Plasma Physics and Controlled Fusion*, IAEA Tokyo. 1974.

# 11 Parametric Instabilities in Inhomogeneous Plasma

## 11.1 Introduction

Phase matching is an important issue in parametric instability. In an inhomogeneous plasma, frequency and wave vector matching conditions are satisfied only locally. As the waves move away from this region, the phase mismatch grows and instability is deterred.[1–6] In direct drive fusion experiments[7,8] the plasma density varies rapidly over a length of a few hundred microns, hence, the localization of the parametric instability and convection losses are strong. In indirect drive fusion,[9] the density scale lengths are longer. However, convection losses decide the threshold pump intensity for the spatial or temporal growth of the parametric instability. Laser interactions with gas jets and thin foil targets one also encounter inhomogeneous plasmas, where, besides the density gradient, a gradient in the plasma flow velocity could also be significant. The parametric instabilities which involve ion motion are influenced by the flow velocity gradient.

For a pump wave of frequency $\omega_0$ launched from the low density side at an angle $\theta_0$ to the density gradient, $\nabla n_0^0 \| \hat{x}$, there are four distinctive regions of interest, viz., the underdense region ($n_0^0 << n_{0cr}$), the quarter critical layer ($n_0^0 \sim n_{0cr}/4$), the turning point ($n_0^0 \sim n_{0cr} \cos^2\theta_0$), and the critical layer ($n_0^0 \sim n_{0cr}$).

In the underdense region, stimulated Raman and Brillouin scattering are the prominent parametric instabilities with electromagnetic sidebands of frequency considerably larger than the local plasma frequency. As long as the decay waves have a significant component of their wave vector along the density gradient, one may employ the Wentzel–Kramers–Brillouin (WKB) approximation to study the scattering processes. For a linear density profile, the $k$ matching condition ($k_{0x}(x) = k_x(x) + k_{1x}(x)$) is satisfied only at a specific value of $x$, say $x = 0$. Beyond this point, the phase mismatch varies as $x^2$ and the region of parametric interaction is localized. The propagation of the decay waves out of this region sets a threshold for the parametric instability. The instability turns out to be convective, where the amplitudes of the daughter waves grow spatially but not temporally after the initial build up.

At large scattering angles with respect to the density gradient, the side-band electromagnetic wave has a turning point in the interaction region, allowing a sufficiently long time for the nonlinear interaction. Consequently the instability could become absolute with temporally growing daughter waves. However, with pump waves of narrow transverse extent, sideways convective losses could be severe.

Near the quarter critical layer, the frequency of the Raman backscattered electromagnetic wave is close to the local plasma frequency ($\omega_1 \sim \omega_p$), that is, it has a turning point that may lead to an absolute instability. In this region the two-plasmon decay is an important parametric instability. The Langmuir waves have turning points and the process turns out to be an absolute instability.

In the region between the quarter critical layer and the turning point, stimulated Brillouin scattering is the only three-wave parametric instability. Near the turning point the pump wave acquires large amplitude and the growth rate of the instability is considerably enhanced. When the angle of incidence $\theta_0$ is small, that is, $\cos^2\theta_0 - 1 << 1$, one may excite the decay instability and the oscillating two-stream instability. Further, if the pump wave is P-polarized and the separation between the turning point and critical layer small, a large amplitude Langmuir wave is produced near the critical layer via linear mode conversion. This wave excites decay and oscillating two-stream instabilities with large growth rates.

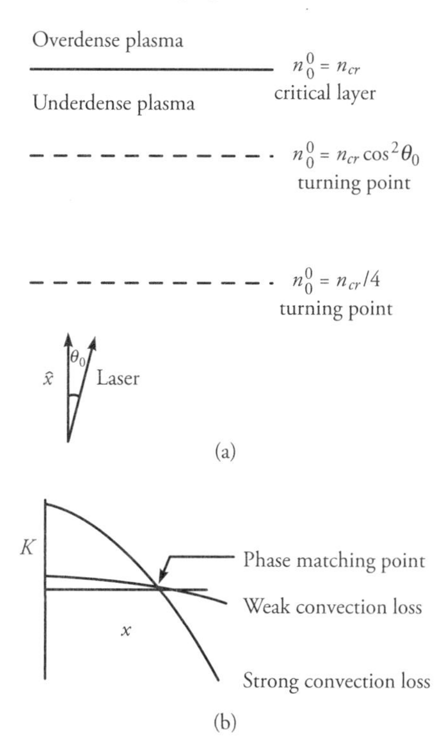


**Fig. 11.1** (a) Various regions of an inhomogeneous plasma, with a laser impinging from vacuum at an angle of incidence $\theta_0$. (b) Wave number mismatch $K = k_x(x) - k_{0x}(x) + k_{1x}(x)$ as a function of $x$; higher rates of variation of $K$ with $x$ lead to higher convection losses of the parametric instability.

In Section 11.2 we will study the convective amplification of stimulated Raman and Brillouin scattering using the WKB approximation. The absolute Raman instability and side-scattering at the quarter critical density will be discussed in Section 11.3. In Section 11.4, we will study the two-plasmon decay. Decay instability will be explained in Section 11.5. In Section 11.6, we will discuss the four-wave process of the oscillating two-stream instability. Finally some concluding remarks are made in Section 11.7.

## 11.2 Convective Raman and Brillouin Instabilities

In an underdense plasma with a gentle density gradient (parallel to the *x*-axis), the field of the electromagnetic pump wave, in the WKB approximation, can be written as

$$\vec{E}_0 = \vec{A}_0 \exp[-i(\omega t - k_{0z}z - \int k_{0x}dx)]. \tag{11.1}$$

Similarly, the potential of the decay electrostatic wave and the electric field of the scattered electromagnetic wave can be written as

$$\phi = Ae^{-i(\omega t - k_z z - \int k_x dx)}, \tag{11.2a}$$

$$\vec{E}_1 = \vec{A}_1 e^{-i(\omega_1 t - k_{1z}z - \int k_{1x}dx)}. \tag{11.2b}$$

The mode coupled equations for the normalized decay wave amplitudes $a' = kA(\varepsilon_0(\partial\varepsilon/\partial\omega)/\hbar)^{1/2}/2$, and $a_1' = A_1(\varepsilon_0/2\hbar\omega_1)^{1/2}$ can be deduced from Eqs. (10.43) and (10.49) by replacing $k_x$ by $k_x - i\partial/\partial x$ and $\omega$ by $\omega + i\partial/\partial t$ in the electrostatic wave dielectric function $\varepsilon(\omega,\vec{k})$, and $k_{1x}$ by $k_{1x} - i\partial/\partial x$ and $\omega_1$ by $\omega_1 + i\partial/\partial t$ in $D_1(\omega_1,\vec{k}_1)$ characterizing the scattered wave, in the same way as in Section 10.5,

$$\frac{\partial a'}{\partial t} + v_{gx}\frac{\partial a'}{\partial x} + \Gamma_d a' = i\Gamma_M a_1'^{*} e^{-iK'x^2/2}, \tag{11.3}$$

$$\frac{\partial a_1'}{\partial t} + v_{g_1x}\frac{\partial a_1'}{\partial x} + \Gamma_{d1} a_1' = i\Gamma_M a'^{*} e^{-iK'x^2/2}, \tag{11.4}$$

where * denotes the complex conjugate, $v_{gx}$ and $v_{g1x}$ are the *x*-components of the group velocities of the decay waves. We have $K = k_x(x) - k_{0x}(x) + k_{1x}(x)$, $K' = (dK/dx)_{x=0}$ at the phase matching point around which we have carried the Taylor expansion of *K*, $\Gamma_M = k|v_0|/(8\omega_1\partial\varepsilon/\partial\omega)^{1/2}$ and we have assumed $x = 0$ as the point at which $K = 0$.

Defining

$$a = a' e^{-i K' x^2/4}, \quad a_1 = a_1' e^{i K' x^2/4}, \tag{11.5}$$

Equations (11.3) and (11.4) can be written as

$$\left(\frac{\partial}{\partial t} + v_{gx}\frac{\partial}{\partial x} - i v_{gx} K' \frac{x}{2} + \Gamma_d\right) a = i\Gamma_M a_1^*, \tag{11.6}$$

$$\left(\frac{\partial}{\partial t} + v_{g1x}\frac{\partial}{\partial x} - i v_{g_1 x} K' \frac{x}{2} + \Gamma_{d1}\right) a_1 = i\Gamma_M a^*. \tag{11.7}$$

Usually $\Gamma_{d1}$ is small and can be neglected.

For the steady-state solution, we set $\partial/\partial t = 0$ in Eqs. (11.6) and (11.7). In a special case when the damping of the electrostatic mode is large, $\Gamma_d > v_{gx}\, \partial/\partial x$, these equations give

$$\frac{da_1}{dx} = \left[\frac{\Gamma_M^2}{\left(\Gamma_d - i\, v_{gx} K' x/2\right) v_{g1x}} + iK' \frac{x}{2}\right] a_1. \tag{11.8}$$

For a back-scattered sideband wave coming from $+\infty$, $v_{g1x}$ is negative with an initial amplitude $a_1 = a_{10}$, (for forward scatter, $v_{g1x}$ is positive). Equation (11.8) can be integrated to obtain the amplitude of the wave after passing through the interaction region (from $+\infty$ to $-\infty$),

$$\left|a_1\right|_{x \to \infty} = a_{10} e^{\pi G}, \tag{11.9}$$

where

$$G = \frac{\Gamma_M^2}{\left|v_{gx} v_{g_1 x} K'\right|}$$

is the convective amplification factor. The linear damping of the low-frequency mode limits the height but broadens the width of the resonant interaction, resulting in an amplification independent of $\Gamma_d$. For significant parametric amplification, one must have $G \geq 1$. This determines the threshold pump power. It can be shown that a temporal growth of the instability, in the WKB approximation considered here, is not possible (cf. Rosenbluth[1]).

### 11.2.1 Stimulated Raman scattering

We now obtain the threshold powers for SRS (stimulated Raman scattering) and SBS (stimulated Brillouin scattering) for a linear density profile with scale length $L_n$,

$$\omega_p^2 = \omega_{p0}^2 \left(1 + \frac{x}{L_n}\right). \tag{11.10}$$

In this case, the decay weaves, viz., the Langmuir wave and the scattered electromagnetic wave, have $v_{gx} = 3v_{\text{th}}^2 k_x / \omega$, $v_{g1x} = c^2 k_{1x} / \omega_1$, $K' \cong \omega_{p0}^2 / 6L_n k_x v_{\text{th}}^2$, and the convective amplification factor $G$ simplifies to

$$G = \frac{k^2 |v_0|^2 L_n}{8c^2 k_{1x}}. \tag{11.11}$$

For backscattering, Eq. (11.11) gives the threshold laser intensity (time-averaged Poynting vector) corresponding to $G = 1$,

$$I_{\text{th}} = \frac{mn_{\text{cr}} c^3}{(\omega_0 L_n / c)}. \tag{11.12}$$

Laser–plasma experiments with exploding foils have confirmed the enhancement of the Raman reflectivity with the scale length of the underdense plasma using 0.53 m laser light at 1 ≥ $10^{14}$ W/cm$^2$. In indirect drive fusion, experiments with 0.351 μm multi-beam lasers have reported strong stimulated Raman back scattering (SRBS)[9] in quasi-homogeneous underdense hohlraum plasma (scale length > 400 μm, $0.03 < n_0^0 / n_{\text{cr}} < 0.1$, $T_e$ = 2.5 keV) at intensities above 2 – 3 × $10^{14}$ W/cm$^2$. In fact Raman reflectivity goes up several orders of magnitude to around 10% as the intensity is varied by a small fraction and then changes only slightly as the intensity is further increased, contrary to convective amplification. Wang et al.[11] have interpreted it as a nonlinear transition from convective to absolute Raman instability as discussed in Chapter 10. In the direct drive fusion, the SRS at the quarter critical density is more dominant and Raman reflectivity correlates well with hot electron generation.[8]

## 11.2.2 Stimulated Brillouin scattering

For $\omega << \omega_{\text{pi}}$ the wave number of the ion-acoustic wave is independent of the electron density, and $K'$ in a stimulated Brillouin scattering process due to the density gradient being rather small, leading to a large amplification. However, the sound speed depends on the electron temperature $T_e$ and the plasma flow velocity, $u\hat{x}$; hence inhomogeneities in $T_e$ and $u$ introduce an $x$-dependence in $k_x$. Here we consider only the inhomogeneity in $u$. The plasma flow modifies the ion susceptibility.

$$\chi_i = -\frac{\omega_{pi}^2}{(\omega - k_x u(x))^2}, \tag{11.13}$$

giving the modified dispersion relation for the ion-acoustic wave

$$\omega = \frac{kc_s}{\left(1 + k^2 c_s^2 / \omega_{\text{pi}}^2\right)^{1/2}} + k_x u(x)$$

$$\cong kc_s + k_x u. \tag{11.14}$$

Then $v_{gx} = u + (k_x/k)c_s$, $K' \sim \left\{k_x u / \left[u + (k_x / k)c_s\right]\right\} / L_u$ and the convective amplification factor turns out to be

$$G = \frac{\Gamma_M^2 L_u k_1}{cuk_x k_{ix}}, \tag{11.15}$$

where $\Gamma_M$ is given by Eq. (10.60) and $L_u$ is the scale length of the $u$ variation. For backscatter,

$$G = \frac{|v_0|^2}{8v_{\text{th}}^2} \frac{\omega_p^2}{\omega_0^2} \frac{\omega_0 L_u}{c} \frac{c_s}{u} \tag{11.16a}$$

$$I_{\text{th}} = 4\frac{n_{\text{cr}}}{n_0^0} \frac{u / c_s}{L_u \omega_0 / c} mv_{\text{th}}^2 n_{\text{cr}} c. \tag{11.16b}$$

Froula et al.[12] have experimentally confirmed the influence of a flow velocity gradient and kinetic effects on SBS. For intensities below $1.5 \times 10^{15}$ W/cm$^2$, they observed that convective amplification was moderated by the flow velocity gradient. Thomson scattering measurements revealed detuning of the SBS by the velocity gradient. At higher intensities, ion trapping in the ion-acoustic wave was seen to be important. The measured two-fold increase in the ion temperature and its correlation with the SBS reflectivity provided quantitative evidence that hot ions were created by ion trapping in the ion-acoustic wave. In later experiments, Froula et al.[13] observed two-fold enhancement in the SBS threshold with polarization smoothing in ignition hohlraum plasma.

### 11.2.3 Brillouin side scattering

At a large scattering angle, the scattered wave, in the stimulated Brillouin scattering process, has a turning point in the interaction region while the ion-acoustic wave does not have a turning point. Consequently, the instability remains convective as demonstrated by Liu and Rosenbluth.[4] The convective amplification factor for Brillouin side-scattering turns out to be the same as obtained in the WKB approximation,

$$G \equiv \frac{\Gamma_M^2}{v_{gx} v_{g1x} \frac{d}{dx}\left(k_x + k_{1x} - k_{0x}\right)}$$

$$\approx \frac{\Gamma_M^2 \omega_0 L_n}{c_s \omega_{p0}^2} \approx \frac{v_0^2}{4v_e^2} \frac{\omega_0 L_n}{c}. \tag{11.17}$$

The major factor responsible for the enhancement of side-scattering is the slowing down of the group velocity of the light wave to

$$v_{g1x} \sim \frac{c\lambda_0}{\lambda_{em}} = c\frac{c/\omega_0}{\left(c^2 L_n / \omega_{p0}^2\right)^{1/3}} \sim c\left(\frac{c}{L_n \omega_{p0}}\right)^{1/3} \frac{\omega_{p0}}{\omega_0}.$$

## 11.3 Absolute Raman Instability at $n_{cr}/4$ and Side-scattering

The WKB approximation explained earlier is valid as long as $(d/dx)ln(k_x) << k_x$ holds for both daughter waves. When the angle between $\vec{k}_1$ and $\nabla n_0^0$ approaches $\pi/2$ or the interaction region is close to the quarter critical layer, $k_{1x} \longrightarrow 0$, the *WKB* approximation fails and one must solve the fluid and Maxwell's equations without replacing $\partial/\partial x$ by $ik_x$ or $ik_{1x}$. The same result can be obtained by replacing $k_x^2\phi$ with $-\partial^2\phi/\partial x^2$ and $k_{1x}^2 E_1$ with $-\partial^2 E_1/\partial x^2$ in the coupled mode equations, Eq. (10.43) and (10.49), resulting in

$$\frac{\partial^2 \phi}{\partial x^2} + \left(\frac{\omega^2 - \omega_p^2}{3v_{\text{th}}^2} - k_z^2 + \frac{2i\omega\Gamma_d}{3v_{\text{th}}^2}\right)\phi = \left|v_0\right| \frac{E_1^* \omega^2 e^{ik_0 x}}{6\omega_1 v_{\text{th}}^2} \tag{11.18a}$$

$$\frac{\partial^2 E_1}{\partial x^2} + \left(\frac{\omega_1^2 - \omega_p^2}{c^2} - k_{1z}^2\right)E_1 = \left|v_0\right| \frac{\omega_1 e^{ik_0 x}}{2c^2}\left(k_z^2 \phi^* - \frac{\partial^2 \phi^*}{\partial x^2}\right). \tag{11.18b}$$

We are interested in the neighborhood of the turning point of $E_1$, where the second term in Eq. (11.18b) vanishes, that is, $k_{1x}^2 \approx \left(w_1^2 - w_p^2\right)/c^2$. For phase matching $k_x \cong k_{0x}$. To solve this coupled set analytically, we will neglect the thermal effects and damping on the Langmuir wave, that is, $\omega^2 - \omega_p^2 > k^2 v_{\text{th}}^2$, $2i\omega\Gamma_d$. Consider $\omega_p^2 = \omega_{p0}^2\left(1 + x/L_n\right)$ and introduce a new variable $\xi' = x/\lambda_{\text{em}}$ where $\lambda_{\text{em}} = \left(c^2 L_n / \omega_p^2\right)^{1/3}$, to obtain from Eqs. (11.18a) and (11.18b),

$$\frac{\partial^2 E_1}{\partial \xi'^2} - \left(\xi' - D - \frac{\lambda}{\xi' - B}\right)E_1 = 0, \tag{11.19}$$

where

$$D = \left(\frac{\omega_1^2 - \omega_{p0}^2}{c^2} - k_{1z}^2\right)\lambda_{\text{em}}^2,$$

$$B = \left( \frac{\omega^2}{\omega_{p0}^2} - 1 \right) \frac{L_n}{\lambda_{\text{em}}},$$

and

$$\lambda = \frac{L_n}{\lambda_{\text{em}}} \frac{|v_0|^2 k^2}{4c^2}.$$

Changing the variable further to $\xi = \xi' - \dfrac{D + B}{2}$, Eq. (11.19) becomes

$$\frac{\partial^2 E_1}{\partial \xi^2} + \frac{\alpha^2 + \lambda - \xi^2}{\alpha + \xi} E_1 = 0, \tag{11.20}$$

where $\alpha = (D - B)/2$. This equation has two turning points at

$$\xi_T = \pm(\alpha^2 + \lambda)^{1/2}$$

As long as $|\xi_T| << \alpha$, as must be verified a posteriori, solutions of Eq. (11.20) are parabolic cylinder functions and the eigenvalues are given by

$$\frac{\alpha^2 + \lambda}{2\sqrt{\alpha}} = n + \frac{1}{2}, \quad n = 0, 1, 2, \ldots \tag{11.21}$$

Solving Eq. (11.21) iteratively for large $\lambda >> 1$ yields

$$\alpha = -i\sqrt{\lambda} \left[ 1 + \left( n + \frac{1}{2} \right) (-\lambda)^{-3/4} \right]. \tag{11.22}$$

The growth rate on solving $\alpha = (D - B)/2$ turns out to be

$$\gamma = \frac{\omega_{p0} k |v_0|}{2\omega_0} \left[ 1 - \frac{(n + 1/2)(\omega_{p0}/kc)^{1/2}}{\sqrt{2}(v_0/c)^{3/2} kL_n} \right]. \tag{11.23}$$

The condition $|\xi_T| << \alpha$ reduces to $\lambda^{1/4} >> 2n + 1$ or

$$|v_0|/c >> (2n+1)^2 \frac{2/k}{(L_n \lambda_{\text{em}})^{1/2}}.$$

The threshold is approximately given by

$$(v_0/c)^{3/2} k_0 L_n > 1. \tag{11.24}$$

In terms of laser intensity in units of $10^{14}$ W/cm$^2$ and wavelength and density scale length in micron, the threshold can be expressed as $I_{14}^{\text{SRS,th}} \approx 2.4\times10^2 / \lambda_\mu^2 L_{n,\mu m}^{4/3}$.

Rosenberg et al.[8] have observed very significant Raman back and side scattering from a direct drive ignition scale plasma, by irradiating a planar CH target simultaneously with a large number of laser beams from the National Ignition Facility (NIF) at 0.351 μm wavelength and 6 – 16 × $10^{14}$ W/cm$^2$ intensity. Strong signals around 0.7 μm, indicative of the SRS instability at the quarter critical density, were observed when the aforementioned threshold condition was exceeded. This was accompanied by the generation of hot electrons with a temperature ~45 – 55 keV. For the SRS back scattering, there is interestingly a gap in the reflected spectrum between $n_{cr}/4$ and the more underdense region.

## 11.4 Two-Plasmon Decay

Near the quarter-critical density an electromagnetic pump wave can decay into two Langmuir waves. The daughter waves may couple to the pump wave to produce scattered electromagnetic waves at ~ 3 $\omega_0/2$. That is observed in many experiments. The Landau damping of the plasma waves heats the plasma, producing hot ELECTRONS. Plasma inhomogeneity has a profound effect on this decay process. Following the procedure reported by Liu and Rosenbluth,[14] we study this parametric instability without resorting to the WKB approximation. Both the decay waves have turning points on the higher density side of the resonant interaction region and propagate at an angle to the density gradient. It turns out that the decay waves, on account of nonlinear coupling, have turning points on the other side (low density side) of the interaction region also and $2\omega_p$ decay becomes an absolute instability.

Consider an electromagnetic pump wave $E_0 = \hat{z}A_0 \exp\left[-i\left(\omega_0 t - k_0 x\right)\right]$ propagating along the density gradient in a plasma with a linear density profile. The pump wave results in an oscillatory electron velocity $\vec{v}_0 = e\vec{E}_0 / mi\omega_0$ and decays into two Langmuir waves of frequencies $\omega$ and $\omega_1$ where $\omega = \omega_0 - \omega_1$. The pump and decay waves exert a ponderomotive force on the electrons, $\vec{F}_p = e\nabla(\phi_p + \phi_{p1})$ with the $\omega$ and $\omega_1$ frequency ponderomotive potentials

$$\phi_p = -\frac{m}{2e}\vec{v}_0\cdot\vec{v}_1^* \approx -\frac{\vec{v}_0\cdot\nabla\phi_1^*}{2i\omega_1}$$

$$\phi_{p1} \cong -\frac{\vec{v}_0\cdot\nabla\phi^*}{2i\omega} \tag{11.25}$$

Using Eq. (11.25), the fluid equations and Poisson's equation for the $w$ frequency component can be written as

$$-i\omega\vec{v} = \frac{e}{m}\nabla\phi - \frac{v_{\text{th}}^2}{n_0^0}\nabla n - \frac{e\nabla\left(\vec{v}_0\cdot\nabla\phi_1^*\right)}{2mi\omega_1}, \tag{11.26a}$$

$$- i\omega n - \nabla \cdot \left[ n_0^0 \left( 1 + \frac{x}{L_n} \right) \vec{v} + \frac{1}{2} n_1^* \vec{v}_0 \right] = 0, \tag{11.26b}$$

$$\nabla^2 \phi = en / \varepsilon_0. \tag{11.26c}$$

The set of equations for the $\omega_1$ frequency component can be written similarly. Writing $\vec{v} = \nabla(\psi(x)\exp[-i(\omega t - k_z z)], \vec{v}_1 = \nabla(\psi_1(x)\exp[-i(\omega_1 t - k_{1z} z)]$, $\partial/\partial z = ik_z$ and Fourier transforming in $x$, we obtain two coupled equations for $u = k^2 \psi_{k_x}$ and $u_1 = k_1^2 \psi_{k_{1x}}$,

$$\frac{i}{L_n}\frac{du}{dk_x} - \frac{\omega^2 - \omega_{po}^2 - 3k^2 v_{\text{th}}^2}{\omega_{po}^2} u = -i\frac{k_z |v_0|}{2\omega_{p0}} \left( \frac{k_1^2}{k^2} + \frac{\omega}{\omega_1} \right) u_1^*, \tag{11.27}$$

$$\frac{i}{L_n}\frac{du_1}{dk_{1x}} - \frac{\omega_1^2 - \omega_{po}^2 - 3k_1^2 v_{\text{th}}^2}{\omega_{po}^2} u_1 = -i\frac{k_z |v_0|}{2\omega_{p0}} \left( \frac{k^2}{k_1^2} + \frac{\omega_1}{\omega} \right) u^*, \tag{11.28}$$

where $k^2 = k_x^2 + k_z^2$, $k_1^2 = \left(k_x - k_0\right)^2 + k_z^2$. Let

$$u = W \exp\left[ -i\frac{L_n}{\omega_{p0}^2} \int (\omega^2 - \omega_{p0}^2 - 3k^2 v_{\text{th}}^2) dk_x \right]$$

$$u_1 = W_1 \exp\left[ -i\frac{L_n}{\omega_{p0}^2} \int (\omega^2 - \omega_{p0}^2 - 3k_1^2 v_{\text{th}}^2) dk_{1x} \right]$$

and $k_x = k_0/2 + K$, $\omega = \omega_0/ 2 + \Delta$. Then we have

$$\frac{i}{L_n}\frac{dW}{dK} = -i\frac{k_z |v_0|}{\omega_{p0}} \frac{k_0 K W_1^*}{\left(K + k_0/2\right)^2 + k_z^2} e^{i\theta}, \tag{11.29a}$$

$$\frac{i}{L_n}\frac{dW_1}{dK} = -i\frac{k_z |v_0|}{\omega_{p0}} \frac{k_0 K W^*}{\left(K - k_0/2\right)^2 + k_z^2} e^{-i\theta}, \tag{11.29b}$$

where

$$\theta = \frac{L_n}{\omega_{p0}^2} \int (\omega^2 - \omega_1^2 - 6k_0 K v_{\text{th}}^2) dK = \frac{L_n}{\omega_{p0}^2} (2\omega_0 K \Delta - 3k_0 K^2 v_{\text{th}}^2)$$

Eliminating $W$ we obtain an equation for $W_1$

$$\frac{d^2W_1}{dK^2} + g\frac{dW_1}{dK} + L_n^2K^2\frac{k_z^2\left|v_0\right|^2 k_0^2W_1}{\omega_0^2\left[(k_0^2/4 + k_z^2 + K^2)^2 - k_0^2K^2\right]} = 0, \tag{11.30}$$

$$g = \left[\frac{2iL_n}{\omega_{po}^2}\left(\omega_0\Delta - 3k_0Kv_{\text{th}}^2\right) + \frac{2\left(K - k_0/2\right)}{\left(K - k_0/2\right)^2 + k_z^2} - \frac{1}{K}\right].$$

Writing $W_1 = v\exp(i\int gdK/2)$ we get

$$\frac{d^2v}{dK^2} + \frac{L_n^2}{\omega_{p0}^2}F(K)v = 0 \tag{11.31}$$

with $F = F_0 + F_1 + F_2$,

$$F_0 = \frac{k_z^2\left|v_0\right|^2 k_0^2K^2W_1}{\left(k_z^2 + K^2 + k_0^2/4\right)^2 - k_0^2K^2} + \frac{(3k_0Kv_{\text{th}}^2 - \omega_0\Delta)^2}{\omega_{p0}^2},$$

$$F_1 = -\frac{i}{L_n}\left[\frac{\omega_0\Delta}{K} - \frac{2\left(K - k_0/2\right)\left(\omega_0\Delta - 3k_0Kv_{\text{th}}^2\right)}{\left(K - k_0/2\right)^2 + k_z^2}\right], \tag{11.32}$$

$$F_2 = \frac{\omega_{p0}^2}{L_n^2}\left[-\frac{3}{4K^2} + \frac{\left(K - k_0/2\right)^3 - k_z^2k_0/2}{K\left(k_z^2 + \left(K - k_0/2\right)^2\right)}\right]$$

in successive orders of $L_n^{-1}$. The solution of Eq. (11.31) can be obtained in the WKB approximation: $v \sim F^{-1/4}\exp[\pm i\int F^{1/2}dK]$. We require the solution to be localized in $k$-space with a finite extent of localization, which implies the localization of the Fourier transformed solution in the $x$ space because of the uncertainly principle. Such a localized solution in $k$-space exists if (i) $F(k)$ possesses a maximum corresponding to a potential well in $k$-space for the equivalent Schrödinger equation, and (ii) the Bohr-Sommerfield quantization condition is satisfied,

$$\frac{L_n}{\omega_{p0}}\int_{K_1}^{K_2}F^{1/2}dK = (n + 1/2)\pi \tag{11.33}$$

where $K_1$ and $K_2$ are the turning points at which $F(K) = 0$, adjacent to the maximum where $dF/dK = 0$. Equation (11.33) is the eigenvalue equation determining the eigenvalues $\Delta$, with Im $\Delta$ giving the growth rate. We obtain an approximate solution treating $L_n^{-1}$ as a small parameter. In the limit $L_n \to \infty$ (homogeneous plasma), the Bohr–Sommerfield quantization condition Eq. (11.33), demands that $\int_{K_1}^{K_2} F_0^{1/2} dK$ be vanishingly small. This implies that $K_1 = K_2 = K_0$ where $K_0$ is the point at which $F_0$ has a maximum. Thus the two turning points surrounding the maximum must coalesce, that is,

$$F_0\left(K_0\right) = \left.\frac{dF_0}{dK}\right|_{K=K_0} = 0. \tag{11.34}$$

Equation (11.34) determines the location of $K_0$ and the lowest order eigenvalue. This condition for the coalescence of the roots with Im $\Delta > 0$ is essentially the condition for the absolute instability in a homogeneous plasma. One may conveniently employ the saddle point method of integration. Let $q = K/k_z$. The eigen-frequency from Eq. (11.32) by setting $F_0 = 0$ turns out to be

$$\omega\left(k_z, K\right) \equiv \frac{\omega_0}{2} + \Delta = \frac{\omega_0}{2} + \left(q + i\lambda \frac{q}{1+q^2}\right)\Omega, \tag{11.35}$$

where $\lambda = 2\left|v_0\right|\omega_{p0} / 3k_z v_{\text{th}}^2$, $\Omega = 3k_0 k_z v_{\text{th}}^2 / \omega_0$. The saddle point is determined by $\partial\omega/\partial q = 0$, or $1 + i\lambda\,(1 - q^2)/\,(1 + q^2)^2 = 0$ which gives

$$q_s^2 = -\left(1 - \frac{i\lambda}{2}\right) - i\left(\frac{\lambda^2}{4} + 2i\lambda\right)^{1/2}, \tag{11.36}$$

where the negative root is picked so that $q = 1$ for $\lambda >> 1$. Setting $q = q_s$ in Eq. (11.35), we find the value of the eigen-frequency at the saddle point

$$\Delta = q_s\Omega\left[1 + \frac{i}{4}(\lambda + (\lambda^2 + 8i\lambda)^{1/2})\right] \tag{11.37}$$

For $\lambda >> 1$ this gives the growth rate of the absolute instability

$$\Gamma_0 = k_0\left|v_0\right| / 4. \tag{11.38}$$

The requirement that $F_0(K_0, \Delta_0)$ is a maximum imposes a condition for the existence of localized modes,

$$\left.\frac{\partial^2 F_0}{\partial K^2}\right|_{K=K_0} = \frac{1}{\omega_{p0}^2}\left[-\frac{k_0^2|v_0|^2}{2k_z^2} + \frac{2(3k_0 v_{\text{th}}^2)^2}{\omega_p^2}\right] < 0$$

or

$$k_z < |v_0|\omega_p / 6v_{\text{th}}^2. \tag{11.39}$$

This is the same condition as that obtained by Liu and Rosenbluth[14] for absolute instability using a Nyquist diagram.

To the next order we include $F_1$. Letting $K = K_0 + K_1$, $\Delta = \Delta_0 + \Delta_1$ and expanding $F$,

$$F = \left.\frac{K_1^2}{2}\frac{\partial^2 F_0}{\partial K^2}\right|_{K_0,\Delta_0} + \Delta_1 \left.\frac{\partial F_0}{\partial \Delta}\right|_{K_0,\Delta_0} + F_1(K_0,\Delta_0)\,. \tag{11.40}$$

With this $F$, Eq. (11.31), for $\left.\partial^2 F_0 / \partial K^2\right|_{K_0,\,\Delta_0} < 0$, becomes the standard Weber equation with a potential well giving

$$\Delta_1 = \left(n + \frac{1}{2}\right)\frac{\left(-2\,\partial^2 F_0 / \partial K^2\right)^{1/2}}{\partial F_0 / \partial \Delta} - \frac{F_1}{\partial F_0 / \partial \Delta}$$

$$\approx -i\left(n + \frac{1}{2}\right)\frac{\omega_{p0}}{2k_x L_n} - \frac{V_{\text{th}}}{2L_n}. \tag{11.41}$$

Therefore, the growth rate including the effect of plasma inhomogeniety is

$$\Gamma = \text{Im}\left(\Delta_0 + \Delta_1\right) = \frac{k_0\,|V_0|}{4} - \left(n + \frac{1}{2}\right)\frac{\omega_{p0}}{2k_z L_n}. \tag{11.42}$$

The threshold condition is thus

$$\sqrt{3}k_z L_n \frac{V_0}{c} \geq 1$$

where $k_0 \sim \omega_{p0}\sqrt{3}/c$ has been used. Substituting from Eq. (11.39), the maximum allowed value of $k_z$, the threshold condition for TPD (two-plasmon decay), can be written as

$$\frac{1}{12}\frac{|V_0|^2}{v_{\text{th}}^2}k_0 L_n \geq 1\,. \tag{11.43a}$$

Adding the Landau damping and collisional damping ($\nu_e$), the growth rate can be written as

$$\Gamma = \frac{k_0|V_0|}{4} - \left(n + \frac{1}{2}\right)\frac{\omega_{p0}}{2k_z L_n} - \nu_e - \sqrt{\pi}\frac{\omega_{p0}^4}{k^3 v_{th}^3} e^{-\omega^2/2k^2 v_{th}^2} \tag{11.43b}$$

where $k^2 = k_z^2 + (k_0/2 + K_0)^2$. Equations (11.43a) and (11.43b) are the same as those obtained by Liu and Rosenbluth,[14] except that their oscillatory velocity is twice that of ours due to the definition of the pump field amplitude.

Many experiments[15-17] on laser produced plasmas have observed TPD. Tanaka et al.[15] observed TPD in a CH plasma produced by 0.351 μm, $10^{13} - 10^{15}$ W/cm$^2$, laser with an electron temperature of the order of 1 keV.

## 11.5 Decay Instability

An electromagnetic pump wave propagating along the density gradient has an Airy function field profile at the critical layer with a scale length $\lambda_{em} = (c^2 L_n/\omega^2)^{1/3}$. It can parametrically drive the decay instability producing ion-acoustic and Langmuir modes of short wavelength, $k\lambda_{em} >> 1$. Since the wave vector of the maximally growing mode is aligned along the field of the pump (i.e., $k \perp \nabla n_0^0$), the Langmuir wave spends a long time near the phase matching point and possesses a turning point in its neighborhood. We examine the problem analytically assuming the ion-acoustic mode to be strongly damped (i.e., $\varepsilon \neq 0$). Then Eq. (10.109) can be cast into the form of a wave equation for $\phi_1$ by replacing $\vec{k}_1$ with $\hat{y}k_1 - i\hat{x}\partial/\partial x$, giving

$$\frac{d^2 A_1}{dx^2} - \frac{x}{\lambda_{es}^3} A_1 + \alpha A_1 = 0, \tag{11.44}$$

where $\lambda_{es} = (3L_n v_{th}^2/\omega_1^2)^{1/3}$, $\alpha = (2\chi_e\,\chi_i/3\varepsilon)\left(k^2|v_0|^2/4v_{th}^2\right)$. We have assumed a linear density profile (cf. Eq. 11.10) with $\omega_{p0}^2 = \omega_1^2 - 3k^2 v_{th}^2$. Since

$$\chi_e \approx \omega_{pi}^2/k^2c_s^2,\ \chi_i \approx -\omega_{pi}^2/\omega^2,$$

$$\varepsilon \approx \left(\omega_{pi}^2/k^2c_s^2\omega^2\right)\left(\omega^2 - k^2c_s^2 + 2i\Gamma_d\omega\right) \approx 2i\Gamma_d \frac{\omega_{pi}^2}{k^2c_s^2\,\omega},$$

Equation (11.44) can be written as

$$\frac{d^2 A_1}{dx'^2} - x'A_1 + i\beta^2 A_1 = 0 \tag{11.45}$$

where $\beta^2 = \left(\omega_{pi}^2 / 3\omega\Gamma_d\right)\left(k^2 |v_0|^2 \lambda_{es}^2 / 4v_{th}^2\right)$, $x' = x/\lambda_{es}$. Following the analysis adopted by Liu,[2] the convective amplification factor can be written as $\exp(\pi G)$ where $G = (\sqrt{2}/3\pi)\beta^3$. The threshold is thus given by

$$\frac{|v_0|}{v_{th}} \geq \frac{(\omega\Gamma_d)^{1/2}}{\omega_{pi}} \frac{1}{k\lambda_{es}}. \tag{11.46}$$

## 11.5.1 Oblique pump

An electromagnetic wave obliquely incident on an inhomogeneous plasma encounters a turning point at $\omega_p = \omega_0 \cos\theta_0$ where $\theta_0$ is the angle $k_0$ makes with $\hat{x}$ at the entry point. When the separation between the turning point and the critical layer is small, the field of the P-polarized light tunnels to the critical layer, acquiring a large value of $E_{0x}$. Following Ginzburg[18] and as done in Chapter 3, $A_{0x}$ may be approximated as

$$A_{0x} = \frac{A_{in}\Phi}{2\pi\left(k_0 L_n\right)^{1/2}} \frac{L_n}{x + i\Delta} \tag{11.47}$$

where $A_{in}$ is the incident wave amplitude, $k_0 = \omega_0/c$. Here, $\Phi$ is a function of $(\omega_0 L_n/c)^{1/3} \sin\theta_0$ and is of the order of unity for a narrow range of $\theta_0$, and $\Delta$ is the width of the resonance $\Delta \sim \nu L_n/\omega_p$ or $\left(L_n v_{th}^2 / \omega_p^2\right)^{1/3}$, depending on whether collisional damping or convection losses are predominant. Such a spatially localized pump may give rise to temporally growing decay instability. Now since $E_0 \| \hat{x}$, the wave vector of the maximally growing decay waves is along the density gradient. We may employ a WKB approximation for both the ion-acoustic and Langmuir waves. From Eqs. (10.78) and (10.81), we thus write, taking $k \to k - i\partial/\partial x$, $\omega \to \omega + ip$,

$$\frac{dA}{dx} + \frac{p}{c_s} A = \frac{ikv_{00}\omega}{4c_s\omega_1} A_1 e^{-iK'x^2/2} \frac{\Delta}{x + i\Delta}, \tag{11.48a}$$

$$\frac{dA_1}{dx} + \frac{2}{3}\frac{\omega_1 p}{k_1 v_{th}^2} A_1 = \frac{iv_{00}^*\omega_1}{6v_{th}^2}\frac{\omega_{pi}^2}{\omega^2} A e^{+iK'x^2/2} \frac{\Delta}{x - i\Delta}, \tag{11.48b}$$

where $K' = d(k - k_1)/dx \approx -\omega_p^2 / 3v_{th}^2 k_1 L_n$, $v_{00} = eA_{in}\Phi L_n / \left[2\pi i\,\Delta\left(k_0 L_n\right)^{1/2} m\omega_0\right]$. Multiplying Eq. (11.61) by $(x - i\Delta)\exp(-iK'x^2/2)$, operating by $(p/c_s + \partial/\partial x)$ and using Eq. (11.48a), we obtain an equation for $A_1$. Introducing a function

$$\psi_1 = A_1 \exp\left[\frac{1}{2}\int\left(\frac{p}{c_s} - \frac{2}{3}\frac{p\omega_0}{kv_{th}^2} - iK'x\right)dx\right], \tag{11.49}$$

and assuming $\Delta p / c_s > 1$, $kv_{\text{th}} > \omega_{\text{pi}}$, we obtain

$$\frac{d^2\psi_1}{dx^2} + \left[\frac{\alpha^2}{x^2+\Delta^2} - \frac{1}{4}\left(\frac{p}{c_s} - iK'x\right)^2\right]\psi_1 = 0 \tag{11.50}$$

where

$$\alpha^2 = \frac{\omega_{pi}^2 \Delta^2}{24 c_s^2}\frac{|v_{00}|^2}{v_{\text{th}}^2}.$$

For $p / c_s > K'x_T$, $\psi_1$ has turning points $x_T = \pm\left(4c_s^2\alpha^2 / p^2 - \Delta^2\right)^{1/2}$. Further, when $x_T << \Delta$, Eq. (11.50) becomes a harmonic oscillator equation with the eigenvalues condition

$$\left(\alpha^2 - \frac{p^2\,\Delta^2}{4c_s^2}\right)\frac{2c_s}{p\Delta} = 2n+1; \quad n = 0,\ 1,\ 2,\ \ldots \tag{11.51}$$

Maximum growth occurs for $n = 0$. Substituting the value of $\alpha$ and taking $\Delta \sim \left(v_{\text{th}}^2\, L_n / \omega_0^2\right)^{1/3}$, the growth rate is approximately given by $\Gamma = \text{Re}(p)$

$$\Gamma = \text{Re}(p) = \frac{\omega_{pi}}{2\pi\sqrt{6}}\left(\frac{L_n\omega_p}{v_{\text{th}}}\right)\frac{eA_{\text{in}}\Phi}{\left(k_0 L_n\right)^{1/2} m\omega_0 v_{\text{th}}}. \tag{11.52}$$

In cases where the turning points are far apart ($x_T \geq \Delta$), Eq. (11.50) can be solved numerically to obtain eigenvalues. Kline et al.[18] have experimentally observed the decay instability and its role in other nonlinear processes.

## 11.6 Oscillating Two-Stream Instability (OTSI)

We have seen that OTSI grows when the frequency mismatch $\Delta \equiv \omega_0 - \omega_p\left(1 + 3k^2 v_{\text{th}}^2 / 2\omega_p^2\right)^{1/2}$ $< 0$. For $\vec{E}_0 \| \hat{x}$, the growth rate maximizes to $\Gamma_{\max}$ for $\Delta = \Gamma_{\max} \equiv \left(k_z^2 / 4k^2\right)\left(|v_0|^2 / v_{\text{th}}^2\right)\omega_p$. In an inhomogeneous plasma with $\nabla n_0 \| \hat{x}$, $k^2 = k_x^2(x) + k_z^2$. As the Langmuir sidebands propagate towards the underdense region $k_x \equiv \left\{\left[\omega^2 - \omega_p^2(x)\right] / \left(3v_{\text{th}}^2\right) - k_z^2\right\}^{1/2}$ increases rapidly leaving $\Delta$ unchanged but diminishing the $\Gamma_{\max}$ severely. Since $\Delta = \Gamma_{\max}$ becomes invalid, the growth rate $\Gamma << \Gamma_{\max}(x)$. This strong localization of the interaction region provides substantial reflections of the sidebands leading to an absolute instability. Here we follow the treatment of Liu[2] to obtain the threshold for the onset of absolute instability.

In order to account for the effects of sideband wave reflection at the critical layer, we start with the full wave equation for the sidebands,

$$\frac{\partial^2 \phi_j}{\partial x^2} - k_z^2 \phi_j = \frac{e}{\varepsilon_0}\left(n_j^L + n_j^{\mathrm{NL}}\right), \quad j = 1, 2, \ldots \tag{11.53}$$

The linear density perturbations $n_j^{\mathrm{L}}$ can be written as

$$n_j^L = \frac{1}{i\omega_j}\nabla\cdot\left[n_0^0\left(-\frac{e\nabla\varphi_j}{mi\omega_j}\right) + \frac{v_{\mathrm{th}}^2}{i\omega_j}\nabla n_j^L\right]$$

$$\cong \frac{e}{m\omega_j^2}\nabla\cdot\left(n_0^0\nabla\phi_j\right) - \frac{v_{\mathrm{th}}^2}{\omega_j^2}\nabla^4\phi_j\frac{n_0^0\, e^2}{m\omega_j^2}. \tag{11.54}$$

Using Eqs. (10.89) and (11.54) in Eq. (11.53) and taking $\phi_j = \Phi_j \exp[-i(\omega_j t - k_z z)]$ we obtain, for the linear density profile,

$$\frac{d^2\Phi_j}{dx^2} - k_z^2\Phi_j = \frac{\omega_{p0}^2}{\omega_j^2}\left(1 + \frac{x}{L_n}\right)\left(\frac{d^2}{dx^2} - k_z^2\right)\Phi_j$$

$$+\frac{\omega_{p0}^2}{\omega_j^2 L_n}\frac{d}{dx}\Phi_j - \frac{3}{2}\frac{\omega_{p0}^2 v_{\mathrm{th}}^2}{\omega_j^4}\left(\frac{d^2}{dx^2} - k_z^2\right)^2\Phi_j$$

$$+\frac{\omega_{p0}^2 k_z^2 |v_0|^2}{4v_{\mathrm{th}}^2\left(1 + T_i/T_e\right)\omega_j}\left(\frac{\Phi_1}{\omega_1} + \frac{\Phi_2}{\omega_2}\right), \tag{11.55}$$

where we have assumed, without any loss of generality $A_0 = i|A_0|$. Expressing

$$\Phi_j = \int_c \Phi_{jk_x} e^{ik_x x}\, dk_x, \tag{11.56}$$

Equation (11.55) takes the form

$$\frac{d}{dk_x}\Phi_{jk_x} = i\frac{\omega_j^2 L_n}{\omega_{p0}^2}\left[1 - \frac{\omega_{p0}^2}{\omega_j^2} - i\frac{\omega_{p0}^2}{\omega_j^2}\frac{k_x}{k^2 L_n} - \frac{3}{2}\frac{\omega_{p0}^2}{\omega_j^2}\frac{k^2 v_{\mathrm{th}}^2}{\omega_j^2}\right]\Phi_{jk_x}$$

$$-\frac{i\omega_j |v_0|^2 L_n}{4v_{\text{th}}^2 (1+T_i/T_e)} \frac{k_z^2}{k^2} \left( \frac{\Phi_{1k_x}}{\omega_1} + \frac{\Phi_{2k_x}}{\omega_2} \right), \tag{11.57}$$

where $k^2 = k_x^2 + k_z^2$. Near the threshold $\omega \simeq 0$, i.e., $\omega_{1,2} = \mp \omega_0$. Further, without any loss of generality, we define $\omega_{p0}^2 \equiv \omega_0^2 + 3k^2 v_{\text{th}}^2 / 2$. Defining $\Phi_\pm = \Phi_{1k_x} \pm \Phi_{2k_x}$, Eq. (11.57) yields

$$\frac{d\Phi_+}{dk_x} = \left( iL_n \frac{k_x^2 v_{\text{th}}^2}{\omega_0^2} - \frac{k_x}{k_x^2 + k_z^2} \right) \Phi_+$$

$$\frac{d\Phi_-}{dk_x} = \left( iL_n \frac{k_x^2 v_{\text{th}}^2}{\omega_0^2} - \frac{k_x}{k_x^2 + k_z^2} \right) \Phi_- \; - i \frac{|v_0|^2 L_n}{2v_{\text{th}}^2 (1+T_i/T_e)} \frac{k_z^2}{k_x^2 + k_z^2} \Phi_- \tag{11.58}$$

giving

$$\Phi_+ = A\left(k_x^2 + k_z^2\right)^{-1/2} e^{ik_x^3 \lambda_{ex}^3/3}$$

$$\Phi_- = B\left(k_x^2 + k_z^2\right)^{-1/2} e^{i\left(k_x^3 \lambda_{ex}^3/3 - \beta\right)}, \tag{11.59}$$

where

$$\beta = \frac{|v_0|^2}{2v_{\text{th}}^2} \frac{L_n k_z}{1+T_i/T_e} \tan^{-1}\frac{k_x}{k_z}, \qquad \lambda_{\text{es}} = \left(v_{\text{th}}^2 L_n / \omega_{p0}^2\right)^{1/3}.$$

From Eq. (11.59), we may write

$$\Phi_1 = \frac{1}{2}\int_c \left(k_x^2 + k_z^2\right)^{-1/2} e^{i\left(k_x^3 \lambda_{es}^3/3 + k_x x\right)} \left(A + Be^{-i\beta}\right) dk_x$$

$$\Phi_2 = \frac{1}{2}\int_c \left(k_x^2 + k_z^2\right)^{-1/2} e^{i\left(k_x^3 \lambda_{es}^3/3 + k_x x\right)} \left(A - Be^{-i\beta}\right) dk_x \tag{11.60}$$

Asymptotically, at large $x$, the integral can be evaluated by the saddle point method. The saddle points are $k_x = \pm\left(-x/\lambda_{\text{es}}^3\right)^{1/2}$. In the underdense region, as $x \to -\infty$, $\tan^{-1}(k_x/k_z) \to \pm \pi/2$. Realizing that $\phi_1 = \Phi_1 e^{+i(\omega_0 t + k_z z)}$ and $\phi_2 = \Phi_2 e^{-i(\omega_0 t - k_z z)}$, the outgoing waves $x \to -\infty$ are

$$\Phi_{1,2} = \left(k_x^2 - \frac{x}{\lambda_{\text{es}}^3}\right)^{-1/2} \left(-\frac{x}{\lambda_{\text{es}}}\right)^{-1/4} e^{\mp i(-x/\lambda_{\text{es}})^3/3} \left[A \pm Be^{\pm i\beta_0}\right], \tag{11.61}$$

where $\beta_0 = |v_0|^2 L_n k_z \pi / 4v_{th}^2 \ (1+T_i/T_e)$. The incoming waves at $x \to -\infty$ are

$$\Phi_{1,2} = \left(k_x^2 - \frac{x}{\lambda_{es}^3}\right)^{-1/2} \left(-\frac{x}{\lambda_{es}}\right)^{-1/4} e^{\pm \frac{i}{3}\left(-\frac{x}{\lambda_{es}}\right)^3} \left[A \pm B\, e^{\mp i\beta_0}\right] \quad (11.62)$$

However, at $x \to -\infty$, there should be no incoming sideband wave, hence, $\beta_0 = \pi/2$, or

$$\frac{|v_0|^2}{v_{th}^2} = \frac{2(1+T_i/T_e)}{k_z L_n}. \quad (11.63)$$

This gives the threshold for the onset of OTSI. The threshold is easily exceeded at $10^{14}$ W/cm$^2$ for 1 μm laser at not too sharp density gradients. Above the threshold, the growth rate is approximately the same as in a uniform plasma, as given in Chapter 10. The instability transfers energy from longer wavelengths to shorter wavelength Langmuir waves and leads to the heating of electrons. Simulations reveal that for $v_0/v_{th} > 1$, electron trapping in the most unstable plasma wave is dominant. Kruer et al.[20] have introduced an effective collision frequency $v^*$ such that $v^* \varepsilon_0 |E_0|^2/2$ represents the anomalous heating rate. The growth rate of energy density of the Langmuir waves due to the OTSI is $2\Gamma \varepsilon_0 |\phi_1|^2$. Equating the two rates one obtains $v^* = 4\Gamma k^2 |\phi_1|^2 / |E_0|^2$. Simulations with $v_0/v_{th} \sim 1$ reveal $v^* \approx 0.06\omega_p$.

## 11.7 Discussion

Experiments with ignition scale indirect drive fusion plasmas have reported high Raman reflectivity at below one-tenth critical density. Inflation (sharp rise) in reflectivity with small increment of pump power is a strong signature of an absolute instability. In shorter scale length low-density plasma, the backscatter SRS and SBS are convective instabilitites. As the scattered electromagnetic wave passes through the interaction region (where the wave number matching condition is satisfied), its amplitude is amplified by a factor that scales exponentially with the density scale-length and the intensity of the laser. In expanding plasmas, the gradient in the plasma flow velocity is the main determining factor for the intensity threshold for the SBS instability. Polarization smoothing raises the threshold for the SBS instability within a factor of 2. The multi-beam feature of the laser introduces cross-beam energy transfer, and its impact on parametric instabilities deserves more investigations.

Near the quarter-critical density, the SRS and two-plasmon decay are absolute instabilities with large growth rates. Side-scatter SRS can also have an absolute instability. Rosenberg et al.[8] have carried experiments on ignition scale direct drive fusion plasmas with 0.351 μm multi-beam laser at intensity 6 – 16 × $10^{14}$ W/cm$^2$ and observed high Raman reflectivity (cf. Fig. 11.2). The spectrum shows a strong half-harmonic component, slightly above 700 nm, and weaker signals between 600 and 650 nm, with a puzzling broad gap between 650 and 700 nm. The fraction of the laser energy converted to hot electrons is 0.7% to 2.9% and correlates well with Raman reflectivity. The TPD is relatively weak. In longer scale-length plasmas, the TPD is stronger. The Langmuir wave, generated in SRS, appears to exceed the threshold for the decay instability when the Langmuir wave phase velocity to electron thermal velocity ratio exceeds 3.5.

Stimulated Raman forward scattering in a very low density plasma converts to modulational instability when the modulation frequency equals the plasma frequency and the phase velocity of the Langmuir wave equals the group velocity of the laser. It is particularly singnificant in high power short pulse laser–plasma interaction (e.g., laser wake-field accelerator) where the back-scattered electromagnetic waves have a much shorter interaction time with the pump while the forward-scattered waves co-move with the laser pump for a long time.

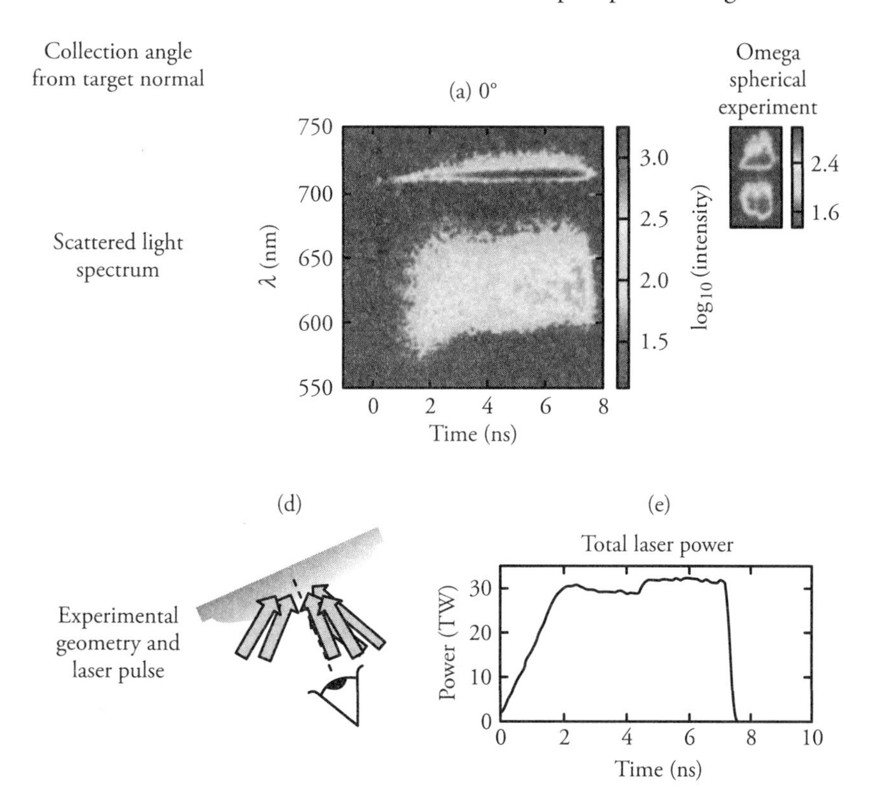


**Fig. 11.2** Time-resolved scattered light spectrum (a) at collection angle 0 degrees relative to the target normal in a CH target experiment. The image corresponds to experiment (d) and ramp flat pulse (e). The streaked spectrum from a spherical-geometry experiment on OMEGA [inset in (a)], same wavelength and time axes] is contrasted to the image in (a). (After Rosenberg et al.[8])

The Filamentation instability is not inhibited by the plasma inhomogeneity. On a shorter time scale the ion motion is unimportant and the spatial growth rate of the filamentation instability is moderate. However, with longer pulses (longer than the ion plasma period), the nonlinearity arising as a result of the ponderomotive force induced depletion of both the electron and ion densities around intensity maxima leads to stronger spatial growth. The enhanced laser intensity in a filament can significantly enhance the growth of parametric instabilities. Liu and Tripathi[21] have studied the impact of filamentation on the SRS instability. While the intensity enhancement of the laser increases the growth rate, the mismatch in the radial mode structures of the Langmuir and electromagnetic eigenmodes in the channel created by the laser filament, tends to reduce the instability.

The OTSI survives the convective losses and remains an absolute instability above the threshold intensity. Studies with computer simulations of parametric instabilities have been reported by Afeyan et al.[22], Albright et al.[23] and Yin et al.[24]. Future studies on controlling parametric instabilities and plasma photonics with fruitful and exciting results are expected. Relevant to laser fusion experiment with spherical holhram or spherical target plasma, the theoretical analysis of parametric instabilities in a spherical plasma was presented in Liu et al.[25].

## References


1. Rosenbluth, Marshall N. 1972. "Parametric Instabilities in Inhomogeneous Media." *Physical Review Letters* 29 (9): 565.

   Rosenbluth, M. N., R. B. White, and C. S. Liu. 1973. "Temporal Evolution of a Three-wave Parametric Instability." *Physical Review Letters* 31 (19): 1190.

2. Liu, C.S. 1976. "Parametric Instabilities in an Inhomogeneous Unmagnetized Plasma." In *Advances in Plasma Physics.* vol. 6, edited by A. Simon and W.B. Thompson. New York: John Wiley, p. 121.

3. Liu, C. S., Marshall N. Rosenbluth, and Roscoe B. White. 1974. "Raman and Brillouin Scattering of Electromagnetic Waves in Inhomogeneous Plasmas." *The Physics of Fluids* 17 (6): 1211–9.

   White, R. B., C. S. Liu, and M. N. Rosenbluth. 1973. "Parametric Decay of Obliquely Incident Radiation." *Physical Review Letters* 31 (8): 520.

4. Drake, J. F., and Y. C. Lee. 1973. "Temporally Growing Raman Backscattering Instabilities in an Inhomogeneous Plasma."*Physical Review Letters* 31 (19): 1197.

5. Afeyan, Bedros B., Albert E. Chou, J. P. Matte, R. P. J. Town, and William J. Kruer. 1998. "Kinetic Theory of Electron-plasma and Ion-acoustic Waves in Nonuniformly Heated Laser Plasmas." *Physical Review Letters* 80 (11): 2322.

6. Campbell, E. M. 1984. "Dependence of Laser Plasma Interaction Physics on Laser Wavelength and Plasma Scale Length" in *Radiation in Plasmas*, vol. 1. edited by B. McNamara. Singapore: World Scientific. p. 579.

7. Craxton, R. S., K. S. Anderson, T. R. Boehly, V. N. Goncharov, D. R. Harding, J. P. Knauer, R. L. McCrory et al. 2015. "Direct-Drive Inertial Confinement Fusion: A Review." *Physics of Plasmas* 22 (11): 110501.

8. Rosenberg, M. J., A. A. Solodov, J. F. Myatt, W. Seka, P. Michel, M. Hohenberger, R. W. Short et al. 2018. "Origins and Scaling of Hot-Electron Preheat in Ignition-Scale Direct-Drive Inertial Confinement Fusion Experiments." *Physical Review Letters* 120 (5): 055001.

9. Montgomery, David S. 2016. "Two Decades of Progress in Understanding and Control of Laser Plasma Instabilities in Indirect Drive Inertial Fusion." *Physics of Plasmas* 23 (5): 055601.

10. Hopkins, L. F Berzak, N. B. Meezan, S. Le Pape, L. Divol, A. J. Mackinnon, D. D. Ho, M. Hohenberger et al. 2015. "First High-convergence Cryogenic Implosion in a Near-vacuum Hohlraum."*Physical Review Letters* 114 (17): 175001.

11. Wang, Y. X., Q. Wang, C. Y. Zheng, Z. J. Liu, C. S. Liu, and X. T. He. 2018. "Nonlinear Transition from Convective to Absolute Raman Instability with Trapped Electrons and Inflationary Growth of Reflectivity." *Physics of Plasmas* 25: 100702.

12. Froula, D. H., L. Divol, D. G. Braun, B. I. Cohen, G. Gregori, A. Mackinnon, E. A. Williams et al. 2003. "Stimulated Brillouin Scattering in the Saturated Regime." *Physics of Plasmas* 10 (5): 1846–53.

13. Froula, D. H., L. Divol, R. L. Berger, R. A. London, N. B. Meezan, D. J. Strozzi, P. Neumayer et al. 2008. "Direct Measurements of an Increased Threshold for Stimulated Brillouin Scattering with Polarization Smoothing in Ignition Hohlraum Plasmas." *Physical Review Letters* 101 (11): 115002.

14. Liu, C. S., and Marshall N. Rosenbluth. 1976. "Parametric Decay of Electromagnetic Waves into Two Plasmons and its Consequences." *The Physics of Fluids* 19 (7): 967–71.

15. Tanaka, K., L. M. Goldman, W. Seka, M. C. Richardson, J. M. Soures, and E. A. Williams. 1982. "Stimulated Raman Scattering from UV-laser-produced Plasmas." *Physical Review Letters* 48 (17): 1179.

16. Baldis, H. A., and C. J. Walsh. 1981. "Experimental Observations of Nonlinear Saturation of the Two-plasmon Decay Instability." *Physical Review Letters* 47 (23): 1658.

17. Seka, W., D. H. Edgell, J. F. Myatt, A. V. Maximov, R. W. Short, V. N. Goncharov, and H. A. Baldis. 2009. "Two-plasmon-decay Instability in Direct-drive Inertial Confinement Fusion Experiments." *Physics of Plasmas* 16 (5): 052701.

18. Ginzburg, V. L. 1970. "The Propagation of Electromagnetic Waves in Plasma." 2nd edition. Oxford: Pergamon.

19. Kline, J. L., D. S. Montgomery, L. Yin, D. F. DuBois, B. J. Albright, B. Bezzerides, J. A. Cobble et al. 2006. "Different k $\lambda$ D Regimes for Nonlinear Effects on Langmuir Waves." *Physics of Plasmas* 13 (5): 055906.

20. Kruer, William L., P. K. Kaw, J. M. Dawson, and C. Oberman. 1970. "Anomalous High-frequency Resistivity and Heating of a Plasma." *Physical Review Letters* 24 (18): 987.

21. Liu, C. S., and V. K. Tripathi. 1996. "Stimulated Raman Scattering in a Plasma Channel." *Physics of Plasmas* 3 (9): 3410–3.

22. Afeyan, Bedros, and Stefan Hüller. 2013. "Optimal Control of Laser Plasma Instabilities Using Spike Trains of Uneven Duration and Delay (STUD Pulses) for ICF and IFE." In *EPJ Web of Conferences*, 59: 05009. EDP Sciences.

23. Albright, B. J., L. Yin, and B. Afeyan. 2014. "Control of Stimulated Raman Scattering in the Strongly Nonlinear and Kinetic Regime Using Spike Trains of Uneven Duration and Delay." *Physical Review Letters* 113 (4): 045002.

24. Yin, L., B. J. Albright, H. A. Rose, D. S. Montgomery, J. L. Kline, R. K. Kirkwood, P. Michel, K. J. Bowers, and B. Bergen. 2013. "Self-organized Coherent Bursts of Stimulated Raman Scattering and Speckle Interaction in Multi-speckled Laser Beams." *Physics of Plasmas* 20 (1): 012702.

25. Liu. C. S. M. N. Rosenbluth and R. B. White. 1975. "Paremetric Instabilities in Spherical, Inhomogeneous Plasmas." In *Proceedings of the Fifth International Conference in Plasma Physics and Controlled Fusion Research*, 1974. Tokyo: IAEA.

# 12 NONLINEAR SCHRÖDINGER EQUATION

## 12.1 Introduction

The nonlinear Schrödinger equation[1–9] (NSE) is a model that describes the propagation of a slowly modulated finite amplitude plane wave. It describes a situation in which the pump wave produces a change in its own dispersion characteristics by changing the parameters of the medium (e.g. the density, cf. Chapter 9). The equation is appropriate for investigating weakly nonlinear effects.

In Section 12.2, we will introduce the nonlinear Schrödinger equation. We will then obtain stationary soliton solutions in a uniform plasma, using the procedure followed by Nishikawa and Liu[1] in Section 12.3. In Section 12.4, we wil study the stability of a one-dimensional soliton to the filamentation instability. We will deduce the criterion for Langmuir wave collapse in Section 12.5. Finally in Section 12.6, we will study resonance absorption, solitons and chaos excited by a laser driver in a non-uniform plasma. Some concluding remarks are given in Section 12.7.

## 12.2 Nonlinear Schrödinger Equation

Consider the propagation of a wave in the absence of modulation,

$$\vec{E} = \vec{a}e^{-i(\omega_0 t - k_0 z)}, \tag{12.1}$$

with complex amplitude $a$. The polarization may be transverse for electromagnetic waves or longitudinal for electrostatic waves. The wave is governed by the nonlinear dispersion relation,

$$D\left(\omega_0, k_0\hat{z}, |a|^2\right) = 0. \tag{12.2}$$

Now we allow a modulation by assuming $a$ to be a slowly varying function of $\vec{r}$ and $t$, around the average value $a_0$ and deducing the equation governing $a$ from Eq. (12.2) as

$$D\left(\omega_0 + i\partial / \partial t, k_0\hat{z} - i\nabla, |a|^2\right)a = 0. \tag{12.3}$$

On carrying out a Taylor expansion around $\omega_0, k_0, a_0^2$, we obtain

$$i\left(\frac{\partial a}{\partial t}+v_g\frac{\partial a}{\partial z}\right)+p\nabla^2 a+q\left(|a|^2-a_0^2\right)a=0, \tag{12.4}$$

where $p=(1/2)\partial^2\omega_0/\partial k_0^2$ is the group dispersion coefficient, $q=(\partial D/\partial a_0^2)/(\partial D/\partial\omega_0)$ is the coefficient of nonlinearity, $v_g=\partial\omega_0/\partial k_0$ is the group velocity, and we have taken $\partial^2 a/\partial t^2\approx v_g^2\partial^2 a/\partial z^2$. We introduce a new set of variables

$$t'=t,\ z'=z-v_g t,\ x'=x,\ y'=y, \tag{15}$$

to refer to quantities in the moving frame, and eventually suppress the primes for brevity, to cast Eq. (12.4) in the form of an NSE[1–4]

$$i\frac{\partial a}{\partial t}+p\nabla^2 a+q\left(|a|^2-a_0^2\right)a=0. \tag{12.6}$$

For a Langmuir wave in an isothermal plasma, one may deduce

$$p=\frac{3}{2}\frac{v_{\rm th}^2}{\omega_0},\quad q=\frac{\omega_p}{8}\frac{e^2k_0^2}{m^2\omega_0^2v_{\rm th}^2}. \tag{12.7}$$

First, we will examine the stability of a plane uniform wave against a one-dimensional modulational perturbation

$$a=a_0+a_1(z,t),\quad a_1=U+iV. \tag{12.8}$$

Using Eq.(12.8) in Eq. (12.6) and separating real and imaginary parts, we obtain

$$-\frac{\partial V}{\partial t}+p\frac{\partial^2 U}{\partial z^2}+2qa_0^2U=0, \tag{12.9}$$

$$\frac{\partial U}{\partial t}+p\frac{\partial^2 V}{\partial z^2}=0. \tag{12.10}$$

Taking $U$ and $V$ to vary as $\exp[-i(\Omega t-kz)]$, Eqs. (12.9) and (12.10) yield the nonlinear dispersion relation

$$\Omega^2=p^2k^4-2pqa_0^2k^2. \tag{12.11}$$

An unstable solution ($\text{Im}\Omega > 0$) occurs when $pq > 0$ and

$$0 < k^2 < 2qa_0^2 / p. \tag{12.12}$$

Maximum growth occurs for $k = (qa_0^2/p)^{1/2}$, with the growth rate

$$\Gamma_{\max} = qa_0^2. \tag{12.13}$$

## 12.3 Stationary Solution

As the modulation grows, higher wave numbers $k$ are produced by harmonic generation, enhancing the dispersion effect and tending to suppress the instability. A stationary state is realized as a balance of the nonlinear bunching effect and the dispersion effect. The nonlinear effect may also produce a frequency shift. Hence, let us attempt to find a stationary solution,

$$a = A(z)e^{-i\omega t}. \tag{12.14}$$

Using this in Eq. (12.6) we get

$$p\frac{d^2 A}{dz^2} + q\left(|A|^2 - a_0^2 + \frac{\omega}{q}\right)A = 0. \tag{12.15}$$

Introducing an eikonal,

$$A = A_0 e^{iS}, \tag{12.16}$$

Equation (12.15) can be separated into its real and imaginary parts,

$$p\left[\frac{d^2 A_0}{dz^2} - \left(\frac{dS}{dz}\right)^2 A_0\right] + q\left(A_0^2 - a_0^2 + \frac{\omega}{q}\right)A_0 = 0, \tag{12.17}$$

$$2\frac{dS}{dz}\frac{dA_0}{dz} + A_0\frac{d^2 S}{dz^2} = 0. \tag{12.18}$$

Equation (12.18) can be integrated to yield

$$A_0^2 \frac{dS}{dz} \equiv M = \text{const.} \tag{12.19}$$

Using Eq. (12.19) in Eq. (12.17), multiplying the latter by $2(dA_0/dz)$ and integrating over $z$, we obtain

$$\left(\frac{dA_0}{dz}\right)^2 + 4M^2 + \frac{4A_0^4}{p}(\omega - qa_0^2 + qA_0^2/2) - EA_0^2 = 0, \tag{12.20}$$

where $E$ is a constant. Equation (12.20) has a general solution in terms of Jacobi's elliptic function $cn$, as

$$A_0^2 = c_2 + (c_1 - c_2)cn^2\left[\left(\left|\frac{q}{p}\right|\frac{c_1 - c_2}{6}\right)^{1/2} z, \chi\right], \tag{12.21}$$

where $c_1, c_2, c_3$ $(c_1 \geq c_2 \geq c_3)$ are the three roots of the equation

$$2\frac{q}{p}c^3 + \frac{4}{p}(\omega - qa_0^2)c^2 - Ec + 4M^2 = 0 \tag{12.22}$$

and

$$\chi = \left(\frac{c_1 - c_2}{c_1 - c_3}\right)^{1/2}.$$

Here, $A_0^2$ is a periodic function of period $2\mathrm{K}(\chi)$, where $\mathrm{K}(\chi_1)$ is a complete elliptic integral of the first kind of modulus $\chi$. It is called a periodic wave train. In the special case when $\chi \to 1$ we get the solitary wave solutions:

$pq \leq 0$:

$$A_0^2 = a_0^2[1 - a'\operatorname{sech}^2(bz)];$$

$$a' = -\frac{2pb^2}{q\phi_{00}^2},\ b = \left(-\frac{\omega}{p} - \frac{q}{2p}\phi_{00}^2\right)^{1/2}, \tag{12.23}$$

called dark envelope solitons, and

$pq \geq 0$:

$$A_0^2 = -\frac{2\omega}{q}\text{sech}^2\left[\left(-\frac{\omega}{p}\right)^{1/2} z\right], M = 0. \tag{12.24}$$

called bright envelope solitons.

## 12.4 Instability of an Envelope Soliton

Now, let us examine the stability of a bright envelope soliton ($pq > 0$) against a perpendicular perturbation, that is, against filamentation,

$$a = [A(z) + a_1(x,z,t)]e^{-i\omega t}, \tag{12.25}$$

where $A = a_0' \text{ sech}((q/2p)^{1/2} a_0' z)], a_0' = (-2\omega/q)^{1/2}$. It may be realized here that the soliton is localized spatially, hence $a_0 = 0$ in Eq. (12.6). Using Eq. (12.25) in Eq. (12.6), one obtains, on linearization,

$$\omega a_1 + i\frac{\partial a_1}{\partial t} + p\frac{\partial^2 a_1}{\partial z^2} + p\frac{\partial^2 a_1}{\partial x^2} = qA^2(2a_1 + a_1^*) = 0. \tag{12.26}$$

Expressing $a_1 = u + iv$, Eq. (12.26) can be separated into real and imaginary parts,

$$-\frac{\partial v}{\partial t} + \left(\omega + p\frac{\partial^2}{\partial z^2} + p\frac{\partial^2}{\partial x^2} + 3qA^2\right)u = 0, \tag{12.27}$$

$$\frac{\partial u}{\partial t} + \left(\omega + p\frac{\partial^2}{\partial z^2} + p\frac{\partial^2}{\partial x^2} + qA^2\right)v = 0. \tag{12.28}$$

Multiplying Eq. (12.27) by $\partial A/\partial z$ and Eq. (12.28) by $A$ and carrying out $z$ integration from $-\infty$ to $+\infty$ we obtain

$$-\frac{\partial}{\partial t}\int dz\frac{\partial A}{\partial z}v + p\frac{\partial^2}{\partial x^2}\int dz\frac{\partial A}{\partial z}u = 0, \tag{12.29}$$

$$-\frac{\partial}{\partial t}\int dzAu + p\frac{\partial^2}{\partial x^2}\int dzAv = 0, \tag{12.30}$$

where an integration by parts of the third terms in Eqs. (12.27) and (12.28) have been carried out.

If one views the perturbation to represent a modification in the amplitude of the soliton, that is, $a_0' \rightarrow a_0' + \delta a_0$, then using the definition of $A$ and $\omega$, $a$ may be written as

$$a \approx Ae^{-i\omega t} + \frac{\partial}{\partial a_0'}(Ae^{iqa_0'^2/2})\delta a_0$$

$$= \left\{ A + \delta a_0 \frac{\partial}{\partial z}\left[ z \text{ sech}\left( \left(\frac{q}{2p}\right)^{1/2} a_0 z \right) \right] \right.$$

$$\left. + iqa_0^2 \text{sech}\left( \left(\frac{q}{2p}\right)^{1/2} a_0 z \right) \int \delta a_0 dt \right\} e^{-i\omega t}. \tag{12.31}$$

The real and imaginary parts of $a_1$ can be written as

$$u = \delta a_0 \frac{\partial}{\partial z}\left[ z \text{ sech}\left( \left(\frac{q}{2p}\right)^{1/2} a_0 z \right) \right], \tag{12.32}$$

$$v = qa_0^2 \text{sech}\left( \left(\frac{q}{2p}\right)^{1/2} a_0 z \right) \int \delta a_0 dt. \tag{12.33}$$

Using these expressions for $u$, $v$ in Eq. (12.31) and differentiating the result with respect to time, we obtain

$$\frac{\partial^2}{\partial t^2}(\delta a_0) + 2pqa_0'^2 \frac{\partial^2}{\partial x^2}(\delta a_0) = 0. \tag{12.34}$$

For $pq > 0$, the solution to this equation is unstable. A similar procedure can be followed to examine a perturbation in the speed of the soliton which turns out to be unstable when $pq < 0$. Thus the envelope soliton is always unstable, whether against the amplitude perturbation (granulation) or against the velocity perturbation (flutter).

## 12.5 Criterion for Collapse

We may deduce some general conclusions about the stability of a multidimensional NSE. Ignoring the $a_0^2$ term, Eq. (12.6) can be written in the form of a Schrödinger equation (writing $a \equiv \phi_0$)

$$i\frac{\partial}{\partial t}\phi_0 = H\phi_0 \tag{12.35}$$

with the Hamiltonian

$$H = -p\nabla^2 - q|\phi_0|^2 - pP^2 - q|\phi_0|^2, \tag{12.36}$$

where equivalent momentum $\vec{P} = -i\nabla$ in the operator form. The localized solution therefore satisfies two conservation laws, density conservation

$$\int|\phi_0|^2 d^3\vec{r} = I_1 = \text{const.} \tag{12.37}$$

and energy conservation

$$\int\left(p|\nabla\phi_0|^2 - \frac{1}{2}q|\phi_0|^4\right)d^3\vec{r} = I_2 = \text{const.} \tag{12.38}$$

We note that $I_1$ is positive definite, while $I_2$ is positive if dispersion dominates over nonlinearity but negative in the opposite case. Let us investigate the time evolution of the spatial variance (or size of the soliton)

$$\left\langle|\vec{r}|^2\right\rangle = \frac{1}{I_1}\int d^3\vec{r}\,|\vec{r}|^2\,|\phi_0|^2.$$

Using the equations of motion

$$\frac{d\vec{r}}{dt} = 2p\vec{P},\ \frac{d\vec{P}}{dt} = q\nabla|\phi_0|^2. \tag{12.40}$$

We can write

$$\frac{d}{dt}\left\langle|\vec{r}|^2\right\rangle = 2\left\langle\vec{r}.\frac{d\vec{r}}{dt}\right\rangle = 4p\left\langle\vec{r}.\vec{P}\right\rangle \tag{12.41}$$

$$\frac{d^2}{dt^2}\left\langle|\vec{r}|^2\right\rangle = 2\left\langle\frac{d\vec{r}}{dt}.\vec{P} + \frac{d\vec{P}}{dt}.\vec{r}\right\rangle$$

$$= 4p\left\langle 2p\left|\vec{P}\right|^2\right\rangle + q\vec{r}.\nabla|\phi_0|^2$$

$$= \frac{4\vec{P}}{I_1} \int \left( 2p \left| \nabla \phi_0 \right|^2 + \frac{1}{2} q\vec{r}.\nabla \left| \phi_0 \right|^4 \right) d^3\vec{r}, \tag{12.42}$$

where we have used $\langle s \rangle = (1/I_1) \int d^3\vec{r} s \phi_0 \phi_0$,

$$\int \phi_0^* \frac{\partial^2 \phi_0}{\partial x^2} dx = -\int \frac{\partial \phi_0}{\partial x} \frac{\partial \phi_0^*}{\partial x} dx = -\int \left| \frac{\partial \phi_0}{\partial x} \right|^2 dx.$$

Integrating the last term of Eq. (12.42) by parts

$$\frac{q}{2} \int \vec{r}.\nabla \left| \phi_0 \right|^4 d^3\vec{r} = -\frac{qd}{2} \int \left| \phi_0 \right|^4 d^3\vec{r},$$

where $d = 3$ for the three-dimensional case ($d$ stands for the dimension of the problem). Equation (12.42) can be written as

$$\frac{d^2}{dt^2} \left\langle \left| \vec{r} \right|^2 \right\rangle = \frac{4p}{I_1} \left[ 2I_2 + q \left( 1 - \frac{d}{2} \right) \int \left| \phi_0 \right|^4 d^3\vec{r} \right].$$

When $pI_2 < 0$, the right-hand side of this equation is always negative for $d \geq 2$. The solution then fulfills

$$\left\langle \left| \vec{r} \right|^2 \right\rangle \leq c_1 + c_2 t + \frac{8p}{I_1} I_2 t^2 \tag{12.43}$$

which collapses to zero at finite time, since the last term on the right-hand side is negative.

The collapse of the plasma wave leads to caviton formation in the plasma and acceleration of charged particles. The dynamics of this collapse has been studied theoretically and computationally with particle simulations.[5-8]

## 12.6 Resonance Absorption, Solitons, and Chaos

A P-polarised laser in an inhomogeneous plasma, at appropriate angle of incidence and density scale length, mode converts into a Langmuir wave near the critical layer, leading to resonance absorption. The electric field of the laser at the critical layer is nearly parallel to the density gradient and acts as a driver to the Langmuir wave. Liu et al.[10] have shown that a high amplitude laser driver excites a train of accelerating solitons, and the field amplitude beyond a critical value leads to chaos. The laser coupling to solitons is governed by an NSE with a driver term and a term accounting for density gradient. One may deduce the equation governing $E_x = -\partial\phi/\partial x$ from Eq. (12.2) (with $D = \omega^2 - (\omega_p^2 + 3k^2 v_{\text{th}}^2) - \omega_p^2 \Delta n / n_0^0$ for the Langmuir wave and $\Delta n / n_0^0$ as the fractional density modification by the plasma wave ponderomotive force), in the presence of a laser driver as

$$\frac{\partial^2 E_x}{\partial t^2} - 3v_{\text{th}}^2 \frac{\partial^2 E_x}{\partial x^2} + \omega_p^2(x)\left(1 + \frac{\Delta n}{n_0^0}\right) = -\omega_p^2 E_{Lx} e^{-i\omega_0 t}. \tag{12.44}$$

On taking the fast time dependence as $E_x = E'(x,t)e^{-i\omega_0 t}$ and $\omega_p^2 = \omega_0^2(1 + x / L_n)$, this equation reduces to

$$i\frac{\partial E}{\partial t} - \left(x + \frac{\Delta n}{n_0^0}\right)E + \frac{\partial^2 E}{\partial x^2} = E_d, \tag{12.45}$$

where $x$ has been normalized to the Airy's length $\lambda_{es} = (3v_{\text{th}}^2 L_n / \omega_0^2)^{1/3}$, time to $2L_n/\lambda_{es}\omega_0$ and $E_d = E_{Lx}L_n/\lambda_{es}$. From the balance between the kinetic and ponderomotive pressures one obtains $\Delta n / n_0^0 = -|v|^2 / 8v_{\text{th}}^2$. Using this relation, and defining $A = E / E_d, P = L_n^2\omega_0^2 v_0^2 / 6v_{\text{th}}^4$, $v_0 = e|E_{Lx}|/m\omega_0$, we obtain

$$i\frac{\partial A}{\partial t} - xA + \frac{\partial^2 E}{\partial x^2} + P|A|^2 A = 1. \tag{12.46}$$

Thus, the process of resonance absorption is represented by an NSE with two additional terms, representing the effect of inhomogeneity and the driver. The inhomogeneity term leads to the acceleration of the soliton into the underdense region, like a wave packet, with the time derivative of group velocity (acceleration)

$$\frac{\partial v_g}{\partial t} = \frac{\partial^2 \omega}{\partial k^2}\frac{\partial k}{\partial t} = -\frac{\partial^2 \omega}{\partial k^2}\frac{\partial \omega}{\partial x} = 3v_{\text{th}}^2/L_n. \tag{12.47}$$

Chen and Liu[4] analytically obtained the exact $n$-solition solutions of an NSE in a nonuniform plasma without a source term. The solitons are now accelerated in an inhomogeneous plasma. The one-soliton solution is

$$E = Ae^{i\psi}, \tag{12.48}$$

$$A = 2\sqrt{2}\eta \operatorname{sech}[2\eta(x + t^2 - 4\xi t - x_0)],$$

$$\psi = (2\xi - t)x - \left[\frac{1}{3}t^3 - 2\xi t^2 + 4(\xi^2 - \eta^2)t + \psi_0\right].$$

Here $x_0$ and $4\xi$ are the initial position and velocity of the wave packet. This represents an accelerating soliton, with velocity increasing toward lower densities.

Liu et al.[10] solved Eq. (12.46) with the driver term numerically for different values of $P$ by a splitting method. The solution reveals the following.

$0 \leq P \leq 0.575$ : The solution is in steady state, much like the Airy function for $P = 0$.

$0.575 < P < 1.088$ : Periodic emission of solitons toward the underdense region is observed (Fig. 12.1 ).

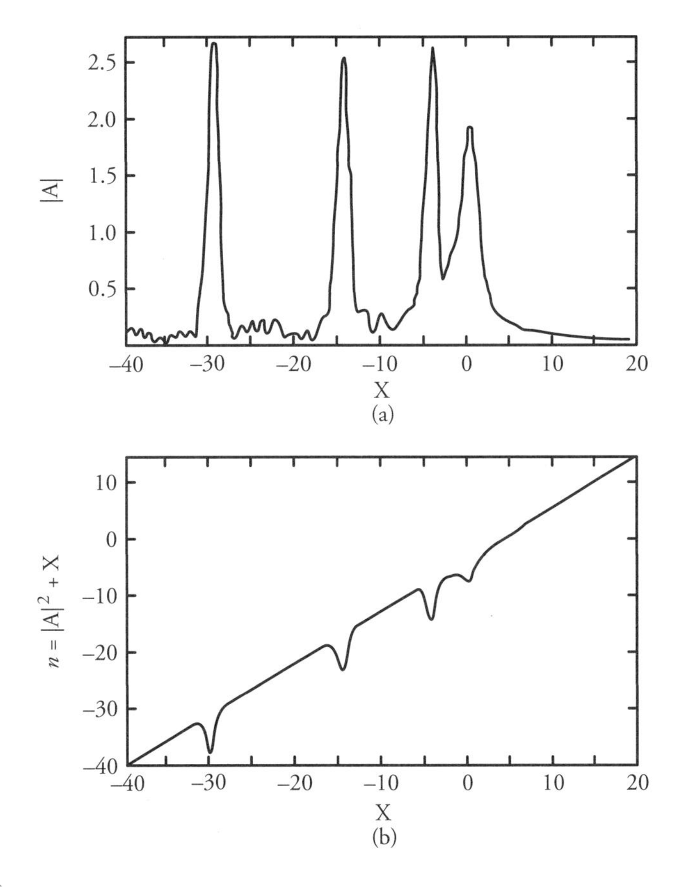


**Fig. 12.1** (a) Wave amplitude vs. $x$. (b) Density profile $n = -|A|^2 + x$ vs $x$. (After Liu et al.[10])

$1.088 < P < 1.175$ : Time evolution of the wave amplitude at $x = 0$ undergoes period doubling (Fig. 12. 2). Figure 12.3 shows a typical period doubled frequency spectrum. The cause of the period doubling is the competition between convection and nonlinear detuning of resonance. The nonlinear propagation tries to push the resonance point inward into the over-dense region

($x > 0$) while the convection effect tries to shift the linear resonance point to the underdense region ($x < 0$). The emission point therefore oscillates between two resonance points, causing period doubling.

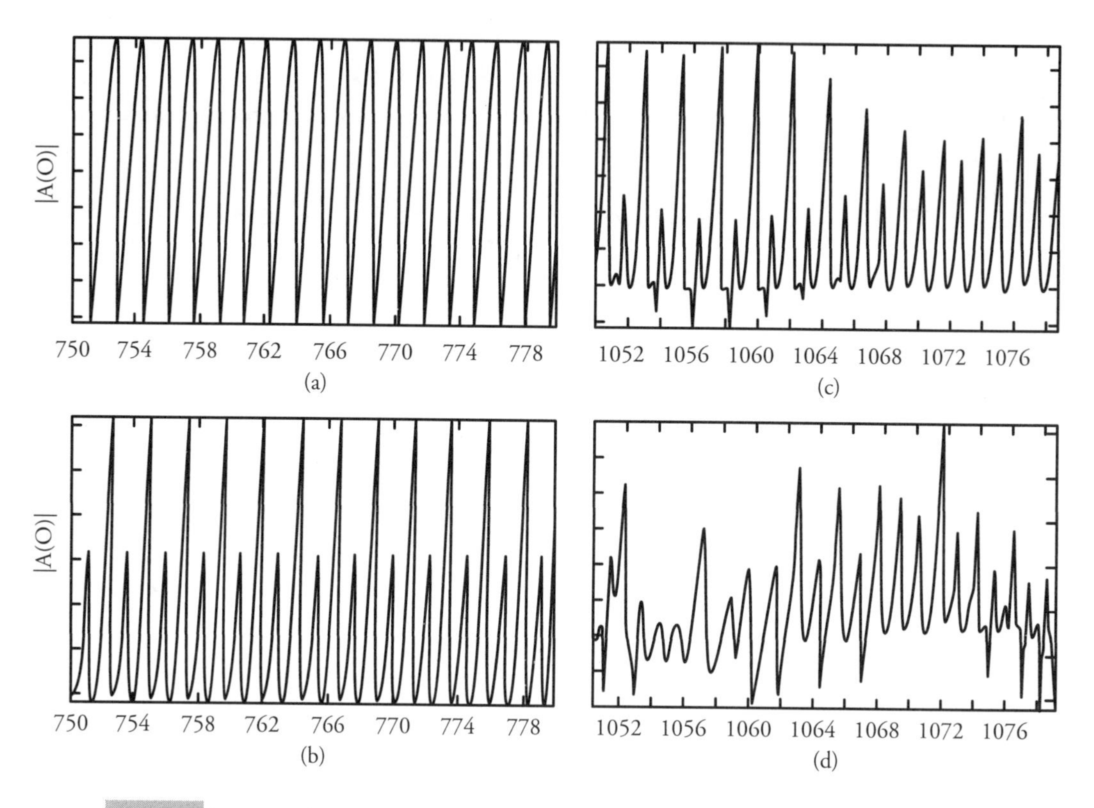


**Fig. 12.2** Time evolution of the wave amplitude at $x = 0$ for (a) periodic state ($P = 0.8$), (b) period doubled ($P = 1.15$), (c) quasi-periodic ($P = 1.2$), and (d) chaotic state ($P = 1.275$). (After Liu et al.[10])

$1.175 < P < 1.255$ : A small frequency modulation beyond the period doubling is observed. The frequency spectrum shows a second frequency $\omega_2$ with $\Delta\omega = \omega_2 - \omega_1 = 0.15$. Other peaks are the linear combinations of these frequencies and their harmonics.

$P > 1.255$ : The emission of solitons shows chaotic behavior. The time evolution of the wave amplitude becomes chaotic. The frequency spectrum shows a very broad distribution. The correlation dimension of the system at $P = 1.3$ is $D = 2.6$; the Lyapunov exponent is (a measure of the growth rate of the instability) 0.3.

Morales and Lee[11,12] also investigated the inhomogeneous Zakharov system taking into account ion-acoustic oscillations to investigate the generation of density cavities and localized electric fields, and where the NLS equation is obtained as a limiting case, for slowly varying ion oscillaions.

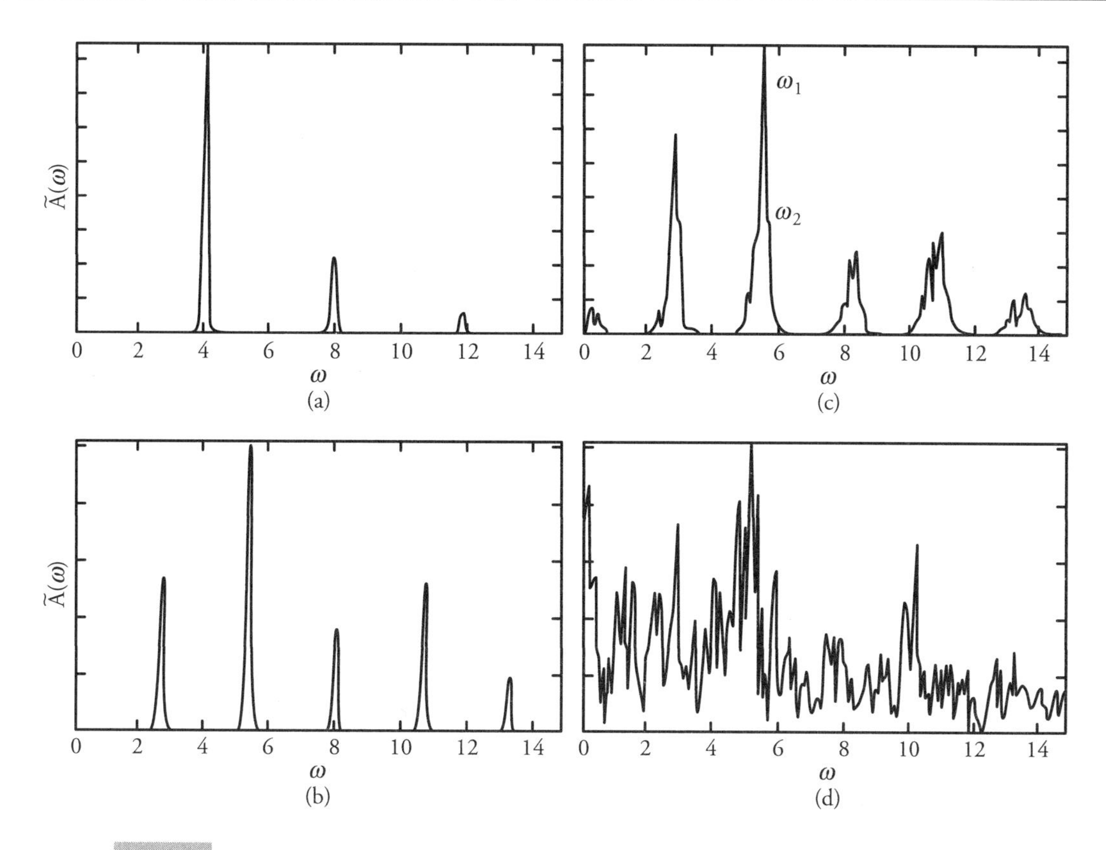


Fig. 12.3 Frequency spectrum at $x = 0$, (a) periodic state ($P = 0.8$), (b) period doubled ($P = 1.15$), (c) quasi-periodic ($P = 1.2$), and (d) chaotic state ($P = 1.275$). (After Liu et al.[10])

## 12.7 Discussion

The nonlinear Schrödinger equation (NSE) is applicable in a wide variety of situations with different kinds of nonlinearities and has been studied extensively. In this chapter, we obtained envelope soliton solutions for the Langmuir wave when nonlinearity arises through the ponderomotive force. Similar solutions exist for electromagnetic waves with relativistic mass nonlinearity. In two space dimensions, the NSE describes the nonlinear stage of the filamentation instability.

## References

1. Kaw, P., W. Kruer, C. Liu, and K. Nishikawa. 1976. *Part I: Parametric Instabilities in Plasma,* in *Advances in Plasma Physics* vol. 6, *edited* by A Simon and W Thompson. New York: John Wiley.

2. Sagdeev, R. Z., and A. A. Galeev. 1969. *Nonlinear Plasma Theory*. New York: Benjamin.
3. Zakharov, V. E., and A. B. Shabat. 1972. "Exact Theory of Two-dimensional Self-focusing and One-dimensional Self-modulation of Waves in Nonlinear Media." *Soviet Physics JETP* 34 (1): 62.

   Zakharov, Vladimir E. 1972. "Collapse of Langmuir Waves." *Soviet Physics JETP* 35 (5): 908–14.
4. Chen, Hsing-Hen, and Chuan-Sheng Liu. 1976. "Solitons in Nonuniform Media." *Physical Review Letters* 37 (11): 693.
5. Galeev, A. A., R. Z. Sagdeev, Iu S. Sigov, V. D. Shapiro, and V. I. Shevchenko. 1975. "Nonlinear Theory for the Modulation Instability of Plasma Waves." *Soviet Journal of Plasma Physics* 1: 10–20.
6. Zakharov, V. E., A. N. Pushkarev, A. M. Rubenchik, R. Z. Sagdeev, and V. F. Shvets. 1988. "Final Stage of 3D Langmuir Collapse." *JETP Letters* 47 (5).
7. Newman, D. L., R. M. Winglee, P. A. Robinson, J. Glanz, and M. V. Goldman. 1990. "Simulation of the Collapse and Dissipation of Langmuir Wave Packets." *Physics of Fluids B: Plasma Physics* 2 (11): 2600–22.
8. DuBois, Don, Harvey A. Rose, and David Russell. 1990. "Caviton Dynamics in Strong Langmuir Turbulence." *Physica Scripta* (T30): 137.
9. Sulem, Catherine, and Pierre-Louis Sulem. 1999. "The Nonlinear Schrödinger Equation." *Applied Mathematical Sciences* 139. New York: Springer Verlag.
10. Liu, C. S.,W. Shyu, P.N. Guzdar, H.H. Chen, and Y.C. Lee. 1992. "Solitons and Chaos in Resonance Absorption of EM Waves in Inhomogeneous Plasmas". in *Research Trends in Physics: Nonlinear and Relativistic Effects in Plasmas*, edited by V. Stefan. American Institute of Physics.
11. Morales, G. J., and Y. C. Lee. 1974. "Ponderomotive-force effects in a Nonuniform Plasma". *Physical Review Letters* 33 (17): 1016.
12. Morales, G. J., and Y. C. Lee. 1977. "Generation of Density Cavities and Localized Electric Fields in a Nonuniform Plasma". *The Physics of Fluids* 20 (7): 1135–47.

# 13 Vlasov and Particle-in-Cell Simulations

## 13.1 Introduction

Particle-in-cell (PIC) and Vlasov simulations both solve the Vlasov equation. The Vlasov equation (cf. Chapter 2) governs the evolution of the distribution function of charged particles (electrons, ions) in the six-dimensional phase space, consisting of three velocity (or momentum) dimensions and three position dimensions, plus time. It offers an accurate description of a plasma in the collisionless limit; that is, when the particles are affected by long-range electric and magnetic fields only, and when short-range fields from their nearest neighbors can be neglected.

PIC simulations[1–6] resolve the distribution function statistically with macro-particles (or super-particles) and follows the solution over trajectories along which the distribution function is constant; the characteristics are given by the equations of motion for the charged particles. This is the Lagrangian description. Many PIC codes have been developed over the years; modern PIC codes include the plasma simulation code[4] (PSC) originally developed by Hartmut Ruhl, the implicit iPIC3D code[5] aimed at connecting kinetic and magnetohydrodynamic time scales, the EPOCH code,[6] partially based on PSC, the VSIM/VORPAL[7,8] code, the OSIRIS code,[9] and QuickPIC.[10] PIC simulations are very adaptive and efficient for many problems, such as high-energy beam–plasma and laser–plasma interactions. On the other hand, they also have limitations; the numerical noise and slow convergence with increasing number of particles are some issues. There is also the need to resolve the Debye length with particles to avoid artificial numerical heating.

A different strategy is followed in Vlasov simulations[11–21] using a Eulerian description. Here, the distribution function is treated as a phase fluid resolved on a fixed numerical grid. Vlasov simulations do not have the statistical noise of PIC simulations; they can also more accurately resolve the high-velocity tail of the particle distribution functions. On the other hand, Vlasov simulations in higher dimensions are very memory demanding due to the need to resolve the six-dimensional phase space on a numerical grid. In some cases, the distribution function can also become oscillatory in phase space, leading to sharp gradients and a need to introduce numerical dissipation in velocity space whilst avoiding artificial numerical heating due to the broadening of the distribution in velocity space. Hence, the choice between PIC and Eulerian Vlasov simulations strongly depends on the physical problem at hand.

## 13.2 The Vlasov Equation

Traditionally, the Vlasov equation in the non-relativistic limit is often written to solve a velocity distribution function; for relativistic treatment, the Vlasov equation is written for the particle momentum distribution. The non-relativistic Vlasov equations governing the velocity distribution functions for ions and electrons are

$$\frac{\partial f_i}{\partial t} + \vec{v} \cdot \nabla f_i + \frac{q_i}{m_i}\left(\vec{E} + \vec{v} \times \vec{B}\right) \cdot \nabla_v f_i = 0 \tag{13.1}$$

and

$$\frac{\partial f_e}{\partial t} + \vec{v} \cdot \nabla f_e + \frac{q_e}{m}\left(\vec{E} + \vec{v} \times \vec{B}\right) \cdot \nabla_v f_e = 0 \tag{13.2}$$

where $q_i = Z_i e$ is the ion charge and $Z_i$ is the ion charge state and $q_e = -e$ is the electron charge. Here $\nabla_v = \hat{x}\partial / \partial v_x + \hat{y}\partial / \partial v_y + \hat{z}\partial / \partial v_z$ is the gradient operator in velocity space. The particles interact via the long-range electric and magnetic fields which accelerate the particles, while neglecting short-range binary interactions between particles. The relativistic Vlasov equations governing the momentum distribution functions for ions and electrons are

$$\frac{\partial f_i}{\partial t} + \frac{\vec{p}}{m_i \gamma_i} \cdot \nabla f_i + q_i \left( \vec{E} + \frac{\vec{p}}{m_i \gamma_i} \times \vec{B} \right) \cdot \nabla_p f_i = 0 \tag{13.3}$$

and

$$\frac{\partial f_e}{\partial t} + \frac{\vec{p}}{m \gamma_e} \cdot \nabla f_e + q_e \left( \vec{E} + \frac{\vec{p}}{m \gamma_e} \times \vec{B} \right) \cdot \nabla_p f_e = 0 \tag{13.4}$$

with the relativistic gamma factors $\gamma_i = \sqrt{1 + p^2 / (m_i^2 c^2)}$ and $\gamma_e = \sqrt{1 + p^2 / (m^2 c^2)}$.

The particle number densities and mean velocities are obtained from the velocity distribution functions as

$$n_i = \iiint f_i d^3 v, \quad \vec{v}_i = \frac{1}{n_i} \iiint \vec{v}\, f d^3 v \tag{13.5}$$

$$n_e = \iiint f_e d^3 v, \quad \vec{v}_e = \frac{1}{n_e} \iiint \vec{v} f_e d^3 v \tag{13.6}$$

and from the momentum distribution functions as

$$n_i = \iiint f_i d^3 p, \quad \vec{v}_i = \frac{1}{n_i} \iiint \frac{\vec{p}}{m_i \gamma_i} f_i d^3 p \tag{13.7}$$

$$n_e = \iiint f_e d^3 p, \quad \vec{v}_e = \frac{1}{n_e} \iiint \frac{\vec{p}}{m \gamma_e} f_e d^3 p \tag{13.8}$$

where $d^3v$ denotes $dv_x dv_y dv_z$ and $d^3p$ denotes $dp_x dp_y dp_z$. The number densities and velocities are used to calculate the electric charge density

$$\rho = q_i n_i + q_e n_e \tag{13.9}$$

and current density

$$\vec{j} = q_i n_i \vec{v}_i + q_e n_e \vec{v}_e . \tag{13.10}$$

The charge and current densities enter into Maxwell's equations as sources for the electric and magnetic fields.

## 13.3 Properties of the Vlasov Equation

The Vlasov equation describes the incompressible flow in the six-dimensional phase space. As an example, we take the Vlasov equation, Eq. (13.1), for the ion distribution function. The distribution function $f_i$ is constant, or

$$\frac{df_i}{dt} = 0 \tag{13.11}$$

along trajectories in the $(\vec{x}, \vec{v}, t)$-space given by

$$\frac{d\vec{x}}{dt} = \vec{v} \tag{13.12}$$

$$\frac{d\vec{v}}{dt} = \frac{q_i}{m_i}\left(\vec{E} + \vec{v} \times \vec{B}\right) \tag{13.13}$$

Equation (13.13) is Newton's second law of motion for a single particle. This property can be seen as an alternative formulation of the Vlasov equation, and forms the basis for PIC simulations.

Hence, standard PIC simulations are numerical solutions of the collisionless Vlasov equation, and even though the PIC simulations sample the particle distributions with the help of macro-particles, there are often no attempts to model binary Coulomb collisions in PIC simulations. A macro-particle represents many real particles. If the macro-particle has mass $M$ and charge $Q$ such that $Q/M = q_i/m_i$, then it represents a collection of ions, and if $Q/M = q_e/m$ then it represents electrons.

The Vlasov equation can be written in a slightly different, conservative form

$$\frac{\partial f_i}{\partial t} + \nabla\cdot(\vec{v}f_i) + \nabla_v\cdot\left[\frac{q_i}{m_i}\left(\vec{E}+\vec{v}\times\vec{B}\right)f_i\right] = 0 \tag{13.14}$$

which has the form of a continuity equation where the distribution function $f_i$ can be seen as the density of a phase space fluid. The phase space fluid is incompressible in velocity space since the phase space divergence is zero,

$$\nabla\cdot\vec{v} + \nabla_v\cdot\left(\frac{q_i}{m_i}\left(\vec{E}+\vec{v}\times\vec{B}\right)\right) = 0, \tag{13.15}$$

combined with Eq. (13.11).

Since the Vlasov equation describes the evolution of an incompressible fluid, the distribution function can become complicated after some time, with large values of the distribution located next to small values. The plots in Fig. 13.1 were produced by solving the one-dimensional electron Vlasov equation coupled with Gauss' law

$$\frac{\partial f_e}{\partial t} + v_x\frac{\partial f_e}{\partial x} - \frac{eE_x}{m_e}\frac{\partial f_e}{\partial v_x} = 0 \tag{13.16}$$

$$\frac{\partial E_x}{\partial x} = \frac{e}{\varepsilon_0}\left(n_0 - \int f_e dv_x\right) \tag{13.17}$$

The oscillations in velocity space seen in Fig. 13.1 can be understood by solving a simpler problem neglecting the electric field

$$\frac{\partial f_e}{\partial t} + v_x\frac{\partial f_e}{\partial x} = 0 \tag{13.18}$$

with the initial condition $f_e = f_0(x,v_x)$ at $t = 0$. The solution at later times is $f_e = f_0(x\text{-}v_x t,\ v_x)$.

For example, if $f_0(x,v_x) = (1 + A\cos(kx))F(v_x)$, where $F(v_x)$ is a Maxwellian distribution, then the solution at later times is $f_e = (1 + A\cos(kx - kv_x t))F(v_x)$. Due to the free-streaming $kv_x t$-term inside the cosine function, the solution becomes more and more oscillatory in velocity space as $t$ increases. In order to calculate the electric field for this solution, we first calculate the electron density. A Maxwellian distribution

$$F(v_x) = \frac{n_0}{\sqrt{2\pi}\, v_{\text{th}}} \exp\left( -\frac{v_x^2}{2v_{\text{th}}^2} \right) \tag{13.19}$$

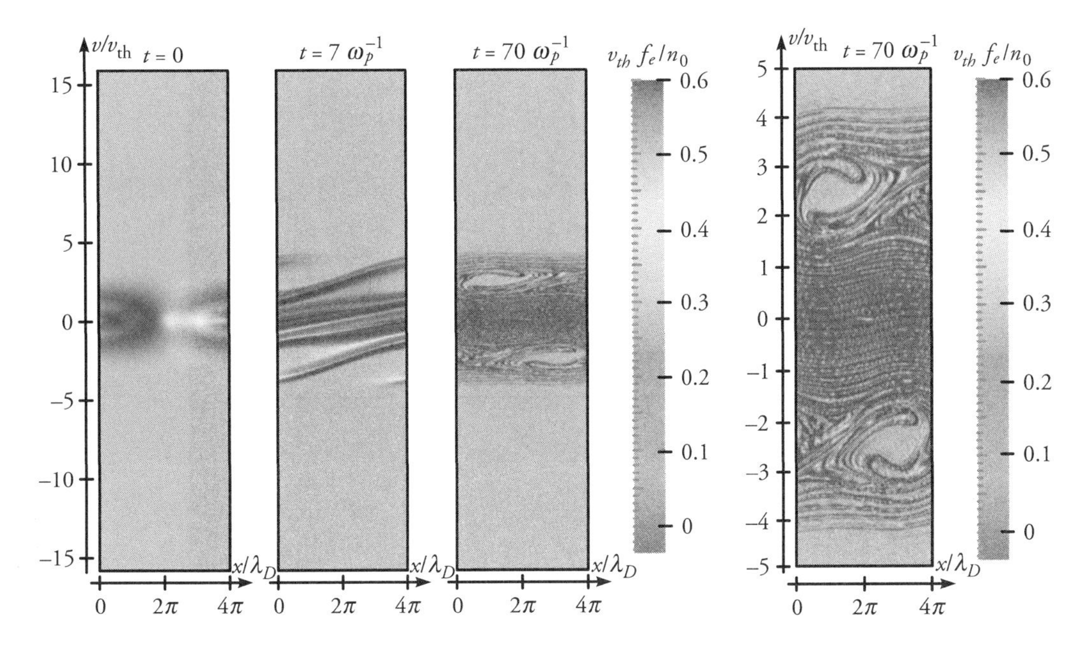


**Fig. 13.1** Nonlinear electron plasma wave (Langmuir wave) at different times. The right-hand panels show a close-up at time $t = 70\ \omega_p^{-1}$. (After Eliasson.[15])

gives

$$f_e = \frac{n_0}{\sqrt{2\pi}\, v_{\text{th}}} \left(1 + A\cos(kx - kv_x t)\right) \exp\left( -\frac{v_x^2}{2v_{\text{th}}^2} \right) \tag{13.20}$$

Integration over velocity space gives the electron density

$$n_e = \int f_e dv_x = n_0 \left( 1 + A\cos(kx) \exp\left( -\frac{k^2 v_{\text{th}}^2 t^2}{2} \right) \right) \tag{13.21}$$

and integrating Gauss' law over space gives the electric field

$$E_x = \int^x \frac{e}{\varepsilon_0} (n_0 - n_e)\, dx = -\frac{en_0}{\varepsilon_0} \frac{A}{k} \sin(kx) \exp\left( -\frac{k^2 v_{\text{th}}^2 t^2}{2} \right) \tag{13.22}$$

The electric field decreases very rapidly as a Gaussian function in time. The solution behaves in the same manner under time-reversal $t \rightarrow -t$ and $v_x \rightarrow -v_x$.

The rapid decay of the electric field is due to the $kv_xt$-term, which leads to the integration of an oscillatory function in velocity space. The oscillatory term can cause problems in numerical solutions of the Vlasov equation. Assume that we want to use the exact solution to calculate the charge density numerically on an equidistant grid in velocity space with grid size $\Delta v_x$. We replace the integral over velocity by a sum,

$$\frac{\partial E_x}{\partial x} = \frac{e}{\varepsilon_0}\left(n_0 - \int f_e dv_x\right) \approx \frac{e}{\varepsilon_0}\left(n_0 - \sum f_{e,j}\Delta v_x\right) \tag{13.23}$$

where we have

$$\sum f_{e,j}\Delta v_x = \sum_j \frac{n_0}{\sqrt{2\pi}v_{\text{th}}}\left(1 + A\cos\left(kx - kj\Delta v_x t\right)\right)\exp\left(-\frac{j^2\Delta v_x^2}{2v_{\text{th}}^2}\right)\Delta v_x \tag{13.24}$$

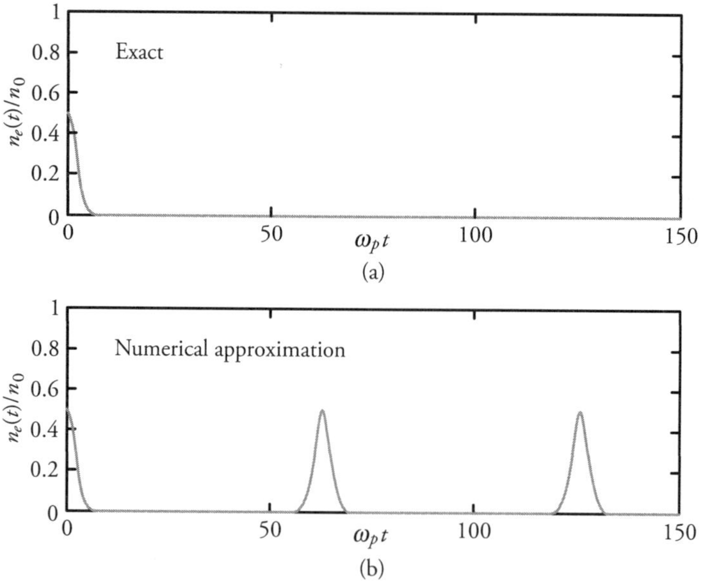

**Fig. 13.2** (a) The exact electron number density perturbation and (b) its numerical approximation. The initial conditions recur on the numerical grid at multiples of the recurrence time $t = 2\pi/(k\,\Delta v_x)$. (After Eliasson.[15])

An interesting effect takes place at time $t = 2\pi/(k\,\Delta v_x)$, called the *recurrence time*. At that time, $\cos(kx - kj\Delta v_x t) = \cos(kx - 2\pi j) = \cos(kx)$ since $\cos(kx)$ is $2\pi$ -periodic. This leads to a recurrence of the initial condition on the numerical grid. The initial condition recurs periodically in time in the numerical approximation of the electron density, which is very different from the exact electron density (Fig. 13.2).

The recurrence phenomenon can easily destroy the results in a numerical simulation. One usually treats this by adding dissipative mechanisms in grid-based simulations.[14,15] This recurrence effect is not a problem in PIC simulations where the macro-particles are sampled randomly in velocity space.

## 13.4 Reduction of Velocity Dimensions

The reduction of velocity dimensions is used in Vlasov simulations to reduce the numerical cost and memory requirements. How do we obtain a 1 × 1-dimensional (in space and velocity space) Vlasov equation

$$\frac{\partial f_e}{\partial t} + v_x \frac{\partial f_e}{\partial x} - \frac{eE_x}{m_e} \frac{\partial f_e}{\partial v_x} = 0 \tag{13.25}$$

from the three-dimensional Vlasov equation? To reduce the number of dimensions in space, one simply assumes that the system varies only in one spatial dimensions (e.g. the $x$-direction) so that space-derivatives are zero in the other space dimensions ($\partial/\partial y = 0$ and $\partial/\partial z = 0$).

However, a similar assumption does not work in velocity space. Here, we instead obtain a distribution function in one velocity dimension (e.g., the $v_x$-dimension) by integrating the distribution function over the other velocity dimensions,

$$f_e\left(x, v_x, t\right) = \iint f_e\left(x, v_x, v_y, v_z, t\right) dv_y dv_z. \tag{13.26}$$

By integrating the Vlasov equation in a similar manner, we can for some cases obtain a Vlasov equation with a reduced number of velocity dimensions. This procedure works for one-dimensional electrostatic waves in un-magnetized plasmas and for electrostatic waves propagating along the magnetic field. For waves propagating perpendicularly to a magnetic field one can integrate the velocity dimension parallel to the magnetic field to obtain a Vlasov equation in the two velocity dimensions perpendicular to the magnetic field.

For cases when the velocity dimensionality cannot be reduced exactly in the Vlasov equation, numerical schemes have been developed where a kinetic description is kept along dominant directions, while fluid moments are developed in the perpendicular directions.[16,17]

As an example, let us consider the propagation of relativistic laser light and electrostatic waves parallel to an ambient magnetic field in plasma. The 3D relativistic Vlasov equation reduced to 1D in space along the $x$-axis reads

$$\frac{\partial f_j}{\partial t} + \frac{\partial}{\partial x}\left(v_{jx} f_j\right) + \nabla_p \cdot \left(q_j \left(\vec{E} + \vec{v}_j \times \vec{B}\right) f_j\right) = 0\,. \tag{13.27}$$

where $f_j = f_j(x, \vec{p}, t)$ is the particle momentum distribution function, $\vec{v}_j = \vec{p}/(m_j \gamma_j)$ is the velocity, $\vec{p} = p_x \hat{x} + p_y \hat{y} + p_z \hat{z}$ is the momentum variable with $\hat{x}, \hat{y}$, and $\hat{z}$, being unit vectors in the $x$, $y$, and $z$ directions, $\gamma_j = \sqrt{1 + |\vec{p}|^2 / \left(m_j^2 c^2\right)}$ is the relativistic gamma factor, $m_j$ and $q_j$ are the particle's mass and charge ($q_e = -e$ for electrons and $q_i = e$ for singly charged ions), $|\vec{p}|^2 = p_x^2 + p_y^2 + p_z^2$, and $\vec{E}$ and $\vec{B}$ are the electric and magnetic fields. The electric field can have any directions in 3D, while in our geometry, the wave magnetic field is in the $y$ and $z$ directions while a constant axial magnetic field is along the $x$ direction.

In the direction perpendicular to the laser propagation direction, the particles mainly perform quivering motion in the laser field, while kinetic effects may be important in the parallel direction. We therefore assume that the particle distribution is "cold" in the perpendicular direction, and write the distribution function as

$$f_j = f_j^{1D}\left(x, p_x, t\right)\delta\left(\vec{p}_\perp - \vec{P}_{j\perp}(x,t)\right) \tag{13.28}$$

where $\delta$ is Dirac's delta function, $\vec{p}_\perp = p_y\hat{y} + p_z\hat{z}$ is the perpendicular momentum variable, and $\vec{P}_{j\perp}(x,t) = P_{jy}(x,t)\hat{y} + P_{jz}(x,t)\hat{z}$ is the perpendicular fluid momentum. The delta function has special properties, for example, that $\iint g(\vec{p}_\perp)\delta(\vec{p}_\perp - \vec{p}_{j\perp}(x,t))dp_y dp_z = g(\vec{P}_{j\perp}(x,t))$, and, via an integration by parts, $\int\int g(\vec{p}_\perp)(\partial\delta(\vec{p}_\perp - \vec{P}_{j\perp}(x,t))/\partial p_y)dp_y dp_z = -\int\int(\partial g(\vec{p}_\perp)/\partial p_y)$ $\delta(\vec{p}_\perp - \vec{P}_{j\perp}(x,t))\ dp_y dp_z = -(\partial g(\vec{p}_\perp)/\partial p_y)_{\vec{p}_\perp = \vec{P}_{j\perp}(x,t)}$. To derive equations for $f_j^{1D}$ and $\vec{P}_{j\perp}$, we take perpendicular moments of the Vlasov equation as

$$\iint\left(\frac{\partial f_i}{\partial t} + \frac{\partial}{\partial x}(v_{jx}f_j) + \nabla_p.(q_j(\vec{E} + \vec{v}_j \times \vec{B})f_j)\right)dp_y dp_z = 0 \tag{13.29}$$

$$\iint\vec{p}_\perp\left(\frac{\partial f_i}{\partial t} + \frac{\partial}{\partial x}(v_{jx}f_j) + \nabla_p.(q_j(\vec{E} + \vec{v}_j \times \vec{B})f_j)\right)dp_y dp_z = 0 \tag{13.30}$$

We can then furnish the 1D Vlasov equation

$$\frac{\partial f_j^{1D}}{\partial t} + \frac{\partial}{\partial x}\left(v_{jx}f_j^{1D}\right) + \frac{\partial}{\partial p_x}\left(q_j\left(E_x + \frac{P_{jy}B_z - P_{jz}B_y}{m_j\gamma_j}\right)f_j^{1D}\right) = 0 \tag{13.31}$$

and the perpendicular momentum equation

$$\frac{\partial(\vec{P}_{j\perp}f_j^{1D})}{\partial t} + \frac{\partial}{\partial z}\left(\vec{P}_{j\perp}v_{jx}f_j^{1D}\right)$$

$$+q_j\left(-\left(\vec{E}_\perp + \frac{\hat{x}\left(P_{jz}B_x - p_xB_z\right)}{m_j\gamma_j} + \frac{\hat{y}\left(p_xB_y - P_{jy}B_x\right)}{m_j\gamma_j}\right)f_j^{1D} + \vec{P}_{j\perp}\frac{\partial}{\partial p_x}\left(f_j^{1D}\left(E_x + \frac{P_{jy}B_z - P_{jz}B_y}{m_j\gamma_j}\right)\right)\right) = 0, \tag{13.32}$$

where $\gamma = \sqrt{1 + (p_x^2 + P_{jy}^2 + P_{jz}^2)/(m_j^2 c^2)}$. Using Eq. (13.30) to eliminate $f_j^{1D}$ in Eq. (13.31) gives

$$\frac{\partial \vec{P}_{j\perp}}{\partial t} + v_{jx}\frac{\partial \vec{P}_{j\perp}}{\partial x} - q_j\left(\vec{E}_\perp + \frac{\hat{y}\left(P_{jz}B_x - p_x B_z\right)}{m_j\gamma_j} + \frac{\hat{z}\left(p_x B_y - P_{jy}B_x\right)}{m_j\gamma_j}\right) = 0. \tag{13.33}$$

Introducing the vector potential through $\vec{B}_\perp = \hat{x} \times \partial \vec{A}_\perp / \partial x$ and $\vec{E}_\perp = -\partial \vec{A}_\perp / \partial t$ in Eq. (13.32) gives

$$\left(\frac{\partial}{\partial t} + v_{jx}\frac{\partial}{\partial x}\right)\left(\vec{P}_{j\perp} + q_j\vec{A}_\perp\right) + \frac{q_j B_x}{m_j\gamma_j}\hat{x} \times \vec{P}_{j\perp} = 0. \tag{13.34}$$

Equation (13.34) still contains the momentum variable $p_x$ in the expressions for $v_{jx}$ and the gamma factor. Multiplying Eq. (13.34) by $f_j^{1D}$ and integrating over $p_x$ gives the fluid momentum equation

$$\left(\frac{\partial}{\partial t} + \tilde{v}_{jx}\frac{\partial}{\partial x}\right)\left(\vec{P}_{j\perp} + q_j\vec{A}_\perp\right) + \tilde{\omega}_{cj}\hat{x} \times \vec{P}_{j\perp} = 0 \tag{13.35}$$

where the parallel fluid velocity and the effective cyclotron frequency are

$$\tilde{v}_{jx} = \frac{1}{n_j}\int \frac{p_x}{m_j\gamma_j} f_j^{1D}\, dp_x, \tag{13.36}$$

$$\tilde{\omega}_{cj} = \frac{1}{n_j}\int \frac{q_j B_x}{m_j\gamma_j} f_j^{1D}\, dp_x, \tag{13.37}$$

and the number density is

$$n_j = \int f_j^{1D}\, dp_x. \tag{13.38}$$

Using the vector potential, the 1D Vlasov equation can be written as

$$\frac{\partial f_j^{1D}}{\partial t} + \frac{\partial}{\partial x}\left(v_{jx} f_j^{1D}\right) + q_j\frac{\partial}{\partial p_x}\left(\left(E_x + \frac{1}{m_j\gamma_j}\vec{P}_{j\perp} \cdot \frac{\partial \vec{A}_\perp}{\partial x}\right) f_j^{1D}\right) = 0 \tag{13.39}$$

Simulation results are shown in Fig. 13.3. One right-hand and one left-hand circularly polarized wave of equal amplitudes $a_\perp^{\mathrm{R}} = a_\perp^{\mathrm{L}} = 0.035$ are injected at the left boundary,

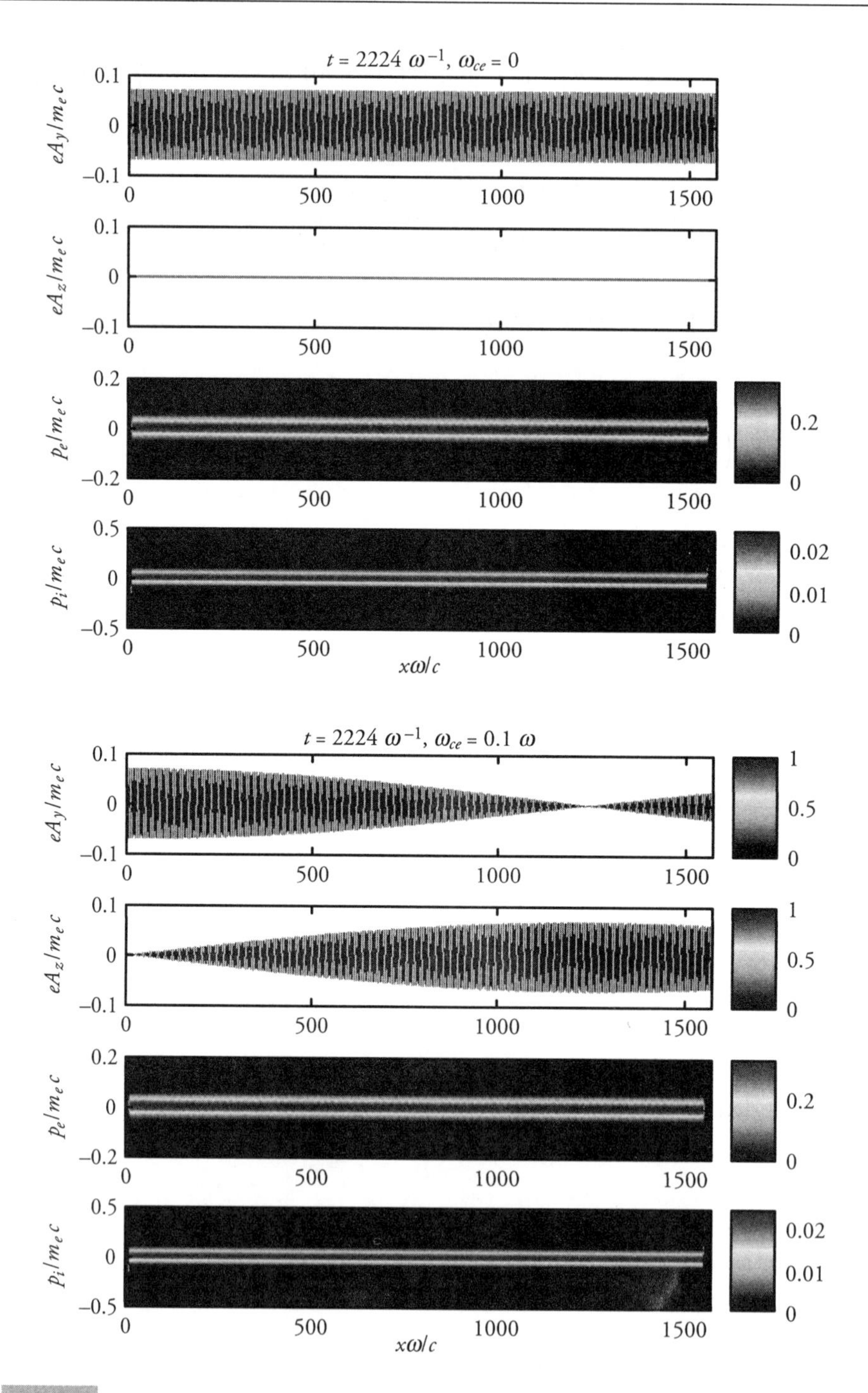


**Fig. 13.3** Simulation result at $t = 2224\omega^{-1}$ without (top) and with (bottom) an axial magnetic field. Faraday rotation of the linearly polarized wave can be seen in the bottom panel.

resulting in a linearly polarized (along the $y$-axis) electromagnetic wave with amplitude $a$ = 0.07. Electromagnetic waves that reaching the boundaries are absorbed. The plasma slab has Maxwell distributed ions and electrons with densities $n_i = n_e = 0.025n_c$ and temperatures $T_i$ = 0.05 keV and $T_e$ = 0.5 keV. One simulation (left panels) is without an axial magnetic field ($\omega_{ce}$ = 0) and one simulation (right panels) with an axial magnetic field $B_x$ such that $|\omega_{ce}|$ = 0.1$\omega$, showing clear signatures of Faraday rotation of a plane polarized wave. The Faraday rotation angle can be estimated as $\theta \approx (1/2)(n_e/n_c)(\omega_{ce}/\omega)(x\omega/c)$. Using $\theta = \pi/2$, gives $x\omega/c \approx$ 1260, in agreement with the simulation where the initial y-polarization at $x$ = 0 has switched to $z$-polarization at that point.

Without an axial magnetic field ($B_x$ = 0), we have from Eq. (13.34), the conservation of perpendicular canonical momentum, $\vec{P}_{j\perp} + q_j\vec{A}_\perp$ = constant, which is used to relate electron momentum to the EM wave amplitude in laser–plasma interactions. Using $\vec{P}_{j\perp} = -q_j\vec{A}_\perp$, the Vlasov equation, Eq. (13.39) can be written as

$$\frac{\partial f_j^{1D}}{\partial t} + \frac{\partial}{\partial x}\left(v_{jx} f_j^{1D}\right) + \frac{\partial}{\partial p_x}\left(\left(q_j E_x - m_j c^2 \frac{\partial \gamma_j}{\partial x}\right) f_j^{1D}\right) = 0 \tag{13.40}$$

with $\gamma_j = \sqrt{1 + \left(p_x^2 + q_j^2\left|\vec{A}_\perp\right|^2\right) / \left(m_j^2 c^2\right)}$. Reduced Vlasov simulations based on the conservation of perpendicular canonical momentum have been used to study Raman instabilities,[18] particle acceleration,[19,20] and relativistic nonlinear structures[21] in laser–plasma interactions.

## References


1. Eastwood, J. W., and R. W. Hockney. 1981. *Computer Simulation Using Particles.* New York: McGraw Hill.
2. Dawson, John M. 1983. "Particle Simulation of Plasmas." *Reviews of Modern Physics* 55 (2): 403.
3. Birdsall, C. K., and A. B. Langdon. 1985. *Plasma Physics via Computer Simulation.* , New York: McGraw Hill. p.331.
4. Germaschewski, Kai, William Fox, Stephen Abbott, Narges Ahmadi, Kristofor Maynard, Liang Wang, Hartmut Ruhl, and Amitava Bhattacharjee. 2016. "The Plasma Simulation Code: A Modern Particle-in-cell Code with Patch-based Load-balancing." *Journal of Computational Physics* 318: 305–26.
5. Arber, T. D., Keith Bennett, C. S. Brady, A. Lawrence-Douglas, M. G. Ramsay, N. J. Sircombe, P. Gillies et al. 2015. "Contemporary Particle-in-cell Approach to Laser-plasma Modelling." *Plasma Physics and Controlled Fusion* 57 (11): 113001.
6. Markidis, Stefano, and Giovanni Lapenta. 2010. "Multi-scale Simulations of Plasma with iPIC3D." *Mathematics and Computers in Simulation* 80 (7): 1509–19.
7. Nieter, Chet, and John R. Cary. 2004. "VORPAL: a Versatile Plasma Simulation Code." *Journal of Computational Physics* 196 (2): 448–73.

8. Tech-X Corporation, Boulder, Colorado, United States. https://www.txcorp.com/vsim. Accessed on: 9 November 2018
9. Fonseca, Ricardo A., Luís O. Silva, Frank S. Tsung, Viktor K. Decyk, Wei Lu, Chuang Ren, Warren B. Mori et al. 2002. "OSIRIS: A Three-dimensional, Fully Relativistic Particle in Cell Code for Modeling Plasma Based Accelerators." In *International Conference on Computational Science*. Berlin, Heidelberg: Springer. Pp. 342–351
10. Huang, Chengkun, Viktor K. Decyk, Chuang Ren, M. Zhou, Wei Lu, Warren B. Mori, James H. Cooley, Thomas M. Antonsen Jr, and T. Katsouleas. 2006. "QUICKPIC: A Highly Efficient Particle-in-cell Code for Modeling Wakefield Acceleration in Plasmas." *Journal of Computational Physics* 217 (2): 658–79.
11. Arber, T. D., and R. G. L. Vann. 2002. "A Critical Comparison of Eulerian-grid-based Vlasov Solvers." *Journal of Computational Physics* 180 (1): 339–57.
12. Filbet, Francis, and Eric Sonnendrücker. 2003. "Comparison of Eulerian Vlasov Solvers." *Computer Physics Communications*150 (3): 247–66.
13. Delzanno, Gian Luca. 2015. "Multi-dimensional, Fully-implicit, Spectral Method for the Vlasov–Maxwell Equations with Exact Conservation Laws in Discrete Form." *Journal of Computational Physics* 301: 338–56.
14. Cheng, Chio-Zong, and Georg Knorr. 1976. "The Integration of the Vlasov Equation in Configuration Space." *Journal of Computational Physics* 22 (3): 330–51.
15. Eliasson, Bengt. 2010. "Numerical Simulations of the Fourier-Transformed Vlasov-Maxwell System in Higher Dimensions—Theory and Applications." *Transport Theory and Statistical Physics* 39 (5–7): 387–465.
16. Newman, D. L., M. V. Goldman, R. E. Ergun, L. Andersson, and N. Sen. 2004. "Reduced Vlasov Simulations in Higher Dimensions." *Computer Physics Communications* 164 (1–3): 122–7.
17. Rose, Harvey A., and William Daughton. 2011. "Vlasov Simulation in Multiple Spatial Dimensions." *Physics of Plasmas* 18 (12): 122109.
18. Bertrand, P., A. Ghizzo, S. J. Karttunen, T. J. H. Pättikangas, R. R. E. Salomaa, and M. Shoucri. 1995. "Two-stage Electron Acceleration by Simultaneous Stimulated Raman Backward and Forward Scattering." *Physics of Plasmas* 2 (8): 3115–29.
19. Grassi, Anna, Luca Fedeli, Andrea Sgattoni, and Andrea Macchi. 2016. "Vlasov Simulation of Laser-driven Shock Acceleration and Ion Turbulence." *Plasma Physics and Controlled Fusion* 58 (3): 034021.
20. Eliasson, Bengt, Chuan S. Liu, Xi Shao, Roald Z. Sagdeev, and Padma K. Shukla. 2009. "Laser Acceleration of Monoenergetic Protons via a Double Layer Emerging from an Ultra-thin Foil." *New Journal of Physics* 11 (7): 073006.
21. Shukla, P. K., and Bengt Eliasson. 2005. "Localization of Intense Electromagnetic Waves in a Relativistically Hot Plasma." *Physical Review Letters* 94 (6): 065002.

# 14 Strong Electromagnetic Field Effects in Plasma

## 14.1 Introduction

While classical electrodynamics (CED) governs the interaction between electromagnetic fields and classical particles,[1] quantum electrodynamics (QED) describes the particles and fields quantum mechanically.[2] In the quantum picture, particles (e.g. electrons) and electromagnetic fields and waves all have both wave and particle properties. QED is one of the most successful theories in precise descriptions of particles and fields and the fine electronic structure of atoms. QED effects are expected to be increasingly important in laser–plasma interactions with the rapid increase of laser power.[3] Important QED effects include vacuum polarization by strong electromagnetic fields leading to photon–photon interactions, and pair creation in extremely strong electric fields.

In Section 14.2 we will discuss vacuum polarization and pair creation and in Section 14.3 radiation reaction. We will also consider oscillating plasma mirror as a source of intense fields in Section 14.4.

## 14.2 Vacuum Polarization and Pair Creation

At very strong electric fields, QED predicts vacuum polarization and the possibility of the creation of electron–positron pairs from vacuum. This is due to the uncertainty principle between time and energy $\Delta W \Delta t < \hbar / 2$. The uncertainty principle states that an energy $\Delta W$ can be "borrowed" from the vacuum for the time $\Delta t \sim \hbar / (2\Delta W)$, after which it has to be returned. In this manner, virtual electron–positron pairs can be created spontaneously from the vacuum, each having the rest mass energy $W = mc^2$ and having to be returned to the vacuum after a time $\Delta t \sim 10^{-21}$ seconds. In a strong electromagnetic field, virtual electrons and positrons are accelerated in opposite directions, leading to non-linear electric polarization and magnetization of the vacuum. This QED nonlinearity can to the lowest order be described through the Euler–Heisenberg Lagrangian density[4,5]

$$\mathcal{L} = \varepsilon_0 \mathcal{F} + \varepsilon_0^2 \kappa (4\mathcal{F}^2 + 7\mathcal{G}^2) \tag{14.1}$$

with

$$\kappa = \frac{2\alpha^2 \hbar^3}{45 m^4 c^5} = \frac{\alpha}{90\pi\varepsilon_0 E_{\text{crit}}^2} \tag{14.2}$$

and the field invariants

$$\mathcal{F} = \frac{1}{2}(c^2 B^2 - E^2) \tag{14.3}$$

$$\mathcal{G} = -c\vec{E} \cdot \vec{B} \tag{14.4}$$

Here $\alpha = e^2/(4\pi\varepsilon_0 \hbar c) \approx 1/137$ is the fine structure constant and $E_{\text{crit}} = m^2c^3 / e\hbar \approx 10^{16}$ V/cm$^{-1}$ is Schwinger's critical field. While the first term on the right-hand side of Eq. (14.1) furnishes the linear Maxwell's equations, the last terms introduce non-linear polarization and magnetization of the vacuum. This may lead to nonlinear interaction such as elastic photon-photon scattering between electromagnetic waves.[3] There are suggestions that QED nonlinearities can be measured in the laboratory using intense lasers.[6–8]

In a very strong electric field, where the amplitude of the electric field approaches $E_{\text{crit}}$, virtual electrons and positrons may gain enough kinetic energy to become real particles, as predicted by Schwinger. The process can be partially understood from a semi-classical picture. Starting at rest, the virtual electron and positron gain speed (neglecting the relativistic mass increase for brevity) $v = eE\Delta t/m$ resulting in the total kinetic energy $\Delta W = mv^2 = m(eE\Delta t/m)^2$. The kinetic energy reaches the total rest mass of an electron–positron pair, $\Delta W = 2mc^2$ in the time $\Delta t = \hbar / (2\Delta W)$ if $E = 4\sqrt{2}E_{\text{crit}}$, thus giving the possibility of a real electron–positron pair. Within QED, the pair creation rate per unit volume per second is predicted from QED to be[5,9-12]

$$q_0 = \frac{c}{(2\pi)^3 \lambda^4} \frac{|\vec{E}|^2}{E_{\text{crit}}^2} \exp\left( -\pi \frac{E_{\text{crit}}}{|\vec{E}|} \right) \tag{14.5}$$

where $\lambda = \hbar / (mc)$ is the Compton wavelength. There are proposals to use plasma-based techniques to focus a laser to approach the Schwinger limit.[13–16]

## 14.3 Radiation Reaction Force

Considering the complex physics of the Reaction Force interacting charged particles in a plasma, it may be surprising that the motion of a single particle in an electromagnetic field is not fully understood. The reason is that an accelerated charged particle radiates electromagnetic waves, leading to an effective friction force on the particle. The equation of an electron, including the radiation reaction force, is

$$\frac{d\vec{p}}{dt} = -e\left(\vec{E} + \vec{v} \times \vec{B}\right) + \vec{F}_{\text{rad}}, \tag{14.6}$$

where the relativistic radiation reaction force, as derived by Lorentz, Abraham, and Dirac,[17–19] is

$$\vec{F}_{\text{rad}} = \frac{\mu_0 e^2}{6\pi mc}\left(\frac{d^2\vec{p}}{dt^2} - \frac{\vec{p}}{m^2c^2}\left(\frac{d\vec{p}}{dt}\right)^2\right). \tag{14.7}$$

As it stands, the second derivative of $\vec{p}$ in Eq. (14.7) leads to unrealistic run-away solutions. A remedy is to consider $\vec{F}_{\text{rad}}$ to be small compared to the electromagnetic force and to replace the time derivative of $\vec{p}$ by $-e(\vec{E} + \vec{v} \times \vec{B})$ in Eq. (14.7). This leads to the Landau–Lifshitz equation[20] which governs the radiation reaction force in the small amplitude limit. While the radiation reaction force is mostly discussed for single-particle motion,[21] plasma effects have also been investigated theoretically and with PIC simulations.[22] Ultimately, QED treatments of the radiation–reaction force[23,24] may be needed to fully predict the effect.

## 14.4 Oscillating Plasma Mirror

An oscillating plasma mirror (OPM) is based on the concept that a moving plasma can act as a perfect mirror if the Doppler-shifted frequency of the incident electromagnetic wave, in the rest frame of the plasma, is below its plasma frequency, and that the frequency of the reflected wave is tunable by the velocity of the mirror. A fast oscillating mirror offers to generate X-ray and gamma ray pulses using intense short pulse lasers.

This mirror can be realized by impinging a linearly polarized intense laser, with normalized amplitude $a \geq 1$, on a thin foil. The laser converts the foil into an overdense plasma. In the skin layer of the plasma, the laser field imparts momentum to the electrons with a component normal to the plasma surface and oscillating at twice the laser frequency (cf. Chapter 5). The oscillating electrons can act as an OPM. Alternatively, one can launch a short pulse laser into a gas jet, creating a wake-field plasma wave moving with a phase velocity equal to the group velocity of the laser. As the plasma wave acquires large amplitude such that the quiver electron velocity equals the group velocity of the laser, the plasma wave breaks, leading to the creation of high density narrow bunches of electrons that may act as plasma mirrors. Bulanov et al.[25] have given an elegant review of relativistic mirrors in plasmas. Here we deduce some salient features of wave reflection from an OPM.

We begin with an overdense plasma slab half-space moving with velocity $v_p\hat{z}$. In the rest frame of the plasma, the plasma frequency is $\omega_p$. An electromagnetic wave of frequency $\omega$ (with respect to the laboratory frame) normally impinges on the plasma. In vacuum, the electric and magnetic fields of the wave are

$$\vec{E} = \hat{x}Ae^{-i(\omega t - \omega z/c)}, \vec{B} = \hat{y}E/c. \tag{14.8}$$

In the moving frame, the electric field, using the Lorentz transformation, can be written as

$$\vec{E}' = \hat{x}\gamma_p(1 - v_p/c)Ae^{-i(\omega' t' - \omega' z'/c)} \tag{14.9}$$

where $\omega' = \gamma_p(1 - v_p/c)\omega$, $k' = \omega'/c$, and $\gamma_p = \sqrt{1 - v_p^2/c^2}$. The reflected wave fields in the moving frame are $\vec{E}'_R = R_A\vec{E}'$, $\vec{B}'_R = -\hat{y}R_A E'/c$ with frequency and wave number $\omega'_R = \omega'$, $k'_R = -\omega'/c$ and amplitude reflection coefficient $R_A = (1 - i\alpha)/(1 + i\alpha)$, where $\alpha = (\omega_p^2/\omega'^2 - 1)^{1/2}$. In the laboratory frame (moving with velocity $-v_p\hat{z}$ with respect to the plasma frame), the electric field and frequency of the reflected wave can be written as

$$\vec{E}_R = \hat{x}\gamma_p^2(1 - v_p/c)^2 R_A A^{-i(\omega_R t - k_R z)}, \tag{14.10a}$$

$$\omega_R = \frac{1 - v_p/c}{1 + v_p/c}\omega, k_R = \omega_R/c, \frac{|E_R|}{\omega_R} = \frac{|E|}{\omega}. \tag{14.10b}$$

For a counter-moving mirror and wave, in the limit $|v_p| \to c$, the reflected wave frequency goes as

$$\omega_R = 4\gamma_p^2\omega. \tag{14.11}$$

As the frequency rises, the amplitude of the reflected wave also rises due to wave compression.

Now we allow the velocity of the mirror to have a time dependence, $v_p(t)$. We may introduce two new variables $\xi = t - z/c$ and $\eta = t + z/c$ to characterize the fields of the incident and reflected waves.[25] The field of the reflected wave can be written as

$$E_R = A\frac{1 + v_p/c}{1 - v_p/c}\cos\psi_R, \tag{14.12}$$

where $\psi_R = \int \omega_R d\eta$ and $\omega_R$ is given by Eq. (14.10b). For an oscillating plasma mirror, $v_p = v_{p0}\cos(2\Omega\eta)$, $t = \eta + (v_{p0}/2\Omega)\sin(2\Omega\eta)$. At large $v_{p0}$, $\psi_R$ has many harmonic components and the reflected wave field has sequence of intense short pulses. Following Bulanov et al.[25], we show in Fig. 14.1 the numerical solution of this equation for $v_{p0}/c = 0.49$, $\omega = \Omega$. One may note that the reflected wave has sequence of short pulses of width $\approx \pi/10\Omega$.

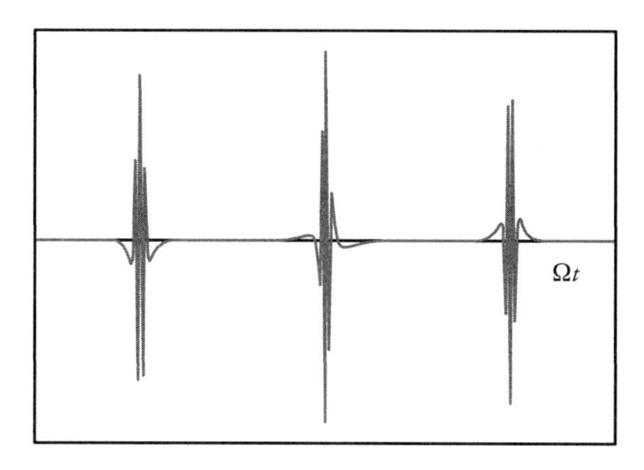


**Fig. 14.1** The electric field of the wave reflected from the oscillating mirror for $v_{p0}/c = 0.49$, $\omega = \Omega$. (Bulanov et al.[25])

In the laser wake-field electron accelerators, we have seen that a large amplitude plasma wave is generated. At the point of wave breaking, the density crests acquire large values and can act as plasma mirrors with very significant reflectivity for a counter-propagating laser. The velocity of the plasma mirror equals the group velocity of the primary laser pulse, $v_p = c(1 - \omega_p^2 / \omega^2)^{1/2}$, giving the Lorentz factor $\gamma_p = \omega / \omega_p$ where $\omega_p$ is the plasma frequency of the equilibrium plasma. For a laser with a Gaussian radial profile of intensity, these plasma mirrors have a paraboidal shape and the second laser, after suffering reflection, gets focused. The focused intensity of the pulse of initial intensity $I_0$ and wavelength $\lambda_0$ after suffering reflection from a paraboidal plasma mirror of diameter $D_0$, is[25]

$$I_R = 32 I_0 (D_0 / \lambda_0)^2 \gamma_p^3. \tag{14.13}$$

It offers to produce ultra-intense fields at much shorter wavelengths.

## 14.5 Conclusions

The rapidly increasing importance of modern laser systems will open up new avenues to investigate strong field effects. The Extreme Light Infrastructure-Nuclear Physics (ELI-NP) facility[26] will enable focused laser intensities above $10^{21}$ W/cm$^2$, and reaching $10^{22}$–$10^{23}$ W/cm$^2$ to be achieved for the first time. This will make possible the experimental investigation of new physical phenomena in plasma, nuclear and particle physics, such as radiation reaction, high-field QED and the production of electron–positron pairs and of energetic gamma rays.

## References

1. Jackson, John David. 1999. *Classical Electrodynamics.* New York: John Wiley & Sons: New York. pp. 841–842.
2. Feynman, Richard P. 2018. *Quantum Electrodynamics.* New York : CRC Press.
3. Marklund, Mattias, and Padma K. Shukla. 2006. "Nonlinear Collective Effects in Photon-photon and Photon-plasma Interactions." *Reviews of Modern Physics* 78 (2): 591.
4. Heisenberg, Werner, and Heinrich Euler. 1936. "Folgerungen aus der Diracschen Theorie des Positrons." *Zeitschrift für Physik* 98 (11–12): 714–32.
5. Schwinger, Julian. 1951. "On Gauge Invariance and Vacuum Polarization." *Physical Review* 82 (5): 664.
6. Brodin, Gert, Mattias Marklund, and Lennart Stenflo. 2001. "Proposal for Detection of QED Vacuum Nonlinearities in Maxwell's Equations by the Use of Waveguides." *Physical Review Letters* 87 (17): 171801.
7. Brodin, Gert, Mattias Marklund, and Lennart Stenflo. 2002. "Influence of QED Vacuum Nonlinearities on Waveguide Modes." *Physica Scripta* 2002 (T98): 127.
8. Eriksson, Daniel, Gert Brodin, Mattias Marklund, and Lennart Stenflo. 2004. "Possibility to Measure Elastic Photon-photon Scattering in Vacuum." *Physical Review A* 70 (1): 013808.
9. Kajantie, Keijo, and Tetsuo Matsui. 1985. "Decay of Strong Color Electric Field and Thermalization in Ultra-relativistic Nucleus-nucleus Collisions." *Physics Letters B* 164 (4–6): 373–8.
10. Gatoff, Gil, A. K. Kerman, and Tetsuo Matsui. 1987. "Flux-tube Model for Ultrarelativistic Heavy-ion Collisions: Electrohydrodynamics of a Quark-gluon Plasma." *Physical Review D* 36 (1): 114.
11. Kluger, Yuval, J. M. Eisenberg, B. Svetitsky, F. Cooper, and E. Mottola. 1991. "Pair Production in a Strong Electric Field." *Physical Review Letters* 67 (18): 2427.
12. Marklund, Mattias, Bengt Eliasson, Padma K. Shukla, Lennart Stenflo, Mark E. Dieckmann, and Madelene Parviainen. 2006. "Electrostatic Pair Creation and Recombination in Quantum Plasmas." *Journal of Experimental and Theoretical Physics Letters* 83 (8): 313–7.
13. Bulanov, S. S. 2004. "Pair production by a Circularly Polarized Electromagnetic Wave in a Plasma." *Physical Review E* 69 (3): 036408.
14. Bulanov, S. S., A. M. Fedotov, and F. Pegoraro. 2005. "Damping of Electromagnetic Waves due to Electron-positron Pair Production." *Physical Review E* 71 (1): 016404.
15. Bulanov, Sergei V., Timur Esirkepov, and Toshiki Tajima. 2003. "Light Intensification towards the Schwinger Limit." *Physical Review Letters* 91 (8): 085001.
16. Shukla, Padma K., Bengt Eliasson, and Mattias Marklund. 2005. "Relativistic Self-compression approaching the Schwinger Limit."*Journal of Plasma Physics* 71 (2): 213–5.
17. Lorentz, H. A. 2007. *The Theory of Electrons.* New York: Cosimo.
18. Abraham, M. 1905. *Theorie der Elektriztät.* Leipzig: Teubner.
19. Dirac, Paul Adrien Maurice. 1938. "Classical Theory of Radiating Electrons." *Proceedings of the Royal Society London A* 167 (929): 148–69.
20. Landau, L. D., and E. M. Lifshitz. 1987. *The Classical Theory of Fields, Course of Theoretical Physics*, Vol. 2. Oxford: Butterworth-Heinemann.

21. Harvey, Chris, and Mattias Marklund. 2012. "Radiation Damping in Pulsed Gaussian Beams." *Physical Review A* 85 (1): 013412.

22. Tamburini, Matteo, Francesco Pegoraro, Antonino Di Piazza, Christoph H. Keitel, Tatyana V. Liseykina, and Andrea Macchi. 2011. "Radiation Reaction Effects on Electron Nonlinear Dynamics and Ion Acceleration in Laser–solid Interaction." *Nuclear Instruments and Methods in Physics Research Section A: Accelerators, Spectrometers, Detectors and Associated Equipment* 653 (1): 181–5.

23. Sokolov, Igor V., John A. Nees, Victor P. Yanovsky, Natalia M. Naumova, and Gérard A. Mourou. 2010. "Emission and its Back-reaction Accompanying Electron Motion in Relativistically Strong and QED-strong Pulsed Laser Fields." *Physical Review E* 81 (3): 036412.

24. Di Piazza, A., K. Z. Hatsagortsyan, and Christoph H. Keitel. 2010. "Quantum Radiation Reaction Effects in Multiphoton Compton Scattering." *Physical Review Letters* 105 (22): 220403.

25. Bulanov, Sergei Vladimirovich, T. Zh Esirkepov, Masaki Kando, Aleksandr Sergeevich Pirozhkov, and Nikolai Nikolaevich Rosanov. 2013. "Relativistic Mirrors in Plasmas. Novel Results and Perspectives." *Physics-Uspekhi* 56 (5): 429.

26. Turcu, I. C. E., C. Murphy, F. Negoita, D. Stutman, M. Zepf, J. Schreiber, C. Harvey et al. 2016. "High Field Physics and QED Experiments at ELI-NP." *Romanian Reports in Physics* 68: S145.

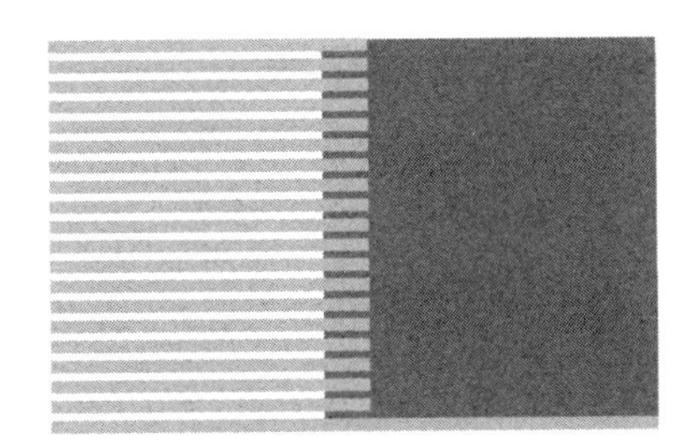

# INDEX